Coarse-Grained Modeling of Biomolecules

Series in Computational Biophysics

Nikolay V. Dokolyan, Series Editor

Coarse-Grained Modeling of Biomolecules
Edited by Garegin A. Papoian

Molecular Modeling at the Atomic Scale: Methods and Applications
in Quantitative Biology
Edited by Ruhong Zhou

Computational Approaches to Protein Dynamics: From Quantum
to Coarse-Grained Methods
Edited by Mónika Fuxreiter

Coarse-Grained Modeling of Biomolecules

Edited by
Garegin A. Papoian

CRC Press
Taylor & Francis Group
Boca Raton London New York

CRC Press is an imprint of the
Taylor & Francis Group, an **informa** business

CRC Press
Taylor & Francis Group
6000 Broken Sound Parkway NW, Suite 300
Boca Raton, FL 33487-2742

First issued in paperback 2021

ISBN-13: 978-0-367-78173-6 (pbk)
ISBN-13: 978-1-4665-7606-3 (hbk)

This book contains information obtained from authentic and highly regarded sources. Reasonable efforts have been made to publish reliable data and information, but the author and publisher cannot assume responsibility for the validity of all materials or the consequences of their use. The authors and publishers have attempted to trace the copyright holders of all material reproduced in this publication and apologize to copyright holders if permission to publish in this form has not been obtained. If any copyright material has not been acknowledged please write and let us know so we may rectify in any future reprint.

Library of Congress Cataloging-in-Publication Data

Visit the Taylor & Francis Web site at
http://www.taylorandfrancis.com

and the CRC Press Web site at
http://www.crcpress.com

Contents

Series Preface, vii

Foreword, ix

Editor, xi

Contributors, xiii

Introduction: A Brief Historical Survey of Coarse-Graining
of Biomolecules, xvii

CHAPTER 1 ■ Inverse Monte Carlo Methods   1

ALEXANDER P. LYUBARTSEV

CHAPTER 2 ■ Thermodynamically Consistent Coarse
Graining of Polymers   27

MARINA G. GUENZA

CHAPTER 3 ■ Microscopic Physics-Based Models of
Proteins and Nucleic Acids: UNRES
and NARES   67

MACIEJ BARANOWSKI, CEZARY CZAPLEWSKI, EWA I. GOŁAŚ,
YUAN-JIE HE, DAWID JAGIEŁA, PAWEŁ KRUPA, ADAM LIWO,
GIA G. MAISURADZE, MARIUSZ MAKOWSKI,
MAGDALENA A. MOZOLEWSKA, ANDREI NIADZVEDTSKI,
ANTTI J. NIEMI, SHELLY RACKOVSKY, RAFAŁ ŚLUSARZ,
ADAM K. SIERADZAN, STANISŁAW OŁDZIEJ, TOMASZ WIRECKI,
YANPING YIN, BARTŁOMIEJ ZABOROWSKI, AND HAROLD A. SCHERAGA

CHAPTER 4 ■ AWSEM-MD: From Neural Networks to Protein Structure Prediction and Functional Dynamics of Complex Biomolecular Assemblies   121

GAREGIN A. PAPOIAN AND PETER G. WOLYNES

CHAPTER 5 ■ Elastic Models of Biomolecules   191

QIANG CUI

CHAPTER 6 ■ Knowledge-Based Models: RNA   221

RAÚL MÉNDEZ, ANDREY KROKHOTIN, MARINO CONVERTINO, JHUMA DAS, ARPIT TANDON, AND NIKOLAY V. DOKHOLYAN

CHAPTER 7 ■ The Need for Computational Speed: State of the Art in DNA Coarse Graining   271

DAVIT POTOYAN AND GAREGIN A. PAPOIAN

CHAPTER 8 ■ Coarse-Grained Modeling of Nucleosomes and Chromatin   297

LARS NORDENSKIÖLD, ALEXANDER P. LYUBARTSEV, AND NIKOLAY KOROLEV

CHAPTER 9 ■ Modeling of Genomes   341

NAOKO TOKUDA AND MASAKI SASAI

CHAPTER 10 ■ Mechanics of Viruses   367

OLGA KONONOVA, ARTEM ZHMUROV, KENNETH A. MARX, AND VALERI BARSEGOV

INDEX, 417

Series Preface

THE 2013 NOBEL PRIZE IN CHEMISTRY WAS AWARDED FOR THE "Development of multiscale models for complex chemical systems." This prize was particularly special to the whole computational community as the role computation has played since the pioneering works of Lifson, Warshel, Levitt, Karplus, and many others was finally recognized.

This Series in Computational Biophysics has been conceived to reflect the tremendous impact of computational tools in the study and practice of biophysics and biochemistry today. The goal is to offer a suite of books that will introduce the principles and methods for computer simulation and modeling of biologically important macromolecules. The titles cover both fundamental concepts and state-of-the-art approaches, with specific examples highlighted to illustrate cutting-edge methodology. The series is designed to cover modeling approaches spanning multiple scales: atoms, molecules, cells, organs, organisms, and populations.

The series publishes advanced level textbooks, laboratory manuals, and reference handbooks that meet the needs of students, researchers, and practitioners working at the interface of biophysics/biochemistry and computer science. The most important methodological aspects of molecular modeling and simulations as well as actual biological problems that have been addressed using these methods are presented throughout the series. Prominent leaders have been invited to edit each of the books, and in turn those editors select contributions from a roster of outstanding scientists.

SERIES IN COMPUTATIONAL BIOPHYSICS would not be possible without the drive and support of the Taylor & Francis series manager, Lu Han. The editors, authors, and I are greatly appreciative of his support and grateful for the success of the series.

Nikolay V. Dokholyan
Series Editor
Chapel Hill, North Carolina

Foreword

F ROM THE PREHISTORIC DRAWINGS ON THE CAVE WALLS OF Lascaux to the paintings hanging in the Louvre in the late nineteenth century, the visual arts seemed to progress monotonically by adding ever more detail seemingly to give greater realism. A new movement then exploded near the turn of the twentieth century. Artists began to explore new ways of expressing truths about the visual world and the visual experience. These new ways involved generally more abstract and superficially less detailed techniques than those taught in the art schools. They required no less intellectual discipline, however. The new "modern art," while not to everyone's taste, nevertheless, in the opinion of many, brought into existence creations of great beauty and insight.

In recent decades the theoretical study of biomolecules has spawned a similar movement. While acknowledging the power of the highly detailed atomistic models of biomolecular structure and dynamics that can now be simulated using the world's most powerful computers, the scientists in this movement have been trying to find new ways of looking at biomolecules and how they move, ways that might be more elegant and efficient. These approaches are often called "coarse grained" because they employ fewer details in picturing molecules and their motion. The goal of this scientific movement is to find computational models that still capture the inner truth and, indeed, beauty of life at the molecular level. The new techniques that this task brings forward are no less intellectually demanding nor intriguing than those traditionally used in computer simulation. New ideas ranging from renormalization group theory to neural networks, machine learning and "Big Data" are key to these new developments.

The chapters in this book aim to survey some of the progress in simulating biomolecular dynamics in this new style. The images conjured up by this work are not yet universally loved, but they are beginning to bring new insights into the study of biological structure and function.

The future will decide whether this scientific movement can bring forth its Picasso or Modigliani.

Peter G. Wolynes
Department of Chemistry, Rice University, Houston, Texas

Editor

Garegin A. Papoian is the first Monroe Martin professor with appointments in the Department of Chemistry and Biochemistry and the Institute for Physical Science and Technology at the University of Maryland, College Park, Maryland. He was born in Yerevan, Armenia, and completed 4 years of undergraduate studies in the Higher College of the Russian Academy of Sciences, followed by graduate work with Professor Roald Hoffmann at Cornell University. He received his PhD in 1999, working on quantum chemistry of intermetallic alloys and heterogenous catalysis. He then joined the research group of Professor Michael Klein at the University of Pennsylvania as a postdoctoral associate in 2000, to study metalloenzymes with Car–Parrinello simulations. This was followed by a National Institutes of Health–sponsored postdoctoral fellowship with Professor Peter Wolynes at the University of California, San Diego, from 2001 to 2004 with research focused on protein folding, binding, and hydration. He started his independent group in 2004 as an assistant professor of chemistry at the University of North Carolina at Chapel Hill, where he received tenure in 2010. He subsequently moved to the University of Maryland in his current position. He is interested in utilizing tools of physical chemistry to shed light on complex biological processes at the molecular and cellular levels. His group is currently working on developing physicochemical models of cytoskeletal dynamics and cell motility, modeling the way DNA packs in cells of higher organisms, and studying protein functional dynamics and allostery.

He is the recipient of numerous awards, including the Phillip and Ruth Hettleman Prize for Artistic and Scholarly Achievement 2010,

ACS Hewlett-Packard Outstanding Junior Faculty Award 2010, National Science Foundation CAREER Award 2009, Camille Dreyfus Teacher-Scholar 2008, Beckman Young Investigator 2007, Camille and Henry Dreyfus New Faculty Award 2004, and National Institutes of Health Postdoctoral Fellowship 2001.

Contributors

Maciej Baranowski
Intercollegiate Faculty of
  Biotechnology
University of Gdansk and Medical
  University of Gdansk
Gdansk, Poland

Valeri Barsegov
Department of Chemistry
University of Massachusetts Lowell
Lowell, Massachusetts

Marino Convertino
Department of Biochemistry and
  Biophysics
School of Medicine
University of North Carolina at
  Chapel Hill
Chapel Hill, North Carolina

Qiang Cui
Department of Chemistry
University of Wisconsin-Madison
Madison, Wisconsin

Cezary Czaplewski
Faculty of Chemistry
University of Gdansk
Gdansk, Poland

Jhuma Das
Department of Biochemistry and
  Biophysics
School of Medicine
University of North Carolina at
  Chapel Hill
Chapel Hill, North Carolina

Nikolay V. Dokholyan
Department of Biochemistry and
  Biophysics
School of Medicine
University of North Carolina at
  Chapel Hill
Chapel Hill, North Carolina

Ewa I. Gołaś
Faculty of Chemistry
University of Gdansk
Gdansk, Poland

and

Baker Laboratory of Chemistry and
  Chemical Biology
Cornell University
Ithaca, New York

Marina G. Guenza
Department of Chemistry and
    Biochemistry
Institute of Theoretical Science
University of Oregon
Eugene, Oregon

Yuan-Jie He
Baker Laboratory of Chemistry
    and Chemical Biology
Cornell University
Ithaca, New York

Dawid Jagieła
Faculty of Chemistry
University of Gdansk
Gdansk, Poland

Olga Kononova
Department of Chemistry
University of Massachusetts Lowell
Lowell, Massachusetts

Nikolay Korolev
Division of Structural Biology and
    Biochemistry
School of Biological Sciences
Nanyang Technological University
Singapore

Andrey Krokhotin
Department of Biochemistry and
    Biophysics
School of Medicine
University of North Carolina at
    Chapel Hill
Chapel Hill, North Carolina

Paweł Krupa
Faculty of Chemistry
University of Gdansk
Gdansk, Poland

and

Baker Laboratory of Chemistry
    and Chemical Biology
Cornell University
Ithaca, New York

Adam Liwo
Faculty of Chemistry
University of Gdansk
Gdansk, Poland

Alexander P. Lyubartsev
Division of Physical Chemistry
Stockholm University
Stockholm, Sweden

Gia G. Maisuradze
Baker Laboratory of Chemistry and
    Chemical Biology
Cornell University
Ithaca, New York

Mariusz Makowski
Faculty of Chemistry
University of Gdansk
Gdansk, Poland

Kenneth A. Marx
Department of Chemistry
University of Massachusetts
    Lowell
Lowell, Massachusetts

Raúl Méndez
Department of Biochemistry and
  Biophysics
School of Medicine
University of North Carolina at
  Chapel Hill
Chapel Hill, North Carolina

Magdalena A. Mozolewska
Faculty of Chemistry
University of Gdansk
Gdansk, Poland

and

Baker Laboratory of Chemistry
  and Chemical Biology
Cornell University
Ithaca, New York

Andrei Niadzvedtski
Faculty of Chemistry
University of Gdansk
Gdansk, Poland

Antti J. Niemi
Department of Physics and
  Astronomy
Uppsala University
Uppsala, Sweden

and

Laboratoire de Mathematiques et
  Physique Theorique
Universite de Tours
Tours, France

and

Department of Physics
Beijing Institute of Technology
Beijing, P.R. China

Lars Nordenskiöld
Division of Structural Biology and
  Biochemistry
School of Biological Sciences
Nanyang Technological University
Singapore

Stanisław Ołdziej
Intercollegiate Faculty of
  Biotechnology
University of Gdansk and Medical
  University of Gdansk
Gdansk, Poland

Gargein A. Papoian
Department of Chemistry and
  Biochemistry
Institute for Physical Science and
  Technology
University of Maryland
College Park, Maryland

Davit Potoyan
Department of Chemistry
Iowa State University
Ames, Iowa

Shelly Rackovsky
Baker Laboratory of Chemistry and
  Chemical Biology
Cornell University
Ithaca, New York

and

Department of Pharmacology and
  Systems Therapeutics
The Icahn School of Medicine at
  Mount Sinai
New York, New York

Masaki Sasai
Department of Computational
  Science and Engineering
Nagoya University
Nagoya, Japan

Harold A. Scheraga
Baker Laboratory of Chemistry and
  Chemical Biology
Cornell University
Ithaca, New York

Adam K. Sieradzan
Faculty of Chemistry
University of Gdansk
Gdansk, Poland

and

Department of Physics and
  Astronomy
Uppsala University
Uppsala, Sweden

Rafał Ślusarz
Faculty of Chemistry
University of Gdansk
Gdansk, Poland

Arpit Tandon
Department of Biochemistry and
  Biophysics
School of Medicine
University of North Carolina at
  Chapel Hill
Chapel Hill, North Carolina

Naoko Tokuda
Department of Computational
  Science and Engineering
Nagoya University
Nagoya, Japan

Tomasz Wirecki
Faculty of Chemistry
University of Gdansk
Gdansk, Poland

and

Baker Laboratory of Chemistry and
  Chemical Biology
Cornell University
Ithaca, New York

Peter G. Wolynes
Department of Chemistry
Rice University
Houston, Texas

Yanping Yin
Baker Laboratory of Chemistry and
  Chemical Biology
Cornell University
Ithaca, New York

Bartłomiej Zaborowski
Faculty of Chemistry
University of Gdansk
Gdansk, Poland

Artem Zhmurov
Department of Chemistry
University of Massachusetts Lowell
Lowell, Massachusetts

Introduction

A Brief Historical Survey of Coarse-Graining of Biomolecules

Garegin A. Papoian

CONTENTS

I.1	Introduction	xvii
I.2	Bottom-Up Coarse-Graining	xviii
I.3	Top-Down Coarse-Graining	xxi
I.4	Summary	xxiv
	Acknowledgments	xxv
	References	xxv

I.1 INTRODUCTION

It is becoming increasingly clear that computer modeling will revolutionize structural and molecular biology. While atomistic simulations of biomolecules are well understood conceptually, coarse-grained models are often viewed both with great interest—because of their promise to cut through length and time scales—but also with certain confusion. The latter primarily arises from the difficulty of assessing the highly fragmented landscape of coarse-grained modeling philosophies and strategies, which should be juxtaposed with the rather straightforward and systematic development of atomistic force fields. However, this diversity of approaches is actually greatly advantageous, providing fertile ground for creative ideas to germinate, compete with each other, and eventually turn into powerful methods for attacking the most difficult problems of modern biology and material science. Confusion, however, has a serious drawback of presenting a barrier of entry for younger scientists, who might be intrigued by the promises of coarse-grained techniques, but are not sure how to compare or evaluate different models. Hence, the main goal of this book is to pool together pedagogical reviews of state-of-the-art coarse-graining strategies for proteins, DNA, RNA, viral capsids, and entire genomes, to help students or other interested scientists achieve a bird's-eye view of various biomolecular coarse-graining approaches.

To help the readers see more clearly the intricate landscape of coarse-graining approaches, I would like to first start by introducing a classification scheme based on the dichotomy of *bottom-up versus top-down* coarse-graining, which can be followed by further grouping methods using additional criteria. Many practitioners tend to rely primarily only on one of these two methodologies. I argue below that both of these strategies are powerful in their own way, and both are likely to stay around for the foreseeable future and contribute to solving many frontier problems in structural and molecular biology. Interestingly, hybrid approaches that leverage and integrate the advantages of both top-down and bottom-up coarse-graining methodologies present yet another powerful paradigm for tackling many specific challenges of biomolecular modeling.

I.2 BOTTOM-UP COARSE-GRAINING

The key idea of bottom coarse-graining is to take advantage of some characteristic time-scale separation within the system's dynamics, and integrate out faster degrees of freedom. For dynamical systems, the Mori–Zwanzig projection operator formalism provides a rigorous basis for such integration [1]; however, it is too cumbersome for most biomolecular applications.

For coarse-graining of systems at equilibrium, Kubo's generalized cumulant expansion [2], published in 1962, is one of the earliest well-known examples of the systematic methodologies underlying bottom-up coarse-graining. This methodology provides the underpinning of the UNRES family of biomolecular force fields, reviewed in Chapter 3 by Scheraga and coworkers. Interestingly, while the functional forms of the UNRES force field, in particular, many-body interactions, are derived in a bottom-up fashion, the force field parameterization relies significantly on knowledge-based (i.e., a type of top-down) approaches. Another celebrated paper in protein modeling was published by Levitt and Warshel (Nobel Prize Laureates in Chemistry in 2013) in 1975, which introduced a one-bead-per-residue polymer model, and showed that computer simulations can be productively used to study protein folding, providing deep conceptual insights [3]. In that work, the interactions between coarse-grained sites were derived in a bottom-up way from an underlying atomistic model by explicitly summing all underlying microscopic interactions, reminiscent of Kubo's cumulant expansion formalism mentioned above [2].

Kadanoff's block-spin construction [4], which was motivated by Widom's scaling hypothesis in near critical systems [5], represents another very prominent early example of bottom-up coarse-graining. Kadanoff proposed to partition spins into groups (blocks), and assign a new spin variable to each group, determined, for example, by the majority rule. This idea was later developed by Wilson into the rigorous framework of renormalization group (RG) theory [6,7], demonstrating the fundamental route for connecting Hamiltonians at neighboring length scales. First, the *real space* RG construction [8] requires an adequate functional basis set of interaction potentials that are applicable at all scales, with only differing coefficients entering specific potentials among the Hamiltonians at different length scales. Then, in a 1-step renormalization, the partition functions (or free energies) of two neighboring scales are required to match, consequently, leading to equations for determining the parameters of a coarse-grained Hamiltonian from a fine-grained one [8]. Iterative application of this algorithm generates a flow in the space of Hamiltonians, revealing interesting features, such as fixed points, which characterize various phase behaviors [7].

However, from the viewpoint of coarse-grained model building, only 1-step renormalization is usually necessary. Following an earlier work by Ma [9], Swendsen suggested [10] a numerical Monte Carlo scheme for numerically estimating critical exponents of Ising models by obtaining and subsequently analyzing the linearized RG transformation matrix from simple correlations functions computed separately at fine-grained and coarse-grained scales. This approach, with some modifications, was applied to coarse-graining atomistic systems by Laaksonen and Lyburatsev [11], and was further generalized for molecular systems with many-body interactions by Savelyev and Papoian [12], who denoted the latter method molecular renormalization group coarse-graining (MRG-CG). In principle, if the chosen basis set of the coarse-grained interaction potentials is nearly complete, in particular including a large number of various many-body terms, the MRG-CG procedure leads to rigorous derivation of the coarse-grained Hamiltonian parameters from analyzing trajectories obtained from fine-grained dynamics. In practice, however, the interaction potentials are far from complete, hence, cumulant expansion approaches can be used to estimate the accuracy of coarse-graining [12]. Chapter 1 by Lyubartsev, Chapter 7 by Potoyan and Papoian, and Chapter 8 by Nordenskiöld, Lyubartsev, and Korolev further elaborate on various flavors of (RG) coarse-graining strategies and their applications to many interesting biomolecular systems, such as DNA, nucleosomes, and chromatin fibers.

In another recent development, Shell and coworkers introduced an elegant coarse-graining framework based on variational minimization of the relative entropy between the distributions obtained from fine-grained and coarse-grained Hamiltonians [13]. Interestingly, the resulting equations that need to be solved turned out to be basically identical with the MRG-CG equations described above. Therefore, the relative entropy coarse-graining might be viewed as a reinterpretation of MRG-CG using an information-theoretic perspective. The latter viewpoint has stimulated exploration of new sets of concepts, however, such as mapping entropy.

Another rigorous perspective on bottom-up coarse-graining is based on potential of mean force (PMF) analyses of the underlying fine-grained Hamiltonian. In particular, faster motions of fine-grained degrees of motion, which will be integrated out, give rise to thermally averaged forces on coarse-grained sites, which are matched in least-square sense within the "force matching" formalism (also called multiscale coarse-graining, MS-CG), developed by Voth and coworkers [14,15]. Subsequently, Noid and coworkers connected this approach to the generalized Yvon–Born–Green equations of statistical mechanics, carefully analyzing various theoretical and practical issues [16]. MS-CG has been used to successfully develop numerous coarse-grained models of different liquids and their mixtures, polymers, lipids, peptides, and other biomolecules [16,17]. In a related strategy, integral equation theories, for example, starting from the Ornstein–Zernike equation, could serve as yet another foundation for rigorous coarse-graining, as elaborated by Guenza in Chapter 2. In that chapter, she sheds light on the remarkable difficulty of reconciling the structural correlations between fine-grained and coarse-grained models, while simultaneously achieving thermodynamic consistency, and also presents approaches for coarse-graining protein dynamics.

Overall, in terms of theoretical rigor and even with regard to practical implementation and computational efficiency, bottom-up coarse-graining currently offers distinct productive routes to accurately integrate out fast degrees of freedom of the underlying fine-grained model. Two fundamental limitations, however, constrain the application of these strategies. First, the quality of the obtained coarse-grained model is limited by the accuracy of the underlying fine-grained model. Although this statement is rather obvious, what is less trivial is how small relative errors of the fine-grained model could impact coarse-graining, because the former encompasses interactions at much higher energy scales. One successful way to address this problem is based on subsequent calibration or fine-tuning of the resulting coarse-grained force field parameters using experimental data [18]; however, this strategy deviates from the strict definition of bottom-up coarse-graining. In any case, while the potential inadequacy of current atomistic force fields for the purpose of coarse-graining is well recognized in principle, this serious pitfall still seems rather overlooked in various practical applications.

Another rather serious limitation of bottom-up coarse graining primarily concerns proteins because of the 20-letter amino acid alphabet of protein sequences. Derivation of pairwise interactions between amino acid pairs leads to 210 unique curves, while the combinatorics of three-body and four-body potentials runs on the order of thousand and ten thousand, respectively. Since prior experience has shown that inclusion of at least some amount of many-body interactions is necessary for developing predictive, transferrable protein force fields, this snowballing of the parameter space and the associated difficulty of learning these parameters presents a rather difficult practical challenge for bottom-up coarse-graining. In contrast, the energy landscape perspective [19], among others, provides a robust route for top-down coarse-graining of proteins. On the other hand, for biopolymers with simpler alphabets, such as DNA and RNA, or smaller biomolecules, such as lipids, this concern is less significant; hence, the adequacy of bottom-up coarse-grained models is primarily determined by the quality of underlying atomistic force fields. Therefore, for such systems, but also in particular for nonbiological soft matter materials and polymers, bottom-up coarse-graining plays an increasingly important role. In particular, in the latter applications, where high-resolution, structure- and data-rich experimental measurements are usually lacking, development of high-quality top-down models is rather challenging. Hence, bottom-up models are nearly indispensable when chemically accurate modeling of these materials is required.

1.3 TOP-DOWN COARSE-GRAINING

Here, I would like to highlight first some of the phenomenal successes of top-down coarse-graining over the last 100 years. Aside from Einstein's modeling of Brownian motion [20], one of the earliest and most impactful coarse-grained models was the Ising model of magnetic spins, where the latter interact with an external magnetic field and each other [21]. On the one hand, as with many top-down models, the Ising model is highly idealized and even unrealistic; however, it ushered a new era of much deeper understanding of critical phenomena in the mid-twentieth century. As a PhD dissertation project of Ising, under supervision of Lenz, the 1D Ising model was developed to study the principles of ferromagnetism, initially disappointing somewhat because it became clear that the 1D model does not undergo a phase transition at finite temperature. However, many important statistical mechanical techniques were developed to solve it, such as the transfer matrix approach [21]. Later, Onsager's tour-de-force solution of the 2D Ising model lead to one of the key breakthroughs of twentieth-century science, namely, the possibility of nonclassical exponents when asymptotically approaching the second-order phase transition point [21]. This insight was later followed by some profound ideas of scaling in strongly interacting systems, culminating in RG theory [4–8,22], which explained many phenomena in statistical physics, condensed matter physics, and particle physics. Interestingly, the RG theory provided one of the early formal routes to bottom-up coarse-graining, discussed above.

Deep physical insights into the fundamental nature of top-down coarse-graining was given by Fisher in a profound review article in 1988 [7]. He credits Landau with the introduction of systematic *effective field theories* and inventing the concept of *the order parameter* to parsimoniously describe the *essential* physics underlying various collective phenomena. Quoting Fisher's supposition of how Landau might have reasoned about this subject [7]: "I may not understand the microscopic phenomena at all, but I recognize that there is a microscopic level and I believe it should have certain general, overall properties especially as regards locality and symmetry: those then serve to govern the most characteristic behavior on scales greater than atomic." Hence, discovering a good order parameter can be one of the most important early steps toward achieving deep physical understanding of complex organization in condensed matter systems [7].

In modeling of biomolecules, top-down coarse-graining also played a critically important role. In 1980s and 1990s, highly idealized lattice description of polymer chains were used to study protein folding [23–26]. While the lattice models are also very much artificial, they nevertheless lead to numerous insights into protein folding thermodynamics, kinetics, and even evolutionary design questions. Starting in the 1990s, off-lattice [27] coarse-grained protein models became dominant, branching into a family of native structure-based models [28], sometimes also called Gō models [23] and so-called *de novo* or transferrable models, where the latter strive to predict structures of proteins with sequences that do not have any known homologues. In this book, Chapter 5 by Cui and Chapter 10 by Barsegov and coworkers rely on native structure-based approaches, while Chapter 4 by Papoian and Wolynes reports on a *de novo* protein force field that can be used to fold unknown sequences or predict the formation of protein complexes where the structures are not yet known.

When studying DNA at larger scales, generic polymer chain models have been used to explain many experimental phenomena, including the Gaussian chain and the worm-like chain models, which are highly successful top-down coarse-grained models describing polymer behaviors [29]. More recently, higher resolution, sequence-specific structural models were developed, for example, a rigid base pair or single base elastic models, where now well-recognized local motion directives, such as tilt, twist, and roll, were introduced [30,31]. Subsequent top-down DNA models allowed formation of bubbles and melting of double-stranded DNA, incorporating also some electrostatic effects [32,33]. In this book, Chapter 6 by Dokholyan and coworkers reports on a top-down coarse-grained RNA model, while Chapter 7 by Potoian and Papoian reviews both top-down and bottom-up approaches to coarse-graining DNA. Finally, taking coarse-graining to a whole other length scale, Tokuda and Sasai in Chapter 9 highlight top-down coarse-grained polymer models of chromatin fibers that comprise genomic DNA in the nuclei of eukaryotic cells.

Top-down coarse-graining of biomolecules can be further subdivided into *physical* models, for example, describing electrostatics via Debye–Hückel interactions or directional hydrogen bonding, and *knowledge-based* potentials, that rely on one, a few, or potentially many thousands of biomolecular structures to infer interaction potentials. I would like to note in passing that some authors distinguish *statistical* potentials as a different subclass from knowledge-based potentials. The main advantage of physical potentials is their simplicity and transferability among very different systems. Knowledge-based potentials, on the other hand, are usually more accurate, especially for proteins. At a fundamental level, one can view knowledge-based potentials as *implicitly* containing many-body, subtly correlated interactions that physical models almost always omit due to cumbersomeness and computational complexity reasons or even more simply because of the lack of awareness. Although it may seem that knowledge-based models are a temporary kludge before more fundamental physical models become feasible, this view most likely will not fully bear out, as evidenced by the recent revolution in machine learning and artificial intelligence that impacts practically all areas of science and technology [34]. In this book, the aforementioned Chapters 4–7, 9, and 10 cover in part knowledge-based top-down coarse-graining strategies, but some of the primarily bottom-up approaches, for example, the UNRES protein force field reviewed in Chapters 3, also rely to some degree on knowledge-based learning of model potentials. Chapter 4 reviews AWSEM-MD, which arose from one of the earliest applications of neural network theories to structural modeling of proteins [35], but evolved over time into a Hamiltonian containing both physical and knowledge-based potentials [36].

Overall, top-down strategies afford researchers almost unlimited creative freedom in proposing novel models of biomolecules and their complex assemblies. The key drawback of these approaches is their non-systematic nature. First, many initially promising top-down models might turn out to be not particularly informative. Second, even if one is fortunate enough to propose a successful coarse-grained model of this type, it is usually rather difficult to further improve on it in some systematic fashion. This is where a hybrid strategy might have most impact: relying on bottom-up coarse-graining to guess or infer additional interaction terms that could improve the top-down model, but subsequently learning the associated force field parameters by relying on knowledge-based approaches. The latter will continue to be particularly important for capturing intricate many-body interactions in proteins and their assemblies with other biomolecules.

I.4 SUMMARY

I hope I provided convincing arguments that bottom-up versus top-down coarse-graining should not viewed as a battle of new ideas versus old ones, or rigorous versus unrigorous statistical mechanics. Instead both approaches can be very powerful and modern, each having remarkable strengths, but also pitfalls, depending on the application area, and both will stay with us for the foreseeable future. In this book, which covers modeling of biomolecules, both approaches are well represented.

Despite covering quite a bit of material in diverse areas of structural biology, this book is not encyclopedic. Additional topics and perspectives have been well reviewed elsewhere, in particular, in an earlier book on coarse-graining, edited by Voth [17] (published in 2008), and a recent comprehensive journal review [16].

Peeking into the near future, my expectation is that protein coarse-grained modeling will continue to rely more on top-down strategies, guided in part, however, by potential forms obtained from bottom-up coarse-graining. On the other hand, the opposite seems plausible for DNA and lipid coarse-grained modeling. RNA presents an interesting borderline case, where both approaches are currently competitive to begin with, and it is likely that powerful hybrid models will emerge as well. Finally, when considering complex biomolecule assemblies, for example, comprised of both proteins and DNA, such as nucleosomes, top-down protein models and bottom-up DNA models will need to be harmoniously blended, which is an area of active research in my laboratory as well as in many others.

ACKNOWLEDGMENTS

I would like to sincerely thank my former mentors and teachers, who showed me how to think deeply about statistical mechanics and quantum chemistry, including Roald Hoffmann, Ben Widom, Michael Klein, and Peter Wolynes. I also learned a lot and am grateful to all members of my laboratory, formerly at the University of North Carolina at Chapel Hill, and currently at the University of Maryland in College Park. I would like to thank Mary Pitman for reading this Introduction and providing helpful feedback. I owe a big debt of gratitude to my wife, Tatiana Meteleva, and my children, Daniel and Karina, for unconditionally supporting me and putting up with my heavy work and travel schedules. Finally, I would like to dedicate this book to my mother and father, Hasmik Galstyan and Haykaz Papoyan, who, despite difficult circumstances, made sure that my brother and myself were provided with high-quality education and opportunities, instilling in us love of knowledge and teaching us how to be good people.

REFERENCES

1. R. Zwanzig. *Nonequilibrium Statistical Mechanics*. Oxford: Oxford University Press, 2001.
2. R. Kubo. Generalized cumulant expansion method. *Journal of the Physical Society of Japan*, 17:1100, 1962.
3. M. Levitt and A. Warshel. Computer simulation of protein folding. *Nature*, 253(5494):694–698, 1975.

4. L. P. Kadanoff. Scaling laws for Ising models near Tc. *Physics*, 2(263–272), 1966.

5. B. Widom. Equation of state in the neighborhood of the critical point. *Journal of Chemical Physics*, 43(11):3898–3905, 1965.

6. K. G. Wilson. Renormalization group and critical phenomena. I. Renormalization group and the Kadanoff scaling picture. *Physical Review B*, 4(9):3174, 1971.

7. M. E. Fisher. Renormalization group theory: Its basis and formulation in statistical physics. *Reviews of Modern Physics*, 70(2):653, 1998.

8. E. Efrati, Z. Wang, A. Kolan, and L. P. Kadanoff. Real-space renormalization in statistical mechanics. *Reviews of Modern Physics*, 86(2):647, 2014.

9. S. Ma. Renormalization group by Monte Carlo methods. *Physical Review Letters*, 37(8):461, 1976.

10. R. Swendsen. Monte Carlo renormalization group. *Physical Review Letters*, 42(14):859–861, Apr 1979.

11. A. Lyubartsev and A. Laaksonen. Calculation of effective interaction potentials from radial distribution functions: A reverse Monte Carlo approach. *Physical Review E*, 52(4):3730–3737, Oct 1995.

12. A. Savelyev and G. A. Papoian. Molecular renormalization group coarse-graining of electrolyte solutions: Application to aqueous NaCl and KCL. *Journal of Physical Chemistry B*, 113(22):7785–7793, Jun 2009.

13. A. Chaimovich and M. S. Shell. Coarse-graining errors and numerical optimization using a relative entropy framework. *Journal of Chemical Physics*, 134(9):094112, 2011.

14. S. Izvekov and G. A. Voth. A multiscale coarse-graining method for biomolecular systems. *Journal of Physical Chemistry B*, 109(7):2469–2473, 2005.

15. W. G. Noid, J.-W. Chu, G. S. Ayton, V. Krishna, S. Izvekov, G. A. Voth, A. Das, and H. C. Andersen. The multiscale coarse-graining method. I. A rigorous bridge between atomistic and coarse-grained models. *Journal of Chemical Physics*, 128(24):244114, 2008.

16. W. G. Noid. Perspective: Coarse-grained models for biomolecular systems. *Journal of Chemical Physics*, 139(9):090901–26, 2013.

17. G. A. Voth, editor. *Coarse-Graining of Condensed Phase and Biomolecular Systems*. Boca Raton, FL: CRC Press, 2008.

18. A. Savelyev and G. A. Papoian. Chemically accurate coarse graining of double-stranded DNA. *Proceedings of the National Academy of Sciences USA*, 107(47):20340–20345, 2010.

19. P. G. Wolynes. Evolution, energy landscapes and the paradoxes of protein folding. *Biochimie*, 119:218–230, 2015.

20. A. Einstein. Investigations on the theory of the Brownian movement. *Annalen der Physik*, 17(8):549–560, 1905.

21. R. J. Baxter. *Exactly Solved Models in Statistical Mechanics (Dover Books on Physics)*. Mineola, NY: Dover Publications, 2007.

22. M. E. Fisher. The theory of condensation and the critical point. *Physics*, 3(5):255–283, 1967.

23. H. Taketomi, Y. Ueda, and N. Gō. Studies on protein folding, unfolding and fluctuations by computer simulation. *Chemical Biology & Drug Design*, 7(6):445–459, 1975.

24. P. E. Leopold, M. Montal, and J. N. Onuchic. Protein folding funnels: A kinetic approach to the sequence–structure relationship. *Proceedings of the National Academy of Sciences*, 89(18):8721–8725, 1992.

25. V. I. Abkevich, A. M. Gutin, and E. I. Shakhnovich. Specific nucleus as the transition state for protein folding: Evidence from the lattice model. *Biochemistry*, 33(33):10026–10036, 1994.

26. K. A. Dill, S. Bromberg, K. Yue, K. M. Ftebig, D. P. Yee, P. D. Thomas, and H. S. Chan. Principles of protein folding—A perspective from simple exact models. *Protein Science*, 4(4):561–602, 1995.

27. D. Thirumalai and D.K. Klimov. Deciphering the timescales and mechanisms of protein folding using minimal off-lattice models. *Current Opinion in Structural Biology*, 9(2):197–207, 1999.

28. C. Clementi, H. Nymeyer, and J. N. Onuchic. Topological and energetic factors: What determines the structural details of the transition state ensemble and "en-route" intermediates for protein folding? An investigation for small globular proteins. *Journal of Molecular Biology*, 298(5):937–953, 2000.

29. M. Rubinstein and R. H. Colby. *Polymer Physics*. Oxford: Oxford University Press, 2003.

30. W. K. Olson, A. A. Gorin, X.-J. Lu, L. M. Hock, and V. B. Zhurkin. DNA sequence-dependent deformability deduced from protein–DNA crystal complexes. *Proceedings of the National Academy of Sciences*, 95(19):11163–11168, 1998.

31. D. Petkevičiūtė, M. Pasi, O. Gonzalez, and J. H. Maddocks. cgDNA: A software package for the prediction of sequence-dependent coarse-grain free energies of B-form DNA. *Nucleic Acids Research*, 42(20):e153–e153, 2014.

32. J. P. K. Doye, T. E. Ouldridge, A. A. Louis, F. Romano, P. Šulc, C. Matek, B. E. K. Snodin, L. Rovigatti, J. S. Schreck, R. M. Harrison, et al. Coarse-graining DNA for simulations of DNA nanotechnology. *Physical Chemistry Chemical Physics*, 15(47):20395–20414, 2013.

33. D. M. Hinckley, G. S. Freeman, J. K. Whitmer, and J. J. de Pablo. An experimentally-informed coarse-grained 3-site-per-nucleotide model of DNA: Structure, thermodynamics, and dynamics of hybridization. *Journal of Chemical Physics*, 139(14):144903, 2013.

34. Y. LeCun, Y. Bengio, and G. Hinton. Deep learning. *Nature*, 521(7553):436–444, 2015.

35. M. S. Friedrichs and P. G. Wolynes. Toward protein tertiary structure recognition by means of associative memory hamiltonians. *Science*, 246(4928): 371–373, 1989.

36. A. Davtyan, N. P. Schafer, W. Zheng, C. Clementi, P. G. Wolynes, and G. A. Papoian. AWSEM-MD: Protein structure prediction using coarse-grained physical potentials and bioinformatically based local structure biasing. *Journal of Physical Chemistry*, 116(29):8494–8503, 2012.

CHAPTER 1

Inverse Monte Carlo Methods

Alexander P. Lyubartsev

CONTENTS

1.1	Introduction	1
1.2	Multiscale Simulations Using IMC	4
	1.2.1 Theoretical Background	4
	1.2.2 IMC: Newton Inversion	6
	1.2.3 IMC: Reconstruction of Pair Potentials from RDFs	7
	1.2.4 Software	10
1.3	Applications of the IMC	11
	1.3.1 Simple Electrolytes	11
	1.3.2 CG Lipid Model	13
	1.3.3 CG DNA Models	17
	1.3.4 Other Systems	19
1.4	Final Remarks	21
	Acknowledgments	22
	References	22

1.1 INTRODUCTION

Modeling of many important biomolecular and soft matter systems requires consideration of length and time scales not reachable by atomistic simulations. An evident solution of this problem is introducing simplified models with lower spacial resolution, which have received a common name: *coarse-grained* (CG) models. In CG models, atoms of (macro)molecules are united into CG sites and solvent atoms are often not considered explicitly. This reduces greatly the number of degrees of freedom of the studied system and allows simulations of much larger systems which are not feasible to simulate at the atomistic level. Studies of models which can be character-ized as "coarse-grained" started at the earlier stages of molecular modeling in the 1960s and 1970s (when the term "coarse-grained" was not used at all). For example, a primitive model of electrolytes represented hydrated ions as charged spheres in a dielectric media (Vorontsov-Velyaminov and Elyashe-vich, 1966; Card and Valleau, 1970), and a simple freely jointed model of a polymer chain (Gottlieb and Bird, 1976) was used to model polymers in solution. Simple rod-like particles with two or three interaction sites were used to describe lipids in lipid bilayers and other self-assembled structures (Noguchi and Takasu, 2001; Farago, 2003; Brannigan and Brown, 2004). Such models were designed to illustrate general physical behavior of the studied systems. In order to relate such models to real physical systems, one needs to find model parameters, such as effective (hydrated) ion radius in the primitive electrolyte model, which can be done empirically by fitting to known experimental data.

Development of more advanced CG models for studies of specific molecular structures including lipids, proteins, DNA, polymers, etc., sets higher requirements for the choice of interaction potentials describing interactions in such systems. In the so-called *top-down* methodology, one is trying to parametrize the model to reproduce experimentally measur-able macroscopic properties of the system. One of the most popular mod-ern CG models of this kind is described in terms of the MARTINI force field (Marrink et al., 2004, 2007), which represents groups of about four heavy atoms by CG sites. The MARTINI force field, originally developed for lipids (Marrink et al., 2007), was later extended to proteins (Monticelli et al., 2008; de Jong et al., 2012), carbohydrates (Lopez et al., 2009), and some other types of molecules (de Jong et al., 2012; Marrink and Tieleman, 2013). Within the MARTINI force field model, CG sites interact by the elec-trostatic and Lennard-Jones potentials with parameters fitted to reproduce experimental partitioning data between polar and apolar media. This func-tional form of the force field is convenient as it coincides with that for the atomistic simulations and is implemented in all major simulation packages, but it may be also a source of problems with overstructuring of molecular coordination and with consistent description of multicomponent systems, which can in principle be solved by using softer (than Lennard-Jones) CG potentials (Marrink and Tieleman, 2013). A complicating circumstance in this respect is that (differently from atomistic models) even a functional form of the effective interaction potentials is in many cases not known *a priori.*

In the alternative *bottom-up* methodology, effective CG potentials are derived from atomistic simulations. Atomistic force fields reflect the real chemical structure of the studied system and they are generally more established than CG force fields. Furthermore, the *bottom-up* methodology can use as a starting point an even deeper, quantum-chemical level of modeling. Within the *bottom-up* methodology, a CG force field is parameterized to fit some important physical properties that result from a high-resolution (atomistic) simulation. Several *bottom-up* approaches to parametrize CG force fields have been formulated recently. Within the force-matching approach (Ercolessi and Adams, 1994; Izvekov et al., 2004), also called multiscale coarse-graining (Izvekov and Voth, 2005; Ayton et al., 2010)), the CG potential is built in a way to provide the best possible fit to the forces acting on CG sites in the atomistic simulation. Within the inverse Monte Carlo (IMC) technique (Lyubartsev and Laaksonen, 1995, 2004) and similar renormalization group coarse-graining (Savelyev and Papoian, 2009b), as well as in the related iterative Boltzmann inversion (IBI) method (Soper, 1996; Reith et al., 2003), the target property is the radial distribution functions (RDFs) as well as internal structural properties of molecules such as distributions of bond lengths, covalent angles, and torsion angles. For this reason, methods based on IMC or IBI techniques are called structure-based coarse-graining. In the relative entropy minimization method (Shell, 2008; Chaimovich and Shell, 2011), the CG potential is defined by a condition to provide minimum entropy change between the atomistic and CG system, which is also equivalent to minimizing information loss in the coarse-graining process. In the conditional work approach (Brini et al., 2011), the effective potentials between CG sites are obtained as free energy (potentials of mean force) between the corresponding atom groups determined in a thermodynamic cycle. More details of different *bottom-up* multiscale methodologies and their analysis can be found in a recent review (Brini et al., 2013).

This chapter is devoted to the systematic coarse-graining methodology based on the IMC method. Here, we discuss the term "inverse Monte Carlo" only in application to multiscale coarse-graining, while there exists a more general definition of "IMC" as a solution of any inverse problem by any type of Monte Carlo (MC) technique (Dunn and Shultis, 2012). In Section 1.2, the theoretical background, practical algorithms, problems, and possible limitations of the approach are considered. Section 1.3 considers various applications of the IMC methodology to the ionic systems, lipid assemblies, DNA, and a few other systems. Finally, perspectives and possible limitations of the methodology are discussed.

1.2 MULTISCALE SIMULATIONS USING IMC

1.2.1 Theoretical Background

Generally, a problem of going from a high-resolution (atomistic) description to a CG one can be formulated as follows. Assume that at the high-resolution level, the system is described by a Hamiltonian (potential energy) $H(\{r_i\})$, where $\{r_i\}, i = 1, ..., n$ are the coordinates of atoms. The potential energy function represents typically the atomistic force field, but in the case of *ab-initio* modeling it can represent the energy surface obtained within any quantum-chemical computation method. A coarse-graining is described in terms of mapping of atomistic coordinates (degrees of freedom) $\{r_i\}_{i=1,...,n}$ to CG coordinates $\{R_j\}_{j=1,...,N}$, which is mathematically expressed in terms of mapping functions $R_j = R_{\text{map}(j)}(\{r_i\})$. Often CG coordinates R_j are center of masses of groups of atoms united in CG site j, but they may just coincide with coordinates of some selected atoms, or other choices can be made. The original Hamiltonian $H(\{r_i\})$ defines all properties of the high-resolution system, and through the mapping functions, all properties of the CG system. The task is to define effective interaction potential for CG sites, which provide the same properties for the CG system as the properties which follows from the CG mapping of the high-resolution system.

If only structural properties are a matter of interest, an exact formal solution of the above formulated problem can be written in terms of N-body potential of mean force. The N-body potential of mean force is obtained by inclusion of the CG degrees of freedom into the partition function of the original high-resolution system and subsequent integration over atomistic coordinates:

$$Z = \int \prod_{i=1}^{n} dr_i \exp(-\beta H(\{r_i\})) =$$

$$= \int \prod_{i=1}^{n} dr_i \int \prod_{j=1}^{N} dR_j \delta(R_j - R_{\text{map}(j)}(\{r_i\})) \exp(-\beta H(\{r_i\}))$$

$$= \int \prod_{j=1}^{N} dR_j \exp(-\beta H_{\text{CG}}(\{R_j\})) \qquad (1.1)$$

with $\beta = 1/k_B T$ and N-body potential of mean force (CG Hamiltonian) $H_{\text{CG}}(\{R_j\})$ defined by

$$H_{\text{CG}}(\{R_j\}) = -\frac{1}{\beta} \ln \int \prod_{i=1}^{n} dr_i \delta(R_j - R_{\text{map}(j)}(\{r_i\})) \exp(-\beta H(\{r_i\})) \qquad (1.2)$$

A CG system with a Hamiltonian (Equation 1.2) has the same structural properties (canonical averages) for the CG degrees of freedom as the underlying high-resolution system with the original Hamiltonian. Since the CG and high-resolution systems have the same partition function, the thermodynamic properties (average energy, free energies, pressure) can also be reconstructed. The important point is, however, that the CG Hamiltonian depends on the thermodynamic conditions (temperature, concentration) and these dependences need to be taken into account while obtaining thermodynamic properties by derivation of the partition functions by the thermodynamics parameters. Reconstruction of the correct dynamics in the CG system is a more challenging task. In addition to renormalization of the Hamiltonian according to Equation 1.2, the dynamic equations of motion need to be changed to the generalized Langevin equation with a memory function (Romiszowski and Yaris, 1991). Some practical approaches to dynamic coarse-graining within dissipative particle dynamics can be found in the literature (Eriksson et al., 2008; Hijon et al., 2010).

As discussed above, the N-body potential of mean force (Equation 1.2) provides an exact solution of the coarse-graining problem; however, in practical terms simulations involving an N-body potential are infeasible. A common way to proceed is to approximate it by a more convenient expression, for example, by a sum of distance-dependent pair potentials:

$$H_{CG}(R_1, ..., R_N) \approx \sum_{i<j} V_{ij}^{eff}(R_{ij}) \tag{1.3}$$

with $R_{ij} = |R_i - R_j|$. It is, however, not necessary to be limited by pair potentials; any practically usable expression can be given in Equation 1.3, for example, angle or torsion potential terms for macromolecular CG models, or some other simple forms expressing three or four body interactions. From the computational point of view, we are first interested in pair-wise approximations: the very aim of coarse-graining is computational speed-up, and extensive use of many-body potentials would greatly hamper this goal.

The task of building a CG force field can be thus reformulated to find an "as best as possible" approximation according to Equation 1.3. Definitions of what "the best approximation" can be, however, differ. Usually one determines a set of target properties that one wish the CG model to keep. These properties can be either of a microscopical character, as forces or instantaneous energies, or canonical averages as RDFs, average energies, or pressure. For example, minimizing the force difference coming from both sides of Equation 1.3 (weighted with the Boltzmann factor) is equivalent to the force-matching method (Ercolessi and Adams, 1994; Izvekov et al., 2004; Izvekov and Voth, 2005). Within the structure-based coarse-graining approach, the target is various structural properties, such as RDFs as well as distributions of internal degrees of freedom in complex molecules. In principle, other properties of interest or any combination of them can be used for parameterization of effective potentials.

1.2.2 IMC: Newton Inversion

The task of determining computationally feasible CG potentials can thus be formulated in the following way (Lyubartsev et al., 2010). Assume our effective potentials $H(\{r_i\})$ (we now remove index "CG" from the notations) are determined by a (finite) set of parameters $\{\lambda_\gamma\}$, and the set of target properties (which we know from the atomistic simulations) is $\{A_\alpha^{ref}\}$. We assume here that the number of potential parameters $\gamma = 1, ..., M$ is equal to the number of target properties $\alpha = 1, ..., M$. If we know the set of $\{\lambda_\gamma\}$, we can always compute average properties $\{\langle A_\alpha \rangle\}$ in direct molecular dynamics (MD) or MC simulations. The inverse problem, finding parameters $\{\lambda_\gamma\}$ from averages $\{\langle A_\alpha \rangle\}$, is less trivial. We can consider the relationship between $\{\lambda_\gamma\}$ and $\{\langle A_\alpha \rangle\}$ as a nonlinear multidimensional equation, and use the Newton inversion method (known also as the Newton–Raphson method) to solve it iteratively.

The method is based on the expression relating small changes of the potential parameters λ_γ and changes of the canonical energies $\langle A_\alpha \rangle$ caused by these changes of the potential parameters:

$$\Delta \langle A_\alpha \rangle = \sum_\gamma \frac{\partial \langle A_\alpha \rangle}{\partial \lambda_\gamma} \Delta \lambda_\gamma + O(\Delta \lambda^2) \qquad (1.4)$$

The matrix of derivatives $\frac{\partial \langle A_\alpha \rangle}{\partial \lambda_\gamma}$ (Jacobian) can, by the use of statistical mechanics expressions for averages in canonical ensemble, be presented in the following form (Lyubartsev et al., 2010; Wang et al., 2013a):

$$\frac{\partial \langle A_\alpha \rangle}{\partial \lambda_\gamma} = \frac{\partial}{\partial \lambda_\gamma} \frac{\int \prod_{i=1}^N dr_i A_\alpha(\{\lambda_\gamma\}, \{r_i\}) \exp(-\beta H(\{r_i\}))}{\int \prod_{j=1}^N dr_j \exp(-\beta H(\{r_j\}))} =$$

$$= \left\langle \frac{\partial A_\alpha}{\partial \lambda_\gamma} \right\rangle - \beta \left(\left\langle \frac{\partial H}{\partial \lambda_\gamma} A_\alpha \right\rangle - \left\langle \frac{\partial H}{\partial \lambda_\gamma} \right\rangle \langle A_\alpha \rangle \right) \qquad (1.5)$$

Now the Jacobian is expressed in terms of canonical averages which all can be evaluated by running direct simulation with a Hamiltonian defined by a given set of parameters $\{\lambda_\gamma\}$. Equations 1.4 and 1.5 can be used to solve the inverse problem iteratively. One starts from some initial potential determined by a trial set of parameters $\{\lambda_\gamma^{(0)}\}$, runs a simulation, and computes the deviation of computed values $\{\langle A_\alpha^{(0)} \rangle\}$ from the target values A_α^{ref}:

$$\Delta \langle A_\alpha \rangle = A_\alpha^{\text{ref}} - \{\langle A_\alpha^{(0)} \rangle\} \qquad (1.6)$$

We also compute a matrix of derivatives (Equation 1.5). Then, the system of linear equations (Equation 1.4) is solved neglecting second-order corrections, resulting in corrected values of parameters λ_γ:

$$\lambda_\gamma^{(n+1)} = \lambda_\gamma^{(n)} + \Delta \lambda_\gamma \qquad (1.7)$$

The procedure is repeated until convergence is reached. If initial approximation $\{\lambda_\gamma^{(0)}\}$ is poor, some regularization of the iterative procedure might be necessary, in which the difference in Equation 1.6 is multiplied by some factor between 0 and 1.

While the Newton inversion procedure was depicted above for the transition from the atomistic to the CG level, it can work in the same way for the connection between *ab initio* and atomistic levels, with the only difference that the quantum-mechanical energy surface is used instead on the N-body potential of mean force in Equation 1.3.

The Newton inversion algorithm can be straightforwardly implemented if the number of potential parameters is equal to the number of target properties. If the number of properties exceeds the number of potential parameters, the problem can be solved in the variational sense, by finding a set of $\{\lambda_\gamma\}$ which provides the least possible deviation of the computed properties from the target values. Optimization using the same equations (Equation 1.5) can in this case be carried out according to the Gauss–Newton algorithm. This approach, under the name the force balance method, has been recently implemented for parametrization of a water model from ab-initio and experimental data (Wang et al., 2013a).

1.2.3 IMC: Reconstruction of Pair Potentials from RDFs

An important case of the general approach described above is when parameters $\{\lambda_\gamma\}$ are values of the pair potential in a regular set of points covering the whole range of distances (i.e., the potential is given in a table form), and the target properties are values of the RDF in the same set of points. Then, the inverse problem is reformulated as finding the pair interaction potential which reconstructs the given RDF. For a multicomponent case, a set of pair interaction potentials is reconstructed from a set of RDFs between the same CG site types. Even intramolecular interactions such as bond, angular, and torsion potentials can be linked to the distribution of the corresponding bond lengths, angles, and torsions and included in the inversion procedure. The equations given in the previous section become equivalent in this case to the IMC algorithm introduced previously (Lyubartsev and Laaksonen, 1995, 2004) and described below.

We consider a class of systems which is described by a Hamiltonian (potential energy) in the following form:

$$H = \sum_{\alpha} V_{\alpha} S_{\alpha} \tag{1.8}$$

The Hamiltonian of any system with pair interaction can be presented in such form when the pair potential is given by a set of tabulated values as a step-wise function of distance:

$$\tilde{V}(r) = V(r_{\alpha}) \equiv V_{\alpha}$$

for

$$r_{\alpha} - \frac{1}{2M} < r < r_{\alpha} + \frac{1}{2M}; \quad r_{\alpha} = (\alpha - 0.5)r_{cut}/M; \quad \alpha = 1, ..., M \tag{1.9}$$

where r_{cut} is some cutoff distance and M is the number of grid points within the interval $[0, r_{cut}]$. The S_{α} values represent the number of particles pairs with the distances found inside α-slice. Evidently, S_{α} is an estimator of the RDF: $\langle S_{\alpha} \rangle = 4\pi r^2 \rho(r) N^2/(2V)$. Thus, the inverse problem is formulated now as finding the values of the interaction potential in grid points V_{α} from RDFs expressed in averages $\langle S_{\alpha} \rangle$.

With these notations, the Jacobian matrix of Newton inversion (Equation 1.5) becomes

$$\frac{\partial \langle S_{\alpha} \rangle}{\partial V_{\gamma}} = -\beta \left(\langle S_{\alpha} S_{\gamma} \rangle - \langle S_{\alpha} \rangle \langle S_{\gamma} \rangle \right) \tag{1.10}$$

Here, indexes α, γ are running over all interaction types (nonbonded, bonded, angular), within each interaction type over all CG types of sites, and for each set of CG types, over the relevant range of distances (or angles). In all cases, average values of $\langle S_{\alpha} \rangle$ and cross-correlation terms $\langle S_{\alpha} S_{\gamma} \rangle$ can be acquired from a simulation of the CG system. In the inverse procedure, one starts from a trial set of potentials (in practical simulations, one can start either from zero or from mean force potentials), runs an MC simulation, computes RDFs expressed in terms of average values of $\langle S_\alpha \rangle$ as well as cross-correlation terms according to Equation 1.10, solves a system of linear equations (Equation 1.4) (with substitution $\lambda_\alpha \to V_\alpha$ and $A_\alpha \to S_\alpha$), obtains new corrections to the interaction potential ΔV_α, and repeats the procedure until convergence.

Another approach to invert RDFs as well as bond or angle distributions was described earlier by Schommers (1983), reintroduced by Soper under the name empirical potential structure refinement method (Soper, 1996), and become most known as the iterative Boltzmann inversion Reith et al. (2003). As in the IMC method, one starts a simulation with some trial values of the potential $V_\alpha^{(0)}$, and corrects at each iteration the potential according to

$$V_\alpha^{(i+1)} = V_\alpha^{(i)} + k_B T \ln \frac{\langle S_\alpha^{(i)} \rangle}{S_\alpha^{\text{ref}}} \qquad (1.11)$$

Correction of potential according to Equation 1.11 is straightforward to implement, and such an approach was used in a number of studies (Shelley et al., 2001; Reith et al., 2003; Harmandaris et al., 2006; Carbone et al., 2008; Wang and Deserno, 2010). In the IBI approach, correction to the potential is determined only by the value of the same distribution function at the same distance point, which is why the IBI approach faces convergence problem in the multicomponent case (Hess et al., 2006a) where different RDFs can be strongly interconnected. In practical calculation of CG potentials by RDF inversion, it might be instructive to start the iterative process using an IBI approach, which brings the system RDFs closer to the reference values, and then switch to IMC, which takes into account correlations between different distribution functions and provides better convergence when the RDFs become close to the reference functions.

It is known that for the relationship between pair potential and RDF, solution of the inverse problem is unique with the precision of an additive constant to the potential. This was proven previously by Henderson for the monocomponent case (Henderson, 1974), and generalized later for a multicomponent case and intramolecular interactions (Rudzinski and Noid, 2011). On the other hand, the inversion problem for the RDF–pair potential relation is often ill-defined; different potentials may in some cases produce RDFs which are very close to each other (Soper, 1996). For this reason, the IMC and IBI approaches can produce different results for effective potentials while having RDFs undistinguishable by eye on a graph (Rühle et al., 2009; Wang et al., 2013b). Still, even a small difference in RDF, especially at large distances, may be of importance in correct reproduction of the Kirkwood–Buff integral (which is determined by expression $\int 4\pi r^2(g(r) - 1)dr$), which is important for consistent description of thermodynamics of mixtures (Mukherji et al., 2012). Within the IMC approach, the target function is proportional to $r^2 g(r)$, which is why the Kirkwood–Buff integrals are well reproduced in the CG simulations, while the IBI method is less sensitive to the behavior of RDF at large distances, and may need corrections in order to reproduce Kirkwood–Buff integrals accurately (Ganguli et al., 2012).

1.2.4 Software

The IMC method is currently implemented in two open source software packages: versatile object-oriented toolkit for coarse-graining applications (VOTCA) (Rühle et al., 2009) and MagiC (Mirzoev and Lyubartsev, 2013). The VOTCA package (which also implements force-matching and IBI methods) uses GROMACS (Lindahl et al., 2001) as a sampling engine for computations of necessary canonical averages at each iteration of the inverse procedure. It analyzes the trajectory obtained by GROMACS with trial potentials, computes necessary averages, and computes the updated tabulated potentials to be used at the next iteration. A problem related to the use of MD to sample system configurations during the IMC procedure is that standard MD software such as GROMACS has certain requirements to the smoothness of the used tabulated potentials while in the IMC procedure the potentials may change unpredictably during the optimization procedure and thus be a source of instability.

The MagiC package was developed specially to implement systematic structure-based coarse-graining of arbitrary molecular models using the IMC or IBI methodology. As input, MagiC uses atomistic trajectories generated by any simulation software, from which it computes necessary reference RDFs between CG sites as well as bond and angle distributions. MagiC has its own MC multithread sampling engine which can run simultaneously many copies of the simulated system on each available processor/core. Simulations are run for a trial set of CG potentials, which may include nonbonded interactions between all CG site types, as well as bonded and angle intramolecular interactions. The electrostatic interactions are taken out of the inversion scheme and are treated by the conventional Ewald summation method (Allen and Tildesley, 1987). The program evaluates the canonical averages necessary for RDF inversion (Equation 1.10) by averaging results generated from all the threads and computes updated potentials which are distributed back to all the threads. The use of multithread methodology improves the quality of sampling, resulting in better stability and faster convergence of the iteration procedure. Recently, MagiC has been used for computations of effective potentials for CG models of lipids (Mirzoev and Lyubartsev, 2014), CG DNA and solvent-mediated DNA–ion interactions (Korolev et al., 2014; Naomé et al., 2014), and a CG model of an ionic liquid (Wang et al., 2013b).

1.3 APPLICATIONS OF THE IMC

1.3.1 Simple Electrolytes

As a first example of application of the IMC methodology, we consider computations of effective solvent-mediated potentials of ions in aqueous solution (Lyubartsev and Laaksonen, 1995, 1997). We recapitulate this study here because of its simplicity and instructive character. One of the typical approximations used in the description of macromolecules and particularly polyelectrolytes is substituting of solvent molecules by a continuum media. For example, in the continuum (called also primitive) electrolyte model, ions in water are substituted by charged spheres moving in dielectric media with a proper dielectric constant. Evidently, this is a serious simplification at a small (a few Å) distance between the ions where it is impossible to define a dielectric constant in a consistent manner. Moreover, an ion radius (in terms of hard sphere, or softer repulsive r^{-12} or r^{-9} potential) within the continuum electrolyte model is an adjustable parameter without clear physical meaning.

A better model of effective ion–ion interactions in aqueous solution must take into account the solvation structure of water around the ions. Practically, effective solvent-mediated ion–ion potentials may be constructed by the IMC method from ion–ion RDFs, generated in high-quality atomistic MD simulations of ions in water. This approach has been already implemented in the first paper describing the IMC technique (Lyubartsev and Laaksonen, 1995). In subsequent publications (Lyubartsev and Laaksonen, 1997; Lyubartsev and Marcelja, 2002; Mirzoev and Lyubartsev, 2011), the effective-solvent mediated potentials for NaCl aqueous solution have been calculated with greater precision as well as for a number of concentrations and temperatures.

An example of computation of solvent-mediated ion–ion potentials for NaCl aqueous solution is illustrated in Figure 1.1. The underlying MD simulations have been performed for the flexible simple point charge (SPC)

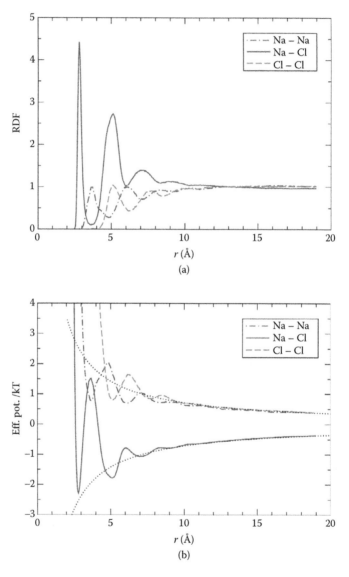

FIGURE 1.1   (a) and (b) Radial distribution functions (RDFs) between Na^+ and Cl^- ions in water computed from atomistic simulations at temperature 298 K and ion concentration 0.5 M (Lyubartsev and Laaksonen, 1997), and corresponding effective potentials derived from these RDFs by the inverse Monte Carlo (IMC) method. Dotted lines show Coulombic potential at dielectric permittivity 80. (Compiled from data in Lyubartsev, A. P. and A. Laaksonen, *Phys. Rev. E*, 55, 5689–5696, 1997.)

water model (Toukan and Rahman, 1985) and Smith–Dang parameters for Na^+ and Cl^- ions (Smith and Dang, 1994) resulting in the ion–ion RDFs shown in Figure 1.1a. They were fed into an IMC procedure resulting in effective potentials between the ions shown in Figure 1.1b. The effective potentials make one to two oscillations, thereby reflecting the molecular nature of the solvent, and then finally approach the primitive model potential with dielectric constant close to 80. With distances more than 10 Å, the effective potentials almost perfectly coincide with the Coulombic potential. These characteristic features of ion–ion solvent-mediated potentials were also observed in other works (Hess et al., 2006a,b; Savelyev and Papoian, 2009a).

An important issue of CG effective potentials is their state-point dependence, which originates in integration over "nonimportant" degrees of freedom (see Equation 1.1). Studies of solvent-mediated ion–ion potentials showed that they do depend both on the ion concentration (Lyubartsev and Laaksonen, 1997) and on the temperature (Mirzoev and Lyubartsev, 2011). It was, however, shown that most of this dependence can be taken care of by introducing concentration (Hess et al., 2006b) and temperature-dependent (Mirzoev and Lyubartsev, 2011) dielectric permittivity. This can be done by considering the short-range part of the effective potential,

$$V_{sh}(r) = V_{tot}(r) - \frac{q_i q_j}{4\pi\epsilon_0 \epsilon r} \tag{1.12}$$

and optimizing dielectric permittivity ϵ by the requirement that the short-range part of all three ion–ion potentials be most close to zero at distances outside 10 Å (or a similar cutoff distance). Besides definition of the transferable short-range part of the effective potential, this approach provides also an alternative definition of effective dielectric permittivity of a solvent. It was demonstrated in a paper (Mirzoev and Lyubartsev, 2011) that such a definition of the dielectric constant is consistent with conventional computations of dielectric permittivity from the dipole moment fluctuations, as well as with experimental data.

1.3.2 CG Lipid Model

Simulations of lipid membranes and other self-assembled structures have attracted much attention during the last decade due to the fact that lipid membranes form the outer shells of living cells (Lyubartsev and Rabinovich, 2011). However, atomistic simulation of even a relatively small piece of membrane consisting of a few hundred lipids and surrounding water is a computational challenge, while many actual biophysical problems, such as studies of inhomogeneous membrane mixtures, membrane mechanical properties, association and partitioning of polypeptides, nanoparticles, other membrane bound compounds, etc., require consideration of substantially larger membrane fragments. For investigation of all these phenomena in molecular simulations, a coarse-grain level of modeling provides practically the only possible choice.

A large variety of various CG models of lipids differing by the level of detail and the way of defining CG potentials have been reported in the last two decades (Drouffe et al., 1991; Goetz and Lipowsky, 1998; Shelley et al., 2001; Murtola et al., 2004; Marrink et al., 2004; Lyubartsev, 2005; Cooke and Deserno, 2005; Izvekov and Voth, 2005; see also the reviews of Pandit and Scott, 2009; Shinoda et al., 2012). In several cases, interaction potentials were determined within a systematic *bottom-up* approach from atomistic simulations, using IBI (Shelley et al., 2001), force matching (Izvekov and Voth, 2005), and IMC (Lyubartsev, 2005). In one work (Lyubartsev, 2005), a dimyristoylphosphatidylcholine (DMPC) lipid was CG to a 10-site model (see Figure 1.2). The CG site–site RDFs as well as CG bond-length distributions were computed from atomistic MD simulations of 16

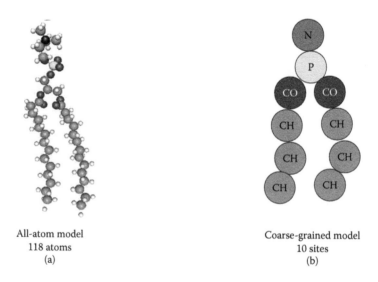

All-atom model
118 atoms
(a)

Coarse-grained model
10 sites
(b)

FIGURE 1.2 Atomistic (a) and coarse-grained (b) lipid models. (With kind permission from Springer Science+Business Media: *Eur. Biophys. J.*, Multiscale modeling of lipids and lipid bilayers, 35, 53, 2005, Lyubartsev, A. P.)

lipids dissolved in water and described by the CHARMM27 force field, and were inverted simultaneously within the IMC procedure, resulting in 10 nonbonded potentials between four different types of CG sites and four bonded potentials. The resulting CG model has showed the ability to reproduce a bilayer structure consistent with atomistic simulations, and to describe self-assembly of lipids into bicell as well as the formation of spherical vesicle structures (Figure 1.3).

In subsequent development (Mirzoev and Lyubartsev, 2014), the effective potentials derived in work (Lyubartsev, 2005) were reparameterized after recomputation of CG site–site RDFs according to recent modification of the CHARMM27 force field described in Högberg et al. (2008). This modification of the CHARMM force field has been done with a primary aim to improve agreement with experiments for atomistic simulations of the lipid bilayer, and to reproduce correctly average area per lipid

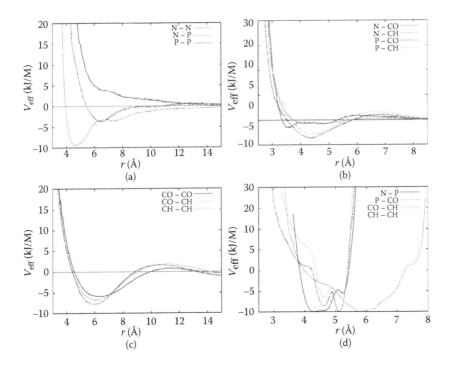

FIGURE 1.3 Potentials for various atom pairs are shown in panels (a)-(d). Effective potentials for the coarse-grained (CG) lipid model displayed in Figure 1.2 computed by the IMC method from RDF and bond-length distributions determined in atomistic simulations. (With kind permission from Springer Science+Business Media: *Eur. Biophys. J.*, Multiscale modeling of lipids and lipid bilayers, 35, 53, 2005, Lyubartsev, A. P.)

in particular. Atomistic simulations of DMPC lipid mixtures with water at different molar ratios were run up to 400 ns in order to ensure reliable converged RDFs. The set of intramolecular potentials of previous work (Lyubartsev, 2005) was complemented by angular potentials determined from distribution of relevant angles between CG sites. It was found that effective potentials obtained from atomistic simulations carried out at different concentrations, and properties of lipid bilayers simulated using these potentials, show nonnegligible concentration dependence. Thus, potentials based on low lipid concentration overestimate the effective hydrophobic attraction of the lipid tails, which favors a more gel-like and more ordered structure of the bilayer. The potentials based on higher lipid concentration in the atomistic simulations provide more fluid-like structure with larger area per lipid. The best agreement with reference data as well with experiment was achieved with a set of potentials derived from atomistic simulations at 1:30 lipid:water molar ratio, which also provides full saturating hydration of the DMPC headgroup in the bilayer. Comparison of some characteristics of the DMPC lipid bilayer obtained in atomistic and CG lipid models is given in Table 1.1.

Despite a certain degree of the state-point dependency of the effective potentials, all the derived potentials obtained in conditions of unordered lipid–water mixture provided a stable bilayer structure with correct partitioning of different lipid groups across the bilayer as well as with acceptable values of the average lipid area, orientational tail ordering, and

TABLE 1.1 Comparison of Some Properties of Lipid Bilayer Obtained in Atomistic and CG Simulations for Four CG Models Derived by IMC from Atomistic Simulations at Different Concentrations.

Model	A (Å^2)	K_A (mN/m)	D (Å)	S1	S2
CG 1:100	49.1	910	38.8	0.73	0.70
CG 1:50	50.1	910	40.2	0.77	0.77
CG 1:30	59.7	370	38.0	0.57	0.52
CG 1:20	65.7	190	37.6	0.52	0.45
Atomistic	60.0	250	39.6	0.58	0.52

Source: Mirzoev, A. and A. P. Lyubartsev, J. Comput. Chem., 35, 1208–1218, 2014.
Note: A, average area per lipid; K_A, compressibility; D, membrane thickness determined from the distribution maxima of N-sites; S1 and S2, order parameters defined by vectors connecting CO and the second CH sites, and the first and the third CH sites of lipid tails, respectively. For details, see Mirzoev and Lyubartsev (2014).

compressibility. They also demonstrated the ability of CG lipids to self-assemble in bilayer, bicell, or vesicle structures depending on simulation condition as is, e.g., displayed in Figure 1.4. This behavior of the CG lipid model, derived using an IMC approach, was reached without use of any additional information except that which was available from atomistic simulations.

1.3.3 CG DNA Models

The DNA molecule was studied by computer simulations methods for many decades. DNA is a strongly charged polyelectrolyte, and understanding of many properties, including DNA packing in the cell nuclei, requires proper description of both long-range electrostatic interactions

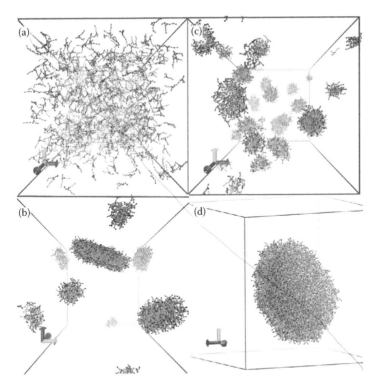

FIGURE 1.4 Trajectory snapshots are shown in panels (a)-(d), as the CG simulation is evolved in time. Formation of a bicell from initially random distribution of CG lipids observed in Langevin molecular dynamics simulations of work. (From Mirzoev, A. and A. P. Lyubartsev. Systematic implicit solvent coarse graining of dimyristoylphosphatidylcholine lipids. *J. Comput. Chem.* 2014. 35. 1208–1218. Copyright Wiley-VCH Verlag GmbH & Co. KGaA. Reproduced with permission.)

and chemical (atomistic) details of DNA interactions with surrounding molecules including ions and histones (Korolev et al., 2012). This necessitates the use of multiscale approaches when continuum solvent CG models are robust enough to catch the atomistic chemically specific details and effect of water hydration.

In one paper (Lyubartsev and Laaksonen, 1999), the IMC approach was used to derive effective solvent-mediated potentials between alkali ions and sites of CG DNA model where each nucleotide was presented as three sites representing phosphate, base, and sugar. The RDFs between ions and these sites used as an input to the IMC procedure were computed in atomistic simulations of a DNA fragment in ionic solution. The DNA periodic fragment used in this study was fixed, and the obtained ion–DNA solvent-mediated potentials were used to model binding affinity of different alkali ions to DNA.

A flexible CG DNA model, presented as two connected helical chains representing DNA phosphate groups, was considered in a series of works from the Papoian group (Savelyev and Papoian, 2009b, 2010; Savelyev et al., 2011). Both solvent-mediated ion–DNA and internal DNA interactions were determined from the structural information obtained in atomistic simulations of DNA in aqueous ion solution using the molecular renormalization group approach, which is in most details equivalent to the IMC method. This model was used primarily to study DNA persistence length at different ion conditions. A similar DNA–ion model has been considered in a recent paper (Naomé et al., 2014) with ion–CG DNA and internal DNA interactions computed by the combined iterative Boltzmann/IMC method from atomistic simulations of a DNA oligonucleotide.

A CG DNA model considered in a recent study (Korolev et al., 2014) included, along with CG sites representing phosphate groups, space filling sites located along DNA axis with each such site representing two base pairs (see Figure 8.5a). The internal structure of DNA was maintained by three bond and three angular potentials between neighboring CG sites. The reference atomistic simulations were carried out for four DNA nucleotides which gave the possibility to extract effective DNA–DNA site–site interactions which are intended to be used in perspective studies of nucleosome folding. It was also found in a paper (Korolev et al., 2014) that the internal DNA potentials can be well fitted to harmonic potentials, which gives the possibility to describe the internal CG DNA interactions in terms of a standard molecular mechanics force field implemented in most MD software.

In another recent work (Biase et al., 2014), the IMC approach was used to parametrize effective potentials for the CG model of a single-strand DNA with CG sites representing phosphate, sugars, and four distinguishable types of DNA bases. The obtained model was used to describe longtime dynamics of a single-strand DNA in a nanopore.

1.3.4 Other Systems

Besides computations of effective potentials for CG models, there were a few attempts to use the IMC approach to derive atomistic potentials from *ab-initio* (Car–Parrinello type) simulations. In on paper (Lyubartsev and Laaksonen, 2000), the RDFs between the O and H atoms of water, obtained in Car–Parrinello molecular dynamics (CPMD) (Car and Parrinello, 1985) simulations of 32 water molecules, were inverted to produce atomistic interaction potentials which turned out to be very similar to potentials of conventional SPC and TIP3P water models. Later unpublished studies have showed, however, that a "ab-initio" derived water model could not compete with the traditional empirically parametrized water models mainly because of deficiencies in the density functional theory (DFT) functionals used in CPMD simulations. In another work (Lyubartsev et al., 2001), CPMD simulations of a Li^+ ion in water were used to extract the interaction potential between the Li^+ ion and water oxygen. The nonelectrostatic part of the Li^+–O potential was found to be well approximated by an exponential function like the one used in the Buckingham potential (see Figure 1.5). The *ab-initio* derived model of Li^+ ion interaction was further used in MD simulations of a LiCl solution (Egorov et al., 2003).

The IMC method in two dimensions was used in a work (Murtola et al., 2007) to study domain formation in a lipid bilayer containing cholesterol. Two-dimensional RDFs of lipids and cholesterol center-of-masses projections to XY plane were determined in atomistic simulations, and were then used to determine effective potentials for a two-dimensional model of a lipid–cholesterol mixture. This two-dimensional model was further used to study formation of cholesterol-rich and cholesterol-poor domains in a mixed lipid–cholesterol bilayer.

In Wang et al. (2013b), the IMC method was used to build potentials for a CG model of $Bmim^+$-PF_6^- ionic liquid. The $Bmim^+$ cation of ionic liquid was presented as three beads and the PF_6^- anion as a single bead as depicted in Figure 1.6. As in other cases, RDFs for the IMC procedure were obtained in atomistic simulations of this system. The obtained effective potentials were used in large-scale simulations of $Bmim^+$-PF_6^- ionic liquid in

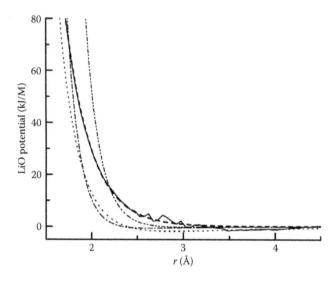

FIGURE 1.5 Short-range part of the effective LiO potential. Solid line—derived from *ab initio* simulation using the inverse Monte Carlo; bold dashed line—exponential fit; and other lines—potentials from other works, see Lyubartsev et al. (2001) for details. (Reprinted with permission from Lyubartsev, A. P. et al., *J. Chem. Phys.*, 114, 3120. Copyright 2001, American Institute of Physics.)

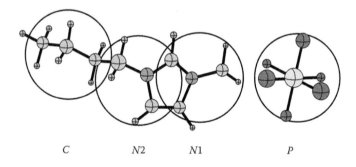

FIGURE 1.6 Coarse-graining scheme of $Bmim^+$-PF_6^- ionic liquid. (From Wang, Y.-L. et al., *Phys. Chem. Chem. Phys.*, 15, 7701, 2013. Reproduced by permission of The Royal Society of Chemistry.)

which, among other properties, experimental x-ray scattering factors were reproduced.

A few other examples of using IMC for reconstruction of interaction potentials from RDF can be mentioned: computation of effective potentials between charged colloids (Lobaskin et al., 2001), single-site water model (Eriksson et al., 2008), proline molecules in dimethyl sulfoxide (DMSO) solvent (Lyubartsev et al., 2010). In Zhang and Berkowitz (2009), atomistic interaction potentials in Ag–Rh were determined by IMC from RDFs obtained from experimental scattering data.

1.4 FINAL REMARKS

In this chapter, the IMC methodology to extract interaction potentials from the structural properties of a molecular system (RDFs, bond and angles distributions) has been presented and its role in multiscale modeling is discussed. Examples given in this chapter show that the method is very general and can find applications for a very wide variety of molecular and macromolecular systems. In principle, the IMC method can serve as a key element in the systematic hierarchical multiscale modeling approach, which starts from *ab-initio* level, produce interaction potentials for classical atomistic simulations, and then proceed to different levels of coarse-graining. It might be, however, unrealistic to believe that the approach can produce CG models suitable for large-scale macromolecular simulations completely in an *ab-initio* manner, so experimental input as well as experimental validation will be always necessary. There are several reasons for this: the approximate nature of DFT functionals or atomistic force fields used at the starting stage; limited sampling of the phase space during RDF calculations; and inherent limitations of the coarse-graining procedure itself which in a typical case neglects high-order terms in the N-body potential of mean force. Transferability of the CG potentials, that is, their dependence on the temperature, concentration, and composition of the system, may be a serious issue. In some cases, like in ionic solutions, temperature and concentration dependence of the effective potentials can be effectively included into effective dielectric permittivity (Hess et al., 2006b; Mirzoev and Lyubartsev, 2011); in other cases, such as in a CG lipid model (Mirzoev and Lyubartsev, 2014), transferability studies need to be carried out. A reasonable strategy to construct effective potentials for CG models might be to determine the general form of the effective potential using IMC or other type of *bottom-up* coarse-graining, and then fine-tune parameters of the potential to fit available experimental data. The IMC methodology presented here would help to increase the fraction of "*ab initio*" derived features in CG molecular models at the expense of "empirically fitted" or "*ad hoc*" ones, which would enhance the predictive character and reliability of large-scale molecular simulations and advance them further to problems not yet covered within today's molecular simulation techniques.

ACKNOWLEDGMENTS

This work has been supported by the Swedish Research Council (Vetenskapsrådet).

REFERENCES

Allen, M. P., and D. J. Tildesley (1987). *Computer Simulations of Liquids* (2nd ed.). Oxford: Clarendon.

Ayton, G. S., E. Lyman, and G. A. Voth (2010). Hierarchical coarse-grained strategy for protein-membrane systems to access mesoscopic scales. *Faraday Discuss. 144*, 347–358.

Biase, P. M. D., S. Markosyan, and S. Noskov (2014). Microsecond simulations of DNA and ion transport in nanopores with novel ion–ion and ion–nucleotides effective potentials. *J. Comput. Chem. 35*, 711–721.

Brannigan, G., and F. L. H. Brown (2004). Solvent-free simulations of fluid membrane bilayers. *J. Chem. Phys. 120* (2), 1059–1071.

Brini, E., E. A. Algaer, P. Ganguly, C. Li, F. Rodriguez-Ropero, and N. F. A. van der Vegt (2013). Systematic coarse-graining methods for soft matter simulations a review. *Soft Matter 9*, 2108–2119.

Brini, E., V. Marcon, and N. F. A. van der Vegt (2011). Conditional reversible work method for molecular coarse graining applications. *Phys. Chem. Chem. Phys. 13*, 10468–10474.

Car, R., and M. Parrinello (1985). Unified approach for molecular dynamics and density functional theory. *Phys. Rev. Lett. 55* (22), 2471–2474.

Carbone, P., H. A. K. Varzaneh, X. Chen, and F. Müller-Plathe (2008). Transferability of coarse-grained force fields: The polymer case. *J. Chem. Phys. 128*, 064904.

Card, D. N., and J. P. Valleau (1970). Monte Carlo study of the thermodynamics of electrolyte solutions. *J. Chem. Phys. 52*, 6232–6240.

Chaimovich, A., and M. S. Shell (2011). Coarse-graining errors and numerical optimization using a relative entropy framework. *J. Chem. Phys. 134*, 094112.

Cooke, I. R., and M. Deserno (2005). Solvent-free model for self-assembling fluid bilayer membranes: Stabilization of the lipid phase based on broad attractive tail potentials. *J. Chem. Phys. 123*, 224710.

de Jong, D. H., G. Singh, W. F. D. Bennett, C. Arnarez, T. A. Wassenaar, L. V. Schäfer, X. Periole, D. P. Tieleman, and S. J. Marrink (2012). Improved parameters for the martini coarse-grained protein force field. *J. Chem. Theory Comput. 9* (1), 687–697.

Drouffe, J. M., A. C. Maggs, and S. Leibler (1991). Computer simulations of self-assembled membranes. *Science 254* (5036), 1353–1356.

Dunn, W. L., and J. K. Shultis (2012). *Exploring Monte Carlo Methods*. Boston, MA: Elsevier.

Egorov, A. V., A. V. Komolkin, V. I. Chizhik, P. V. Yushmanov, A. P. Lyubartsev, and A. Laaksonen (2003). Temperature and concentration effects on Li$^+$-ion hydration. A molecular dynamics simulation study. *J. Phys. Chem. B 107*, 3234–3242.

Ercolessi, F., and J. B. Adams (1994). Interatomic potentials from first-principles calculations: The force-matching method. *Europhys. Lett. 26*, 583–588.

Eriksson, A., M. N. Jacobi, J. Nyström, and K. Tunstrøm (2008). Using force covariance to derive effective stochastic interactions in dissipative particle dynamics. *Phys. Rev. E 77*, 016707.

Farago, O. (2003). "Water-free" computer model for fluid bilayer membranes. *J. Chem. Phys. 119* (1), 596–605.

Ganguli, P., D. Mukherji, C. Junghans, and N. F. A. van der Vegt (2012). Kirkwood–Buff coarse-grained force fields for aqueous solutions. *J. Chem. Theory Comput. 8*, 1802–1807.

Goetz, R., and R. Lipowsky (1998). Computer simulations of bilayer membranes: Self-assembly and interfacial tension. *J. Chem. Phys. 108* (17), 7397–7409.

Gottlieb, M., and R. B. Bird (1976). A molecular dynamics calculation to confirm the incorrectness of the random-walk distribution for describing the Kramers freely jointed bead-rod chain. *J. Chem. Phys. 65* (8), 2467–2468.

Harmandaris, V. A., N. P. Adhikari, N. F. A. van der Vegt, and K. Kremer (2006). Hierarchical modeling of polystyrene: From atomistic to coarse-grained simulations. *Macromolecules 39*, 6708–6719.

Henderson, R. L. (1974). A uniqueness theorem for fluid pair correlation functions. *Phys. Lett. A 49A* (3), 197–198.

Hess, B., C. Holm, and N. van der Vegt (2006a). Modeling multibody effects in ionic solutions with a concentration dependent dielectric permittivity. *Phys. Rev. Lett. 96* (14), 147801.

Hess, B., C. Holm, and N. van der Vegt (2006b). Osmotic coefficients of atomistic NaCl (aq) force fields. *J. Chem. Phys. 124* (16), 164509.

Hijon, C., P. Espanol, E. Vanden-Eijnden, and R. Delgado-Buscalioni (2010). Mori–Zwanzig formalism as a practical computational tool. *Faraday Discuss. 144*, 301–322.

Högberg, C.-J., A. M. Nikitin, and A. P. Lyubartsev (2008). Modification of the CHARMM force field for DMPC lipid bilayer. *J. Comp. Chem. 29*, 2359–2369.

Izvekov, S., M. Parrinello, C. J. Burnham, and G. A. Voth (2004). Effective force fields for condensed phase systems from ab initio molecular dynamics simulation: A new method for force-matching. *J. Chem. Phys. 120*, 10896–10913.

Izvekov, S., and G. A. Voth (2005). A multiscale coarse-graining method for biomolecular systems. *J. Phys. Chem. B 109*, 2469–2473.

Korolev, N., Y. Fan, A. P. Lyubartsev, and L. Nordenskiöld (2012). Modelling chromatin structure and dynamics: Status and prospects. *Curr. Opin. Struct. Biol. 22* (2), 151–159.

Korolev, N., D. Luo, A. P. Lyubartsev, and L. Nordenskiöld (2014). A coarse-grained DNA model parameterized from atomistic simulations by inverse Monte Carlo. *Polymers 6*, 1655–1675.

Lindahl, E., B. Hess, and D. van der Spoel (2001). Gromacs 3.0: A package for molecular simulations and trajectory analysis. *J. Mol. Model. 7*, 306–317.

Lobaskin, V., A. P. Lyubartsev, and P. Linse (2001). Effective macroion-macroion potentials in asymmetric electrolytes. *Phys. Rev. E 63*, 020401.

Lopez, C. A., A. J. Rzepiela, A. H. de Vries, L. Dijkhuizen, P. H. Hunenberger, and S. J. Marrink (2009). Martini coarse-grained force field: Extension to carbohydrates. *J. Chem. Theory Comput. 5*, 3195–3210.

Lyubartsev, A. P. (2005). Multiscale modeling of lipids and lipid bilayers. *Eur. Biophys. J. 35*, 53–61.

Lyubartsev, A. P., and A. Laaksonen (1995). Calculation of effective interaction potentials from radial distribution functions: A reverse Monte Carlo approach. *Phys. Rev. E 52* (4), 3730–3737.

Lyubartsev, A. P., and A. Laaksonen (1997). Osmotic and activity coefficients from effective potentials for hydrated ions. *Phys. Rev. E 55* (5), 5689–5696.

Lyubartsev, A. P., and A. Laaksonen (1999). Effective potentials for ion–DNA interactions. *J. Chem. Phys. 111* (24), 11207–11215.

Lyubartsev, A. P., and A. Laaksonen (2000). Determination of effective pair potentials from ab-initio simulations: Application to liquid water. *Chem. Phys. Lett. 325*, 15–21.

Lyubartsev, A. P., and A. Laaksonen (2004). On the reduction of molecular degrees of freedom in computer simulations. *Lect. Notes Phys. 640*, 219–244.

Lyubartsev, A. P., K. Laasonen, and A. Laaksonen (2001). Hydration of Li^+ ion. An *ab initio* molecular dynamics simulation. *J. Chem. Phys. 114* (7), 3120–3126.

Lyubartsev, A. P., and S. Marcelja (2002). Evaluation of effective ion-ion potentials in aqueous electrolytes. *Phys. Rev. E 65*, 041202.

Lyubartsev, A., A. Mirzoev, L. J. Chen, and A. Laaksonen (2010). Systematic coarse-graining molecular models by the Newton inversion method. *Faraday Discuss. 144*, 43–56.

Lyubartsev, A. P., and A. L. Rabinovich (2011). Recent development in computer simulations of lipid bilayers. *Soft Matter 7*, 25–39.

Marrink, S. J., A. H. de Vries, and A. E. Mark (2004). Coarse grained model for semiquantitative lipid simulations. *J. Phys. Chem. B 108*, 750–760.

Marrink, S. J., H. J. Risselada, S. Yefimov, D. P. Tieleman, and A. H. de Vries (2007). The MARTINI force field: Coarse grained model for biomolecular simulations. *J. Phys. Chem. B 111* (27), 7812–7824.

Marrink, S. J., and D. P. Tieleman (2013). Perspective on the Martini model. *Chem. Soc. Rev. 42*, 6801–6822.

Mirzoev, A., and A. P. Lyubartsev (2011). Effective solvent-mediated interaction potentials of Na^+ and Cl^- in aqueous solution: Temperature dependence. *Phys. Chem. Chem. Phys. 13*, 5722–5727.

Mirzoev, A., and A. P. Lyubartsev (2013). MagiC: Software package for multiscale modeling. *J. Chem. Theory Comput. 9*, 1512–1520.

Mirzoev, A., and A. P. Lyubartsev (2014). Systematic implicit solvent coarse graining of dimyristoylphosphatidylcholine lipids. *J. Comput. Chem. 35*, 1208–1218.

Monticelli, L., S. K. Kandasamy, X. Periole, R. G. Larson, D. P. Tieleman, and S. J. Marrink (2008). The MARTINI coarse-grained force field: Extension to proteins. *J. Chem. Theory Comput. 112*, 819–834.

Mukherji, D., N. F. A. van der Vegt, K. Kremer, and L. D. Site (2012). Kirkwood–Buff analysis of liquid mixtures in an open boundary simulation. *J. Chem. Theory Comput. 8*, 375–379.

Murtola, T., E. Falck, M. Karttunen, and I. Vattulainen (2007). Coarse-grained model for phospholipid/cholesterol bilayer employing inverse Monte Carlo with thermodynamic constraints. *J. Chem. Phys. 126* (7), 075101.

Murtola, T., E. Falck, M. Patra, M. Karttunen, and I. Vattulainen (2004). Coarse-grained model for phospholipid/cholesterol bilayer. *J. Chem. Phys. 121* (18), 9156–9165.

Naomé, A., A. Laaksonen, and D. P. Vercauteren (2014). A solvent-mediated coarse-grained model of DNA derived with the systematic Newton inversion method. *J. Chem. Theory Comput. 10* (8), 3541–3549.

Noguchi, H., and M. Takasu (2001). Self-assembly of amphiphiles into vesicles: A Brownian dynamics simulation. *Phys. Rev. E 64*, 041913.

Pandit, S. A., and H. L. Scott (2009). Multiscale simulations of heterogeneous model membranes. *Biochim. Biophys. Acta 1788*, 136–148.

Reith, D., M. Pütz, and F. Müller-Plathe (2003). Deriving effective mesoscale potentials from atomistic simulations. *J. Comp. Chem. 24*, 1624–1636.

Romiszowski, P., and R. Yaris (1991). A dynamic simulation method suppressing uninteresting degrees of freedom. *J. Chem. Phys. 94*, 6751–6761.

Rudzinski, J. F., and W. G. Noid (2011). Coarse-graining entropy, forces, and structures. *J. Chem. Phys. 135*, 214101.

Rühle, V., C. Junghans, A. Lukyanov, K. Kremer, and D. Andrienko (2009). Versatile object-oriented toolkit for coarse-graining applications. *J. Chem. Theory Comput. 5* (12), 3211–3223.

Savelyev, A., C. K. Materese, and G. A. Papoian (2011). Is DNA's rigidity dominated by electrostatic or nonelectrostatic interactions? *J. Am. Chem. Soc. 133*, 19290–19293.

Savelyev, A., and G. A. Papoian (2009a). Molecular renormalization group coarse-graining of electrolyte solutions: Applications to aqueous NaCl and KCl. *J. Phys. Chem. B 113*, 7785–7793.

Savelyev, A., and G. A. Papoian (2009b). Molecular renormalization group coarse-graining of polymer chains: Application to double-stranded DNA. *Biophys. J. 96*, 4044–4052.

Savelyev, A., and G. A. Papoian (2010). Chemically accurate coarse graining of double-stranded DNA. *Proc. Natl. Acad. Sci. USA 107*, 20340–20345.

Schommers, W. (1983). Pair potentials in disordered many-particle systems: A study for liquid gallium. *Phys. Rev. A 28*, 3599–3605.

Shell, M. S. (2008). The relative entropy is fundamental to multiscale and inverse thermodynamic problems. *J. Chem. Phys. 129*, 144108.

Shelley, J. C., M. Y. Shelley, R. C. Reeder, S. Bandyopadhyay, and M. L. Klein (2001). A coarse grained model for phospholipid simulations. *J. Phys. Chem. B 105*, 4464–4470.

Shinoda, W., R. DeVane, and M. L. Klein (2012). Computer simulation studies of self-assembling macromolecules. *Curr. Opin. Struct. Biol. 22*, 175–186.

Smith, D. E., and L. X. Dang (1994). Computer simulations of NaCl association in polarizable water. *J. Chem. Phys. 100*, 3757–3766.

Soper, A. K. (1996). Empirical potential Monte Carlo simulation of fluid structure. *Chem. Phys. 202*, 295–306.

Toukan, K., and A. Rahman (1985). Molecular dynamics study of atomic motions in water. *Phys. Rev. B 31*, 2643–2648.

Vorontsov-Velyaminov, P. N., and A. M. Elyashevich (1966). Theoretical investigations of thermodynamic properties of strong electrolyte solutions by the Monte Carlo method. *Elektrokhimia (Russ. J. Electrochem.) 2*, 708–716.

Wang, Z.-J., and M. Deserno (2010). A systematically coarse-grained solvent-free model for quantitative phospholipid bilayer simulations. *J. Phys. Chem. B 114*, 11207–11220.

Wang, L.-P., T. Head-Gordon, J. W. Ponder, P. Ren, J. D. Chodera, P. K. Eastman, T. J. Martinez, and V. S. Pande (2013a). Systematic improvement of a classical molecular model of water. *J. Phys. Chem. B 117*, 9956–9972.

Wang, Y.-L., A. Lyubartsev, Z.-Y. Lu, and A. Laaksonen (2013b). Multiscale coarse-grained simulations of ionic liquids: Comparison of three approaches to derive effective potentials. *Phys. Chem. Chem. Phys. 15*, 7701–7712.

Zhang, Z., and M. L. Berkowitz (2009). Orientational dynamics of water in phospholipid bilayers with different hydration levels. *J. Phys. Chem. B 113*, 7676–7680.

CHAPTER 2

Thermodynamically Consistent Coarse Graining of Polymers

Marina G. Guenza

CONTENTS

2.1 Bottom-Up and Top-Down Models   31
2.2 Modeling Synthetic Polymers as Chains of Coarse-Grained Units   33
    2.2.1 Fine-Graining Approaches   35
2.3 The Problem of Representability and Transferability of CG
    Models   36
2.4 Integral Equation Theory of CG   37
2.5 Derivation of the Effective Coarse-Grained Potential in a
    Liquid of Polymers Represented as Chains of Multiblobs   38
2.6 Thermodynamics   42
2.7 A Universal Equation of State for a Variable-Level
    Coarse-Grained Representation of Polymer Melts   45
    2.7.1 Methods to Evaluate the Compressibility Parameter c_0 47
2.8 Testing the Theoretical Predictions of the Coarse-Grained
    Model against Atomistic Simulations   48
    2.8.1 Potential Energy   51
    2.8.2 Entropy   52
2.9 Insights in CG from the Thermodynamically Consistent
    Coarse-Grained Model   53
2.10 Reconstruction of the Dynamics: Why the Dynamics are Too
    Fast in Coarse-Grained Models   56
2.11 A Coarse-Grained Method for Protein Dynamics: The
    Langevin Equation for Protein Dynamics   59

2.12 Summary   60

 Acknowledgments   62

 References   62

M OLECULAR DYNAMICS simulations have become an essential tool in developing and testing theories in complex systems of either biological or synthetic origins. However, explicit-atom molecular dynamics simulations are computationally costly, which limits their applicability to a restricted range of length and timescales. A way to overcome this limitation is to design specific coarse-graining (CG) models. These models can be highly efficient with respect to atomistic descriptions because they represent the system at a lower resolution, thus greatly reducing the degrees of freedom that need to be sampled during the simulation [1–6].

However, reducing the degrees of freedom comes with the consequence that some thermodynamic properties and the dynamical quantities are modified in a way that depend on the extent of CG. This can be intuitively understood by considering that when a system is coarse grained, a number of atomistic units are combined together into a new, effective, fictitious unit. Those CG units interact by means of an effective potential that is simplified and, in general, is smoother than the original atomistic potential. While the atomistic units spend time sampling a myriad of local configurations, the effective CG units can slide freely on a smoother free-energy landscape.

A CG procedure is formally equivalent to applying the Mori–Zwanzig projection operator technique to a molecular liquid, originally described at the level of the Hamiltonian by a Liouville equation. The projection simplifies the system to a number of coarse-grained units interacting through an effective free-energy-type potential in the reduced representation. The dynamics of the coarse-grained units is then described by a Langevin equation where dissipation emerges from the procedure. Because the potential between units in the coarse-grained description is a free energy, it is parameter dependent and changes with the thermodynamic settings and the type of molecular system. We should expect a different potential to act between the coarse-grained units when the value of the temperature, density, molecular concentration, or number of atoms in the coarse-grained unit, or even the molecular weight of the molecule, is varied. This is a problem for the proper design of a coarse-grained description if the potential is numerically derived, i.e., an analytical solution is not possible, leading to representability problems for the potential.

Most of the effective potentials in traditional molecular dynamics simulations, such as the Lennard–Jones potential or the united atom (UA) potential [7], are dependent on thermodynamic and molecular parameters in a trivial way. They depend on temperature and on the chemical identity of the particles involved but do not depend on other parameters, such as the number of atoms in the molecule or the molecular concentration; effective coarse-grained potentials usually present a more complex behavior. We argue that the capability for traditional potential, such as the Lennard–Jones potential, of providing a reasonable estimate of the properties by molecular dynamics simulations, even in regions of the phase diagram where they perhaps should not be applied, is due to the small error typical of CG models that have a very limited level of CG. This work shows that the possible error in the potential due to lack of transferability increases with increasing the degree of CG.

Much of the difficulty in designing consistent CG approaches stems from the fact that it is not always clear how various many-body effects are incorporated into simple two-body interactions, and how errors in the numerical optimization of a pair potential are propagated to thermodynamic properties [8]. The CG approach that we developed, and that we summarize in this chapter, is called the Integral Equation Coarse-Grained (IECG) model. This approach has the advantage of providing an analytical solution of the potential, structural properties, and thermodynamic properties as a function of the parameters in the system, which is represented with a variable level of resolution [9].

In most CG models, which are numerically solved, the step of running detailed atomistic simulations is necessary to be able to parameterize the model. However, running detailed atomistic simulations in part defeats the purpose of developing accurate CG models, because a detailed atomistic description is already available for the system under study in the thermodynamic region of interest. The hope is that the models optimized against detailed atomistic simulations can perform well in regions of the phase space that is close to the point where the coarse-grained model has been optimized. This is a problem of transferability of the potential.

In general, numerically optimized CG potentials do not apply to regions in configurational space where they have not been optimized, and atomistic simulations should be performed at any condition of interest to test the applicability of the potential. However, after atomistic simulations are performed at any conditions of interest, it is not clear what would be the purpose of also running mesoscale simulations in the coarse-grained description.

One parameter that seems reasonable to vary is the time covered by the simulation, which in the CG description can be orders of magnitude longer than in the atomistic simulation. The underlying assumption is that the effective potential that acts between coarse-grained sites does not change with time. It is sometimes the case that increasing the time of the simulation can hardly bring new information, but the opposite is also often true, where the longtime prediction of the CG model could be incorrect. In general, it is difficult to know if a model parameterized in a short simulation contains information from all the relevant free-energy barriers, some of which could come into play only at a timescale that exceeds the length of the initial atomistic simulation. The process of crossing those high energy barriers could exceed the timescale sampled in the atomistic simulation and the coarse-grained description derived from the atomistic simulations would likely predict incorrect longtime behavior of the system. This is a problem of ergodicity of the system.

In general, the key point in analyzing the advantages and challenges of the different CG models is to understand which properties should be conserved, which properties should be modified in the process, and how they will be modified as a consequence of CG. As a starting step, with the purpose of understanding the microscopic motivations of the observed thermodynamic and dynamical inconsistencies, it is useful to consider an intuitive representation of the effects of CG. Here, we look at a very simple model given by the comparison between the free-energy landscape for the rotation of the dihedral angle in an ethane molecule in the coarse-grained and atomistic representations. In the atomistic representation, the energy presents three minima corresponding to one trans and two gauche configurations. In the mesoscale representation of the UA model, each methyl unit, CH_3, is an effective CG sphere, the whole molecule is a dumbbell, and the rotation of the two units with respect to each other is free, leading to a completely flat free-energy landscape.

Several effects typical of coarse-grained descriptions arise from site-averaging, as in the following example. First, the dynamics speeds up as the CH_3 units, which in the atomistic description employed a finite amount of time in crossing the configurational barriers that separate states of low energy, are fully free to rotate in the coarse-grained description. The speeding up of the dynamics gives the computational advantage to the coarse-grained description; the more extensive the CG, the larger the gain in the computational time. The dynamic properties of the coarse-grained representations are, however, unrealistically accelerated. For example, the diffusion coefficient measured in mesoscale simulations can be orders of magnitude too high. One could be tempted to look into slowing down the coarse-grained model so that the simulation can reproduce the correct dynamics; however that choice would also slow down the computational time, losing the gain of having a CG description. Different strategies to recover the correct dynamics in a fast computational framework have been implemented. For example, it is possible to directly modify the equations of motion for the CG units to reproduce the correct dynamics in a coarse-grained description, as is done in dissipative particle dynamics [10]. Another strategy is to use reduced dynamical models such as a generalized Langevin representation of a restricted number of molecules moving slowly in a field of fast-moving molecules (solvent or polymers). This strategy was applied in a model that we developed to describe protein dynamics [11,12].

In general, the CG representation has fewer configurational states than in the atomistic representation. As a result, the partition function of the system is modified, as discussed below. Even when the structure of the system is correctly represented by the pair correlation function in both the atomistic and coarse-grained representations, some thermodynamic properties are different. Because the number of configurational states that the system samples in the CG representation is reduced, the entropy of the CG system as measured in the simulation is also reduced with respect to the one in the related atomistic simulation. The free energy, on the other hand, is consistent in the two representations, leading to the conclusion that the internal energy also has to change during the process of CG. In order to develop a precise understanding of how these properties are modified during CG, it is essential to build models that are based on a solid statistical mechanics foundation and analyze CG procedures by means of the tools of statistical mechanics.

2.1 BOTTOM-UP AND TOP-DOWN MODELS

The coarse-grained models that have been proposed so far are mostly divided into two groups: (1) bottom-up and (2) top-down. The *bottom-up* approaches start from the atomistic description and group atoms into new coarse-grained units. This procedure, if rigorously done, should provide information on the effective interaction potential between CG units. The potentials are used as an input to the molecular dynamics simulations of the system represented by the CG description.

In the bottom-up models, the coarse-grained description is, in most cases, derived by matching specific physical quantities between the two levels of descriptions: atomistic and coarse grained. Motivated by Henderson's theorem, which states that an isotropic potential that reproduces the correct pair structure of a fluid is unique up to a constant, the physical quantity that is most commonly matched is the pair distribution function [13]. This is the basis of the iterative Boltzmann inversion (IBI) procedure [14]. Statistical mechanics directly relates the pair distribution function to the thermodynamic properties [15], so that their proper agreement should also be ensured. Other approaches reproduce the data from atomistic simulations by matching the internal energy [16], forces [17], or conditional interaction free energy between two groups of atoms as a function of the distance separating them [18], or simply by minimizing the relative entropy [19].

The second type of approach is *top-down*, in the sense that the model is built to reproduce specific global properties of the system on the large scale (mesoscale) that characterizes the coarse-grained description. These methods are needed when there is not an atomistic simulation that can provide a valid reference description for the numerical optimization. This happens when (1) the system to be simulated is quite complex, for example, in simulations of large biological objects, or (2) when the properties that are of interest are on a scale that is too large to be reached in atomistic simulations, for example, dynamical properties close to a second-order phase transition where the concentration fluctuation lengthscale diverges, or (3) when the atomistic description is itself imprecise, as can be the case in atomistic simulations of RNA and DNA, where, in some thermodynamic conditions, the potential is known to produce overstacking of the base pairs and incorrect melting temperature [20–22]. Belonging to the category of top-down approaches are all the mean-field theories of polymer structure and dynamics [23,24], and models of membranes [25].

Necessarily top-down models are less precise than the bottom-up approaches for which a rigorous statistical mechanics procedure can guide the CG of the atomistic description. Top-down approaches are, however, quite reliable in reproducing the properties that are used in the construction of the model [26], but are less reliable as far as other physical properties are concerned. For systems that are very complex, such as biological macromolecules and their complexes, for which the coarse-grained units cannot be straightforwardly defined, top-down models are the only feasible way to approach CG.

For some biological macromolecules, like proteins, a number of bottom-up approaches have been designed, mostly to study the dynamic of fluctuations around minima of energy and folding [27–29], while for nucleic acids, the approaches have been mostly top-down because the atomistic forcefields need to be further developed [21,30].

Our approach to the structure and dynamics of polymer liquids belongs to the group of bottom-up methods, where the pair distribution function is reproduced across variable levels of CG. However differently from the methods discussed above, the CG approach is largely analytical, and not numerical, and does not require the numerical fitting to atomistic simulations. We also proposed a fine-grained model for the dynamics of proteins in solution based on a Langevin equation for a chain of amino acids represented as effective spheres. This model, which can be parameterized using either atomistic simulations or experiments, shows an excellent agreement with the time-correlation functions measured in nuclear magnetic resonance (NMR) relaxation experiments [11].

2.2 MODELING SYNTHETIC POLYMERS AS CHAINS OF COARSE-GRAINED UNITS

In recent years, we have proposed a series of models to coarse-grain the structure of macromolecular liquids [31–37]. Our models concern liquids of polymers that are isotropic and composed of n molecules in a volume V, with each chain including a total number of monomer N. The density of chains is $\rho_{ch} = n/V$ and is related to the liquid monomer density $\rho = \rho_{ch}N$. Every chain in the liquid is partitioned in a variable number of coarse-grained units or blobs, n_b, each containing a number $N_b = N/n_b$ of monomers, with the blob density $\rho_b = \rho/N_b$.

Starting from the Ornstein–Zernike integral equation, we calculated the pair distribution function of the coarse-grained model and the effective potential acting between units. Using the potential, we performed simulations of the coarse-grained systems, and then compared thermodynamic quantities and structural quantities of interest from these coarse-grained simulations with UA simulation data. The agreement between CG and atomistic descriptions is quantitative, whereas the direct correlation contribution at large distances, $c(k \rightarrow 0) = c_0$, is the only nontrivial parameter, which is evaluated either from experiments or from theory. Notice that atomistic simulations are not needed to parameterize the CG model.

Having an analytical solution brings some advantages to the method. The theory produces formal solutions of the static and dynamic quantities of interest and a formal analysis of how different properties are modified during the process of CG becomes feasible. Furthermore, the approach is useful in multiscale simulations of dense polymer systems with specific chemical structures because it is system specific and reproduces the correct equation of state across various levels of CG.

The model applies to any type of polymer, because the lengthscale of CG is assumed to be larger than its local persistence length. By selecting a lengthscale larger than the persistence length, which is specific to the polymer considered, the coarse-grained units are statistically uncorrelated and follow a random walk in space. The chain of blobs can then be modeled as freely jointed, which affords an analytical formalism for the blob chain structure and interacting potentials. This model has unique characteristics because being analytical is fully transferable; it applies to different points in the phase diagram and represents well any type of homopolymer liquid, independent of the molecular structure of the monomer.

In our method, atomic scale simulations can be used only as a test and not to provide information to the CG approach. The purpose of developing CG methods is to have a mesoscale description that can be directly used in molecular dynamics simulations without the need of performing atomistic simulations to parameterize them.

Figure 2.1 shows a schematic outline of our approach. We perform two different types of simulations: UA and coarse grained. The coarse-grained potentials, obtained analytically by solving the Ornstein–Zernike equation, are used in coarse-grained simulations with variable numbers of effective sites. Structural and thermodynamic quantities are then compared directly to UA simulations. Tests have been done for polyethylene melts at a variety of CG levels and for systems in different thermodynamic conditions and variable chain length.

The model has been extended not only to treat polymeric liquids where phase separation occurs, such as mixtures of polymers with different monomeric units and local semiflexibility, but also mixtures of polymers with the same molecular form but different degrees of polymerization, and also melts of diblock copolymers with variable composition of the blocks. Because the model has been solved analytically, it is possible to perform a precise analysis of the advantages and the challenges of using a CG model. Some considerations emerge from this study that are more general than the model presented and relate to all CG models, not just ours. In this way, the discussion has a value beyond the simulation of liquids of polymers.

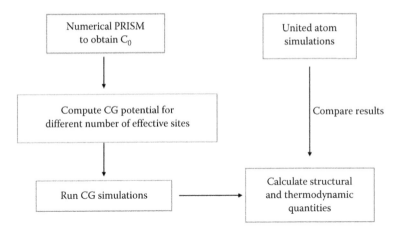

FIGURE 2.1 Schematic outline of integral equation theory coarse-graining (CG) approach. There are two different types of simulations: united atom (UA) and coarse grained. The coarse-grained potentials are obtained from the analytical theory and used in coarse-grained simulations with variable numbers of effective sites. Structural and thermodynamic quantities are then compared directly to UA simulations. (Reprinted with permission from J. McCarty et al., *J. Chem. Phys.*, 140. Copyright 2014, American Institute of Physics.)

2.2.1 Fine-Graining Approaches

The model that we have studied and developed is of the type where the degree of CG is larger than the polymer persistence length, and belongs to the family of ultra CG models or molecular scale models, whereas other CG models target a finer scale. These are the models that group together a small number of atoms to create a unit, for example, they replace an aromatic ring with a bead [5], or they group together the atoms in the side chain of each amino acid [38], or they represent each base in DNA with a reduced number of beads [39].

Among these models, one that is general enough to be fully transferable to different polyolefins is the UA model, which assembles together hydrogens and the related carbon atom in one bead. Units such as CH_3, CH_2, and CH all have different optimized values of the parameters that define their potential. The success of the UA model is due in large part to the small error that one can make when adjusting the parameters of the potential [7]. As discussed later in this review, the error is small because the level of CG is minimal. The numerical optimization of the potential against atomistic simulations is feasible with negligible negative consequences on the precision of the calculations.

Because this complex energy landscape can be influenced by the surrounding atoms, which can exert forces on the hidden atoms of the coarse-grained unit to the parameterization of the potential is hardly uniquely defined; it depends on the position of the coarse-grained unit inside the molecule. In a nutshell, the coarse-grained unit is not statistically independent of its surrounding atoms and it can hardly be treated as an independent statistical unit. This generates the problems that most CG approaches have in relation to the transferability and representability of the potential.

2.3 THE PROBLEM OF REPRESENTABILITY AND TRANSFERABILITY OF CG MODELS

It is often noticed that CG models at given state conditions of temperature and density that are optimized to reproduce correctly, for example, a physical quantity like the pair distribution function, cannot reproduce correctly other quantities, such as the pressure or the free energy of the system. In most cases, CG models would need to be optimized for each given set of thermodynamic parameters, and for every type of molecular system that one wants to study.

The lack of thermodynamic consistency between various representations is in fact a limitation that has delayed the widespread use of coarse-grained models in engineering and material science. For many numerical CG schemes, there is no guarantee that the resulting behavior of the coarse system will be consistent with what would have been observed by using a more detailed and expensive model [16,40,41].

There is a second and more subtle reason that hampers the application of CG models. Numerically optimized CG models are built to reproduce faithfully the properties of the atomistic model against which the CG parameters were optimized. Even when considering all the possible physical properties of interest, and when the model is capable of reproducing those with very small errors, the quality of the CG model is limited by the quality of the atomistic simulation and by the possible lack of ergodicity of the system simulated. Because it is not known if the atomistic simulation is able to sample efficiently the free-energy landscape, the quality of the CG model is limited by the quality of the atomistic simulation [42–44].

Motivated by these considerations, some recent studies have focused on understanding how well a less-than-optimal coarse-grained description, that is, a CG model that contains errors, is still able to provide some useful guidance to the related atomistic simulations. In these studies, CG models are used for sampling the configurational space and uncover deep energy minima in the atomistic potential, which would be otherwise not accessible to the atomistic simulation [45]. The coarse-grained description can guide the system to sample a larger region in the configurational space than the one accessible to the atomistic simulation in the same time window.

Simulations that run on the mesoscale are periodically stopped to rebuild the atomistic structure around the coarse-grained one and then the simulation is carried on with atomistic resolution. Mesoscale and atomic scale simulations are alternated for a full and consistent sampling of the free-energy landscape. The quality of the sampling depends on the reliability of the CG procedure and on the quality of the atomistic description.

2.4 INTEGRAL EQUATION THEORY OF CG

In constructing a coarse-grained model, it is common to start from an atomistic representation and build the new coarse-grained description in a reduced representation by matching the quantitative value of a physical property in the two descriptions, atomistic and reduced. Different theories match different quantities; specifically the most common quantities are the matching of the structural distribution, that is, the pair distribution function, using the IBI procedure [46,47], the force matching procedure [17], to map the derivative of the mean-force potential, the mapping of the internal energy [18], and finally the search for a minimized information entropy [19]. These methods are successful in reproducing quantitatively the physical properties that are initially fitted, but in most cases, they cannot ensure representability and transferability of the potential so derived.

We approached the problem from a different perspective. It is known that the pair distribution function is uniquely defined [13] and that the real pair distribution function, when known, allows for the calculation of all the thermodynamic quantities of interest [48]. We derived the pair distribution function from the solution of the integral equation theory for the coarse-grained description, starting from the integral equation theory in the atomistic representation, which is the PRISM approach [49].

The solution of the pair distribution function from the integral equation theory does not require performing atomistic simulations. Instead the atomistic pair distribution function is defined starting from the molecular and the thermodynamic parameters of the system. Once the pair distribution function is derived, the effective potential between the coarse-grained units is calculated using the appropriate closure.

The derived potential is an input to mesoscale simulations of the coarse-grained system, which, once they are performed, provide all the molecular and thermodynamic quantities of interest. Those are, for example, the pair distribution function of the coarse-grained description, pressure, excess free energy, as well as internal energy and entropy. The structural and thermodynamic quantities are then compared with the ones measured in atomistic simulations for the system under study. Atomistic simulations are used as a test of the correctness of the proposed coarse-grained potential, and not to optimize the coarse-grained description.

2.5 DERIVATION OF THE EFFECTIVE COARSE-GRAINED POTENTIAL IN A LIQUID OF POLYMERS REPRESENTED AS CHAINS OF MULTIBLOBS

In our multiblob coarse-grained model, each polymer is described as a chain of n_b soft blobs [8,32–34]. The number of soft spheres can be varied, starting from the single soft-sphere representation [35–37], and progressing to the multiblob description. The only constraint is that the size of the blob has to be larger than the persistence length of the polymer; for polyethylene, each blob needs to have at least 30 monomers. This allows us to adopt a freely jointed chain model, as the chain of blobs will follow Markov statistics. By adopting this coarse-grained description, the simulation can include many more polymer chains than would otherwise be possible to simulate.

Given N monomers with a chain density ρ_{ch}, and an effective segment length $\sigma = \sqrt{6/N}R_g$ with R_g being the polymer radius of gyration and $R_{gb} = Rg/\sqrt{n_b}$ the blob radius of gyration, the number of underlying monomers per blob is given as $N_b = N/n_b$. Coarse-grained or fictitious interacting sites are taken to be located at the center of mass of the polymer chain for the soft-sphere model or at the centers of mass of several monomers along the same chain for the connected blob model. The relation between center-of-mass fictitious sites and real monomer sites is derived by solving a generalized matrix Ornstein–Zernike equation [33].

A single soft-sphere representation and three- and five-blob representations have been formally derived. For the three-blob and five-blob representations, the blobs are not all equivalent, as the blobs at the end of the chains are different than the one(s) in the internal part of the chain. For chains with a large number of blobs, more specifically, when there are more than five coarse-grained sites per chain, end effects become negligible and it is possible to formally derive a blob-averaged description [8].

The intramolecular distributions in the blob-averaged limit are normalized as $\hat{\Omega}(k) = \omega(k)/N$, for the blob–blob (bb), blob–monomer (bm), and monomer–monomer (mm) distributions. The normalized bm and the bb distributions are given by

$$
\hat{\Omega}^{bm}(k) = \frac{1}{n_b}\left[\frac{\sqrt{\pi}}{kR_{gb}} Erf\left(\frac{kR_{gb}}{2}\right) e^{-\frac{k^2 R_{gb}^2}{12}} \right.
$$

$$
\left. - 2\left(\frac{e^{-n_b k^2 R_{gb}^2} - n_b e^{-k^2 R_{gb}^2} + n_b - 1}{k^2 R_{gb}^2 n_b (e^{-k^2 R_{gb}^2} - 1)}\right) e^{-k^2 R_{gb}^2/3} \right] \tag{2.1}
$$

and

$$
\hat{\Omega}^{bb}(k) = \frac{1}{n_b} + 2\left[\frac{e^{-n_b k^2 R_{gb}^2} - n_b e^{-k^2 R_{gb}^2} + (n_b - 1)}{n_b^2 (e^{-k^2 R_{gb}^2} - 1)^2}\right] e^{-2k^2 R_{gb}^2/3} \tag{2.2}
$$

The monomer distribution $\Omega^{mm}(k)$ is normalized as

$$
\hat{\Omega}^{mm}(k) = \hat{\omega}^{mm}(k)/N = \frac{2}{n_b^2 k^4 R_{gb}^4}\left(k^2 R_{gb}^2 n_b - 1 + e^{-n_b k^2 R_{gb}^2}\right) \tag{2.3}
$$

Given the Ornstein–Zernike relation for the coarse-grained blob representation

$$
\hat{h}^{bb}(k) = n_b \hat{\Omega}^{bb}(k)\hat{c}^{bb}(k)\left[n_b\hat{\Omega}^{bb}(k) + \rho_b\hat{h}^{bb}(k)\right] \tag{2.4}
$$

the direct correlation function is given as

$$
\hat{c}^{bb}(k) = \frac{\hat{h}^{bb}(k)}{n_b\hat{\Omega}^{bb}(k)\left[n_b\hat{\Omega}^{bb}(k) + \rho_b\hat{h}^{bb}(k)\right]} \tag{2.5}
$$

From the direct correlation function, the interaction potential is calculated by evaluating the hyper-netted chain potential from the Fourier transform of Equations 2.4 and 2.5 as

$$
\frac{v^{bb}(r)}{k_B T} = -\ln\left[h^{bb}(r) + 1\right] + h^{bb}(r) - c^{bb}(r) \tag{2.6}
$$

The solution can be performed either analytically or numerically.

When $|h^{bb}(r)| << 1$, which always holds at large separations ($r >> 1$ in units of R_{gb}) and at any separation for representations with large N_b and high densities, the potential further simplifies to

$$V^{bb}(r) \approx -k_b T c^{bb}(r) \tag{2.7}$$

In the literature, this formula is referred to as the mean spherical approximation (MSA) and applies to low compressible polymer liquids. If this formalism is improperly used to treat low density liquids, where the MSA does not hold, this approximation would lead to unphysical behavior.

We now focus on the effective direct correlation function and the MSA potential for large separations in real space. In this limit, $r >> 1$ (in R_{gb} units), the inverse transform integral is sufficiently dominated by $\hat{c}^{cc}(k)$ (in the small wave vector limit) that the large wave vector contribution can be entirely neglected. Furthermore, since the expansion for small wave vectors is bounded at large k, the error incurred in using the small k form with the integral bounds extended to infinity is small. This approximation leads, for $r >> 1$, to

$$
\begin{aligned}
c^{bb}(r) &\approx \frac{-N_b \Gamma_b}{2\pi^2 \rho_m R_{gb}^3 r} \int_0^\infty \left(k \sin(kr) \left[\frac{45}{45 + \Gamma_b k^4} + \frac{5k^2}{28} \frac{13\Gamma_b k^4 - 3780}{(\Gamma_b k^4 + 45)^2} \right] \right) dk \\
&= \left[-\left(\frac{45\sqrt{2} N_b \Gamma_b^{1/4}}{8\pi\sqrt{3}\sqrt[4]{5}\rho_m R_{gb}^3} \right) \frac{\sin(Q'r)}{Q'r} e^{-Q'r} + \left(\frac{\sqrt{5} N_b}{672\pi\rho_m \Gamma_b^{1/4} R_{gb}^3} \right) \right. \\
&\quad \times \left[(13Q^3(Q'r - 4)) \cos(Q'r) + \left(\frac{945 + 13Q^4}{\Gamma_b^{1/4}} \right) r \sin(Q'r) \right. \\
&\quad \left. \left. + \frac{945r}{\Gamma_b^{1/4}} \cos(Q'r) \right] \frac{e^{-Q'r}}{Q'r} \right]
\end{aligned}
\tag{2.8}
$$

where $Q' = 5^{1/4}\sqrt{3/2}\,\Gamma_b^{-1/4}$ and $Q \equiv Q'\Gamma_b^{1/4}$. The key quantity of interest is $\Gamma_b = N_b \rho |c_0|$, which is defined once the molecular and thermodynamic parameters are known, but also depends on the direct correlation function at $k = 0$, which is unknown.

The range of the potential, in units of the radius of gyration of the blob, scales as $N_b^{1/4}$. This scaling behavior describes how the interaction between effective units propagates through the atomistic sites in the macromolecular liquid. This pathway follows a random walk in the space defined by the lengthscale of the blob–blob interpenetration, which also scales with the degree of polymerization as $N_b^{1/2}$.

Interestingly, the range of the potential increases with the number of monomers comprised in the coarse-grained unit, that is, blob or soft sphere, whereas the potential at contact decreases. However, the interblob potential does not vanish even when the length of the polymer chain becomes infinity, indicating that intermolecular interactions between polymers are important even for infinitely long chains. This result disagrees with the conventional assumption in polymer physics that intramolecular interactions are dominant over intermolecular contributions, and a polymer melt can simply be described by mean-field approaches of the single chain [24].

The potential becomes longer ranged when increasing the lengthscale of CG, improving the gain in computational time. However, the presence of long-ranged forces in the molecular dynamic simulations makes the use of a large box necessary, as the simulation box is usually chosen to be at least twice the range of the effective potential. The inconvenience of having long-range interactions can be alleviated by the use of simulations in reciprocal space, as is conveniently done in the case of electrostatic interactions using the Ewald summation [15], and by other methods.

The potential has a long-ranged, slowly decaying repulsive component and a second attractive part that is smaller in absolute value than the repulsive part. This attractive contribution is important when one evaluates the thermodynamic properties of the system and cannot be discarded. Higher order terms, which are present in the equation of the potential, tend to give increasingly more negligible contributions; the potential in our simulations is often truncated after the first attractive well, or more rarely the second repulsive contribution, depending on error minimization.

It is interesting to note that the attractive contribution is present in the effective potential even when the intermolecular atomistic potential, from which the coarse-grained potential is derived, is purely repulsive. This indicates that the attractive contribution to the intermolecular potential is, at least partially, a consequence of CG and the propagation of the interactions through the liquid. Being the resultant of the projection of many-body interactions onto the pair of coarse-grained units, the attractive component of the potential is, at least in part, entropic in nature.

When the atomistic-scale interaction already includes an attractive part, that is, for example, the monomer–monomer interaction is a Lennard–Jones potential or a finite extensible nonlinear elastic (FENE) potential, the latter provides a contribution to the total attractive component of the CG potential. In that case, the attractive CG component is enhanced with respect to one arising from a "pure" hard-sphere monomer–monomer interaction.

The CG potential becomes longer ranged with increasing level of CG, so that for models of the fine-graining type, the potential is still short ranged and the error that can possibly occur in the thermodynamics when there are imperfections in the calculations of the pair distribution functions, $g(r)$, and the related potential, $v(r)$, is still small. It is for these models that the IBI procedure becomes a promising strategy to calculate the interacting potential, which is optimized to reproduce the structure but also can predict the thermodynamics with small error. Other methods, which optimize the potential by optimizing the forces or by minimizing the information entropy, should be most precise when the number of atoms that are grouped together into the effective coarse-grained unit is small.

In general, fine-grained models are complex because they are very specific to the local structure and geometry of the unit that is being coarse-grained. Dihedral angles, branching, local angles are in most cases different in different monomeric units, and the potential derived from the optimization against atomistic level simulations is hardly applicable to similar units belonging to polymers that have different chemical composition [50]. A typical example of this problem is CG models for proteins, where, even if the number of building blocks is reduced to only 20 amino acids, not only the position of each amino acid inside the primary structure of the protein but also the chemical nature of its near-neighbor and next-near neighbor amino acids is important to correctly predict the properties of the protein, such as, its folding. In this way, a simple potential that is specific to pairs of amino acid is limited in predicting quantitative and experimentally consistent physical quantities, when they are related to a precise evaluation of the energy of the system.

2.6 THERMODYNAMICS

From the pair distribution function, the equation of state and the related thermodynamic quantities of interest are derived for our multiblob coarse-grained description. The normalized pressure is given by the virial expansion

$$\frac{P}{\rho_{ch} k_B T} = 1 - \frac{N \rho c_0}{2} \tag{2.9}$$

with ρ_{ch} the number of chain density and $c_0 < 0$ the $k = 0$ limit of the total correlation function. The isothermal compressibility is

$$\rho k_B T \kappa_T = \frac{N}{1 - \rho N c_0} \tag{2.10}$$

The excess Helmholtz free energy per molecule, calculated relatively to the energy of the system in its gas phase, is

$$\frac{F - F_0}{n k_B T} = -\frac{N c_0 \rho}{2} \tag{2.11}$$

These equations are derived from the approximated analytical solution of the effective potential, which is accurate in the mean-field limit of a nearly incompressible liquid [51]. This equation of state holds for any level of CG in the multiblob description, while all the nonideal contributions that arise from system-specific interactions are contained in the nontrivial parameter c_0. When deriving thermodynamic properties from the equation of state, it is necessary to account for the state dependence of the parameter c_0 for the system under study.

The *internal* energy per chain, defined as the potential energy plus the kinetic energy, $U = E + K$, is found to have a more complex behavior. In the soft-sphere representation, where $n_b = 1$ and $N_b = N$, the internal energy per chain is

$$\frac{U}{n k_B T} = \frac{3}{2} - \frac{\rho N c_0}{2} \tag{2.12}$$

where the first term in the right-hand side of the equation is the kinetic energy of the n classical point particles, whereas the second term is the ensemble average of the potential energy arising from the intermolecular interaction contribution, which is identical in the soft-sphere representation to the excess free energy. The related entropy per chain for the liquid of soft particles is given by the simple identity $U = F + TS$ as

$$\frac{S}{n k_B} = \frac{3}{2} + F_0 \tag{2.13}$$

In the limit that the whole macromolecule is represented as a single-point particle, the entropy is only translational; no intramolecular configurational entropy is present in each coarse-grained site.

In the underlying atomistic system, however, the entropy and the internal energy have additional contributions from the chain configurations, which are not accounted for in the soft-sphere model. In the same way, those thermodynamic quantities in the multiblob description contain information from the multiblob chain configurations. Both entropy and internal energy are expected to depend on the level of CG.

The *potential* energy in the multiblob description is composed of an *intra*molecular and an *inter*molecular contribution. The intermolecular component is calculated by a simple generalization of the soft-sphere procedure as

$$\frac{E_{inter}^{bb}}{nk_B T} = \frac{2\pi\rho_b}{k_B T} \int_0^\infty v^{bb}(r) g^{bb}(r) r^2 dr \approx -\frac{\rho N c_0}{2} \tag{2.14}$$

The dependence on the coarse-grained model emerges instead from the intramolecular contributions to the potential energy,

$$\frac{E_{intra}^{bb}}{nk_B T} = \frac{3}{2}(n_b - 1) + \frac{1}{2}(n_b - 2) \tag{2.15}$$

where $n_b = N/N_b$ is the number of blobs.

At a given temperature, while the excess free energy is constant, the number of degrees of freedom and the related entropy depend on the extent of CG. In this way, the internal energy and the potential energy also change with the number of internal degrees of freedom. Furthermore, the entropy correlates with the lengthscale of CG; the structure defined at a lengthscale that is larger than the lengthscale of CG is conserved, whereas the structure defined at a smaller lengthscale is averaged out.

As the internal energy changes with the level of CG so does the specific heat, defined as $C_V = (\partial U/\partial T)_V$. The specific heat directly depends on the number of degrees of freedom that are available to the system to store energy. When a molecule is represented with two different levels of CG, the number of degrees of freedom changes and so does C_V.

The emergence of a phase transition, however, is determined by the free energy and the discontinuity in one of its derivatives with respect to the related thermodynamic variable. The value of the free energy as a function of the thermodynamic parameters does not change when the level of CG is modified, so that the phase diagram predicted by our CG model is identical, independent of the level of CG that is selected.

We find it interesting to notice that all the quantities that relate to the global/macroscopic properties of the liquid, such as the free energy and the pressure, do not depend on the choice of the lengthscale of the coarse-grained unit, but depend only on the thermodynamic parameters that are defined at the monomer level, namely the number of monomers in a chain, N, the monomer liquid density, ρ, and the direct correlation function at the macroscopic scale, c_0. This is correct, as the "bulk" properties of a liquid should not depend on the level of detail employed to describe the molecules if the coarse-grained description is consistent.

2.7 A UNIVERSAL EQUATION OF STATE FOR A VARIABLE-LEVEL COARSE-GRAINED REPRESENTATION OF POLYMER MELTS

In our model, it is possible to take advantage of the fact that the excess free energy, compressibility, and pressure do not depend on the number of coarse-grained units in which the molecule is partitioned. We derived an equation of state for the polymer liquid starting from the simplest, most reduced, representation, where the whole molecule is described as a point particle interacting through an effective long-ranged potential. This is the so-called "soft-sphere model," where the interaction potential is given by the solution of the integral equation for the center of mass of the molecule.

As mentioned earlier, the only nontrivial parameter in our theory is the thermodynamic parameter c_0, which is related through the Ornstein–Zernike equation to the compressibility of the system and to the equation of state. This parameter is also independent of the degree of CG and can be conveniently calculated in the soft-sphere representation.

The resulting equation of state is of the form of a Carnahan–Starling expression but includes numerical prefactors that reflect the chain connectivity and the fact that the real potential is not of the simple hard-sphere form

$$\frac{P}{\rho k_B T} = \frac{4(\eta_{\text{eff}} + c_1 \eta_{\text{eff}}^2 + c_2 \eta_{\text{eff}}^3)}{(1 - \eta_{\text{eff}})^3} \tag{2.16}$$

The pressure is given as a function of the soft-sphere packing fraction

$$\eta_{\text{eff}} = \frac{\pi}{6} \rho d^3 \tag{2.17}$$

and three parameters: an effective soft-sphere diameter, d, and two other parameters, c_1 and c_2, which are specific for the polymer under study.

Figure 2.2 shows data for the normalized pressure as a function of the effective packing fraction for a number of UA simulations of polyethylene

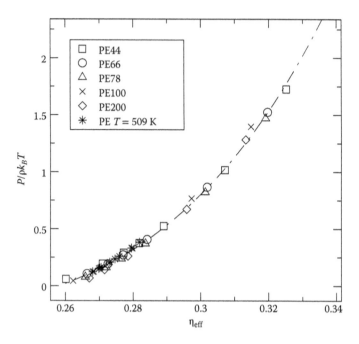

FIGURE 2.2 Pressure as a function of the packing fraction for UA simulations. Simulations carried out at constant temperature $T = 509$ K and increasing polymer chain length are depicted with stars. All the other points are simulations at $T = 400$ K and variable densities and chain lengths. The dotted-dashed line is Equation 2.16, which does not depend on the degree of polymerization, N. (Reprinted with permission from J. McCarty et al., *J. Chem. Phys.*, 140. Copyright 2014, American Institute of Physics.)

performed at $T = 400$ K and increasing degree of polymerization ($N = 44$, 66, 78, 100, and 200) and variable density, as well as samples at $T = 509$ K, density $\rho = 0.03153$ sites/A^3, and degree of polymerization $N = 36, 44,$ 66, 78, 100, 192, 224, and 270. All the samples fall onto a universal curve, which is well represented by the equation of state for soft spheres, Equation 2.16.

In the same theoretical framework, the direct correlation function at $k \rightarrow 0$ is expressed as a function of the parameters as

$$c_0 = -\frac{4\pi d^3}{3} \frac{1 + c_1 \eta_{\text{eff}} + c_2 \eta_{\text{eff}}^2}{(1 - \eta_{\text{eff}})^3} \qquad (2.18)$$

By plotting the normalized pressure as a function of the density for a number of atomistic simulations of polyethylene melts at varying temperature and degree of polymerization, we see that the data follow an equation of state if plotted as a function of an effective packing fraction, η_{eff}. Once we assume that the parameters c_1 and c_2 are independent of N and temperature, the optimized effective sphere diameter is found to be $d = 2.5$ Å, while the other two parameters are $c_1 = -11.9$ and $c_2 = 31.11$.

The simulations are found to reproduce quantitatively the trend of pressure as a function of density of the equation of state without any postoptimization scheme or fitting procedure, and across variable levels of CG.

2.7.1 Methods to Evaluate the Compressibility Parameter c_0

By using Equation 2.18 we can estimate c_0 for any chain length at any temperature for polyethylene. c_0 is the only parameter that does not describe in a trivial way physical or molecular quantities. Other parameters, besides the direct correlation c_0, are the thermodynamic properties of temperature, T, and density, ρ, as well as the structural properties of N, the number of monomers, and the effective segment size, σ. These parameters allow the method to be readily applied to a variety of polymers in variable experimental conditions.

A second possible procedure to calculate the monomer direct correlation parameter, c_0, is to solve the numerical solution of the PRISM equations with a realistic representation of the polymer chain. The solution of the PRISM equation provides results that are consistent with the equation of state method described above [9].

An analytical equation for the c_0 parameter was obtained earlier using the Gaussian thread model, which relies on the description of the polymer chain as infinitely long and infinitely thin, and this is the model used to represent polymers in field theory [52,53]. While the PRISM thread model represents an idealized limiting case, it is not expected to give quantitative predictions for real chains of finite length and thickness. The analytical solution of the potential that has been discussed here does not rely on the use of the thread model.

A third method to evaluate the direct correlation function at large distances is to directly use the isothermal compressibility of the liquid under study, experimentally determined. The isothermal compressibility, κ_T, which is also preserved in CG, is related to the static structure factor $\hat{S}(k = 0)$ as

$$\rho k_B T \kappa_T = \hat{S}(k = 0) = [N + \rho \hat{h}^{mm}(0)] \tag{2.19}$$

with $\hat{S}(k = 0)$ the $k \to 0$ limit of $\hat{S}(k) = [\hat{\omega}^{mm}(K) + \rho \hat{h}^{mm}(K)]$. The isothermal compressibility in the blob description is identical to the isothermal compressibility in the monomer description. The value of c_0 is then determined, for example, in the monomer description, as

$$c_0 = \frac{\hat{h}^{mm}(0)}{\rho N \hat{h}^{mm}(0) + N^2} \tag{2.20}$$

with $\hat{h}^{mm}(0)$ related to the isothermal compressibility through Equation 2.19.

2.8 TESTING THE THEORETICAL PREDICTIONS OF THE COARSE-GRAINED MODEL AGAINST ATOMISTIC SIMULATIONS

To test the quality of the theoretical predictions, we performed a series of simulations of soft-blob coarse-grained models with variable levels of CG and UA models, under the same set of molecular and thermodynamic conditions, for a variety of chain lengths and densities at two different temperatures. The UA simulations do not provide information to the coarse-grained model, but they are used to test the consistency of the coarse-grained models. The calculations start from the equation of state, from which all the thermodynamic properties of interest are derived and then compared to the results from simulations following the scheme of Figure 2.1.

As expected, we see that macroscopic properties of the polymeric liquid do not depend on the lengthscale that we select to coarse-grain the macromolecules, whereas the internal energy and entropy are model dependent [9]. See Figure 2.3 for an example of the calculation of the pressure as a function of the degree of polymerization for a set of simulations with variable degrees of CG. The pressure shows consistency for all the samples; the data are also in agreement with the analytical expression of the equation of state.

The Helmholtz free energy per monomer is obtained through thermodynamic integration of the pressure as a function of the packing fraction

$$\frac{F^{ex}}{Nnk_B T} = \int_{\eta_1}^{\eta_2} \left(\frac{P}{\rho_{ch} k_B T} \right), \tag{2.21}$$

$$= \frac{-2(1 - c_1 - 3c_2)\eta_{eff}^2 + 4(1 - c_2)\eta_{eff}}{(1 - \eta_{eff})^2} - 4c_2 \ln(1 - \eta_{eff})$$

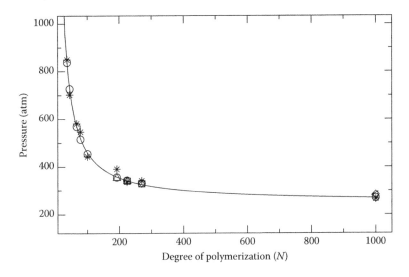

FIGURE 2.3 Pressure as a function of the degree of polymerization for polymer liquids represented at different levels of CG. A hierarchy of soft-blob simulations are compared to atomistic simulations and to the analytical theory, Equation 2.9 with c_0 from Equation 2.18. The simulations were carried out at constant temperature, $T = 509$ K, and constant density, $\rho = 0.733$ g/mL. UA simulations are represented by circles, the soft-sphere model by asterisks, tri-blob by triangles, penta-blob by squares, 10-blob by diamonds, and 20-blob by left-oriented triangles. (Reprinted with permission from J. McCarty et al., *J. Chem. Phys.*, 140. Copyright 2014, American Institute of Physics.)

The excess Helmholtz free energy, associated to the liquid as distinct from the ideal gas, is shown to be independent of the degree of CG, as is also illustrated by the example in Figure 2.4 [9].

The excess Gibbs free energy in the canonical ensemble is calculated in the mean-field approximation, which holds at liquid density, as

$$G^{ex} = \frac{nn_b\rho}{2N_b} \int v^{bb}(r)g^{bb}(r)d\mathbf{r}, \qquad (2.22)$$

with N_b the number of monomers per blob, from which we obtain the excess free energy per monomer

$$\frac{G^{ex}}{nNk_BT} = -\frac{\rho c_0}{2} \qquad (2.23)$$

It is worth noticing that the expressions for the Helmholtz and the Gibbs excess free energies are different because their calculations account for the different thermodynamic parameters that are controlled. The Helmholtz

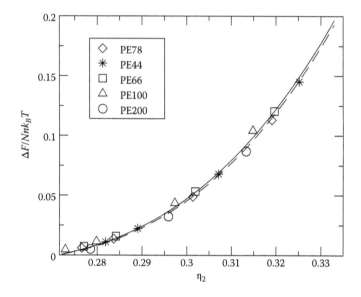

FIGURE 2.4 Helmholtz free energy changes as a function of the packing fraction compared to a reference packing fraction of $\eta_1 = 0.27$, which is the lowest packing fraction that was simulated. All systems collapse to a universal curve, within numerical precision and independent of the degree of polymerization. Solid and dashed lines are Equation 2.22 for $N = 44$ and $N = 200$, respectively. (Reprinted with permission from J. McCarty et al., *J. Chem. Phys.*, 140. Copyright 2014, American Institute of Physics.)

free energy is calculated at constant volume by integration over the packing fraction and so the density; these calculations account for the density dependence of the direct correlation function, c_0. The Gibbs free energy, instead, is calculated in the canonical ensemble, where the density is constant, and so is c_0. For this quantity, we evaluated the pressure dependence, input to

$$\frac{\Delta G}{nNk_BT} = \frac{1}{\rho k_BT}\int_{P_1}^{P_2} dP \tag{2.24}$$

by calculating the change in free energy related to the change in pressure observed when simulations at constant volume and temperature are performed as a function of the number of monomers per chain, and then compared to our analytical expression

$$\frac{\Delta G}{nNk_BT} = \left[\frac{1}{N} - \frac{\rho c_0}{2}\right]_2 - \left[\frac{1}{N} - \frac{\rho c_0}{2}\right]_1 \tag{2.25}$$

The excess free energy in both ensembles is a constant quantity and does not depend on the coarse-gained model adopted.

2.8.1 Potential Energy

The internal energy, and the related potential energy, displays a more subtle dependence on the number of CG units selected to partition the molecule. For the *soft-sphere* description, for which $n_b = 1$ and $N_b = N$, the potential energy includes only intermolecular contributions and is equivalent to the excess free energy

$$\frac{E^{\text{soft-sphere}}}{nk_BT} = \frac{2\pi\rho_{\text{ch}}}{k_BT}\int_0^\infty v^{\text{ss}}(r)g^{\text{ss}}(r)r^2dr \approx -\frac{\rho Nc_0}{2} \tag{2.26}$$

In the multiblob description, the potential energy has both inter- and intramolecular contributions. The intermolecular part is calculated as an extension of the formula for the soft-sphere

$$\frac{E^{\text{bb}}}{nk_BT} = \frac{2\pi\rho_b}{k_BT}\int_0^\infty v^{\text{bb}}(r)g^{\text{bb}}(r)r^2dr \approx -\frac{\rho Nc_0}{2} \tag{2.27}$$

and gives a contribution that even in the multiblob description is a constant. The potential energy, however, also contains, in this case, contributions from the intramolecular structure, such as bond stretching, angle bending, torsional rotation, and nonbonded pair interactions, which are representation dependent. Since the bond energy is a harmonic potential with a Gaussian probability distribution, the average bond energy is simply the equipartition result,

$$\left\langle\frac{E_{\text{bond}}}{nk_BT}\right\rangle = \frac{3(n_b - 1)}{2} \tag{2.28}$$

For the angular contribution to the energy, we add an additional $E_{\text{angle}} \approx (n_b - 2)/2k_BT$ contribution per chain. The total predicted energy is shown in Figure 2.5 and represented by the line.

From the simulations, the potential energy is calculated as the average total energy minus the kinetic energy contribution. In Figure 2.5, we show the potential energy per molecule for two different systems, PE200 at $\rho = 0.8$ g/mL at $T = 400$ K and PE1000 at $\rho = 0.733$ g/mL and $T = 509$ K as a function of the number of effective sites, n_b. The UA simulations are represented by the last data point on the right of the figure, where the number of effective sites is equal to the number of UAs.

The figure shows that in both sets of simulations, the potential energy changes as the number of sites is increased, and that the agreement between theoretical expressions and simulations is quantitative up to the

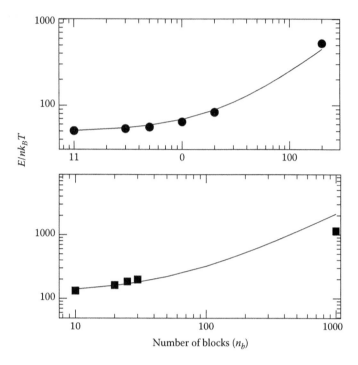

FIGURE 2.5   Potential energy changes as a function of the number of blobs for two samples. The first is polyethylene with $N = 200$ (top) and the second is polyethylene with $N = 1000$ (bottom). The last point is the potential energy from UA simulations. The solid line is the theoretical prediction and is extrapolated to the last point in the limit of large n_b. (Reprinted with permission from J. McCarty et al., *J. Chem. Phys.*, 140. Copyright 2014, American Institute of Physics.)

atomistic description where a small error is observed. In this case, however, the small disagreement observed between theory and simulations is due to the approximation of using the mean-field equation, $g^{bb}(r) \to 1$, which becomes increasingly less accurate as the local structure becomes important.

2.8.2 Entropy

The basic procedure of any CG formalism is the averaging of the microscopic states that are then represented by effective units, with the consequence that the entropy of the system in a given coarse-grained representation is different with respect to the atomistic description. This is a direct consequence of the fact that the CG reduces the dimensionality of the configuration space and smooths the probability distributions.

The extent of the change in entropy depends on the level of detail maintained in the coarse-grained representation, which determines the number of atomistic configurations that can be mapped into a single coarse-grained configuration. It can be quite large when the level of CG is extreme and the underlying chain is flexible. This is commonly called the "mapping entropy" and is simply the difference in entropy of the atomistic model when viewed from the atomistic configurations and the coarse-grained configurations [54].

If the chain is assumed to have a statistical distribution of monomers in space that follows a Gaussian form, the entropy associated with increasing the number of blobs in the CG procedure is given as

$$\frac{S_{bb}}{nk_B} \approx \frac{3}{2}n_b + \frac{3}{2}(n_b - 1) - \frac{3}{2}(n_b - 1)\ln\left(\frac{3n_b}{8\pi R_g^2}\right) + \ln\left(\frac{Ve}{\Lambda^3 n}\right) \quad (2.29)$$

The first two terms in Equation 2.29 arise from the kinetic energy and bond potential energy, whereas the final two terms are the ideal translational and vibrational free energy. Importantly, there is no contribution in Equation 2.29 from the potential or c_0, since the increasing entropy with the number of blobs n_b is due solely to the increasing configurational degrees of freedom and not the interaction potential itself.

Another type of entropy of interest is the relative entropy [19]. This function is based on the "information" that is lost during CG, which has to be minimized to optimize the coarse-grained model. Our coarse-grained formalism, based on liquid state theory, is devised to reproduce the correct distribution function, so that the relative entropy between the coarse-grained sites and monomer sites is minimized, and the potential is optimized, without need for any variational approach. This is equivalent to saying that the relative entropy, which is based on the information function that discriminates between coarse-grained configurations sampled in the two levels of representation, is zero.

2.9 INSIGHTS IN CG FROM THE THERMODYNAMICALLY CONSISTENT COARSE-GRAINED MODEL

One of the advantages of having an analytical method of CG is that it is possible to study how physical inconsistencies can arise from a selected CG procedure. More precisely, it is possible to see how errors in the procedure can lead to errors in the resulting simulated quantities on the mesoscale [8].

An issue often reported when performing simulations of coarse-grained systems is that the mesoscale simulation describes a liquid that is too compressible in comparison to the more realistic modeling of the related atomistic simulations. This is not a problem for our model, which is largely analytical, but it affects numerically optimized CG methods such as the IBI procedure.

The conventional IBI procedure optimizes the CG potential by minimizing the disagreement between the atomistic and the mesoscale pair distribution functions, $g(r) = h(r) - 1$, or equivalently the total correlation function, $h(r)$. When the effective potential is calculated from $g(r)$, the IBI procedure rapidly converges to a total correlation function indistinguishable from the one which the procedure is started with. The pressure, however, resulting from the mesoscale simulation that uses the derived potential, is found to be reduced from the correct value of the atomistic simulation.

To study the reason for the observed disagreement, we started by comparing our analytical total correlation function for the soft-sphere representation to data from simulations. The agreement between analytical theory and atomistic simulations is quantitative (see Figure 2.6).

In the IBI, the total correlation function against which the potential is optimized is defined up to a given interparticle distance, r_{cut}, which is a fraction of the box size in the atomistic simulation. To mimic the IBI, we simply set the total correlation function to be identical to zero at a distance r' that is larger than a given r_{cut}, where we select for the value of the cutoff distance 3.5 R_g, where R_g is the radius of gyration of the polymer chain. Outside the radius of gyration of a polymer, for $r > R_g$, the probability to successfully find another polymer chain is ≈100% favorable and $h^{bb}(r) \approx 0$ is a valid approximation, even more so at a distance as large as the one we selected.

Once the total correlation function is optimized with data up to $r = 3.5\ R_g$, we derive the potential and run molecular dynamic simulations for the soft-sphere liquid. The total correlation function obtained in the mesoscale simulation is indistinguishable within numerical error from the results of the atomistic simulations. However, the pressure in the atomistic and mesoscale simulations is different. Figure 2.7 presents the calculation of the pressure with and without truncation of $h^{cc}(r) = g^{cc}(r) - 1$. The error due to the truncation of the total correlation function at large distance leads to large errors in the pressure, because in the virial equation, Equation 2.30, the pair correlation function is weighted by the distance elevated to the third power; small errors in the tail of a long-ranged potential strongly affects the precision of the equation of state. By truncating the total

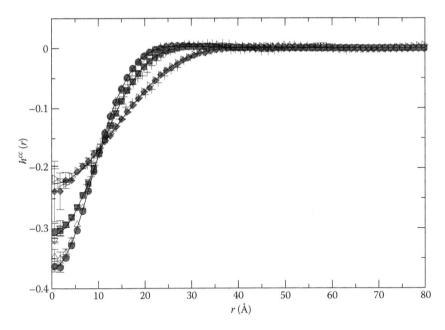

FIGURE 2.6 Total correlation function for the soft-sphere representation as a function of polymer center-of-mass distance, for chains of increasing degree of polymerization: $N = 78$ (filled gray circles), $N = 100$ (filled gray squares), and $N = 192$ (filled gray diamonds). The total correlation function from atomistic simulations (star symbols) is compared to the theory (lines) and to the results from mesoscale simulations of the coarse-grained system, where the soft-sphere potential is derived by cutting $h^{cc}(r)$ at $r = 3.5\,R_g$. $N = 78$ (unfilled up-pointing triangles), $N = 100$ (unfilled down-pointing triangles), and $N = 192$ (unfilled right-pointing triangles). All produce the same total correlation function.

correlation function, the potential that results from the procedure is also truncated and the pressure of the coarse-grained simulation is underestimated. As the equation of state is different when the pair distribution function is truncated, all the thermodynamic quantities that are derived from the equation of state will also be plagued by errors. The virial equation is

$$\frac{P}{\rho_{ch}k_BT} = 1 - \frac{2\pi\rho_{ch}}{3k_BT}\int_0^\infty g(r)\frac{dv(r)}{dr}r^3dr \qquad (2.30)$$

The error is larger the longer the range of the potential, the larger the lengthscale of CG, and the higher the density of the liquid. Unfortunately, those are the conditions where CG is most useful. The range of the long repulsive tail of the potential increases with the level of CG and large-scale precision becomes important for polymer liquids when the chains are highly coarse grained. Fine-gained models have smaller errors in the thermodynamics, but for these models, the gain in computational time is limited.

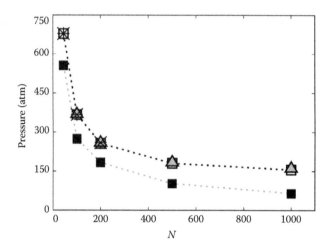

FIGURE 2.7 Pressure measured from molecular dynamics simulations of the coarse-grained system performed either using the full tail of $g(r)$ in the effective potentials (open squares) or the $g(r)$ set to be equal to one for $r > 3.5\,R_g$ (filled squares). Also shown are UA simulations (stars) for systems where they are available ($N \leq 200$). Data are for systems consistent with Figure 2.6. Despite both potentials reproducing the structure with high accuracy, the pressure is clearly affected by cutting the pair distribution function at large distances. (Reprinted with permission from A. J. Clark et al., *J. Chem. Phys.*, 139. Copyright 2013, American Institute of Physics.)

Interestingly, even if the potential incorrectly predicts the thermodynamic properties the structure of the liquid in the form of the total correlation function is correctly reproduced. This is because the pair distribution function and its related total correlation function are notoriously insensitive to small differences in the potential. It is known that different potentials can lead to the same pair distribution function. For example, the total correlation function of a Lennard–Jones liquid is very similar to the total correlation function of a liquid of hard spheres of the appropriate diameter. In fact, pair distribution functions are mostly defined by the repulsive part of the potential, whereas the attractive contribution has little or no influence.

2.10 RECONSTRUCTION OF THE DYNAMICS: WHY THE DYNAMICS ARE TOO FAST IN COARSE-GRAINED MODELS

While structure and thermodynamic properties are well described across multiple levels of CG, the dynamical properties measured in mesoscale simulations of coarse-grained systems are always too fast when compared to the related atomistic simulations. For example, the mean-square displacement and diffusion coefficient of a coarse-grained representation can be several orders of magnitude faster than in the atomistic simulations.

In the coarse-grained model, local degrees of freedom are averaged out and the molecules move rapidly over the simplified free energy landscape. While the system explores the reduced configurational landscape, the measured dynamics is artificially sped up by the smoothness of the potential. The speed up in the dynamics is proportional to the level of CG, so that the largest computational gain is obtained for the most coarse-grained system.

Given that the measured dynamics is unrealistically fast, it needs to be properly rescaled *a posteriori* to recover the correct dynamics of the atomistic description. We have proposed an analytical procedure to rescale the mesoscale dynamics. Because all the structural and thermodynamic quantities are known in our model, it has been possible to derive from first principles a procedure for the reconstruction of the dynamics that is based on the solution of the generalized Langevin equations for the atomistic and the coarse-grained representations. We identified two steps in the rescaling procedure of the coarse-grained dynamics to reconstruct the atomistic description. A first rescaling aims at including the dissipation of energy due to the internal vibrational degrees of freedom in a time rescaling. Those degrees of freedom are averaged out in the soft-colloid representation during the CG process. The second step accounts for the change in shape, of the molecular surface exposed to the solvent, and friction coefficient of the polymeric units as a consequence of CG [12,55,56].

The procedure of dynamical reconstruction that we proposed is different from the usual one because it does not require performing atomistic simulations, and once the only parameter, that is the effective hard-sphere diameter of the monomer, is defined, the theory is fully predictive. The common strategy is based, instead, on the application of a "calibration curve" previously obtained through the numerical fitting of dynamical quantities; in the calibration curve procedure, parameters are optimized until one achieves agreement for the dynamical properties of interest calculated in the atomistic and mesoscale simulations [5,57]. However, the same reasoning that applies for static properties also applies in the case of the dynamics; once the atomistic simulations are performed, it is not obvious the need of performing also mesoscale simulations.

The numerical calculation of optimized calibration curves for the dynamics is quite difficult to achieve for macromolecular systems because the dynamics is mode dependent; there are, in principle, N internal modes in any molecule formed by N units and the degree of polymerization of a long chain can be of the order of 1 million monomers. Numerically optimized parametric quantities are, in general, not transferable between systems in different thermodynamic conditions or with different chemical structure or increasing degree of polymerization. To overcome this problem, it is common to select coarse-grained units that are very close in size to the atomistic units, so that the needed corrections to reach consistency in static, thermodynamic, and dynamic properties is minimal. In this case, corrections to the measured dynamics can be evaluated through a perturbative formalism, which should rapidly converge to the desired value. The resulting computational gain is, however, limited.

Our procedure has been developed so far for the soft-sphere representation, which affords the largest dynamical gain. The same principles hold for variable levels of multiblob representations. In the soft-sphere representation, the internal dynamics cannot be studied, but the coarse-grained simulation can provide information on the center-of-mass diffusion. The mesoscale simulations of the coarse-grained system provide data that, once they are properly rescaled have shown to predict center-of-mass dynamics in quantitative agreement with experiments and atomistic simulations. Because the soft-sphere representation is the one with the highest level of CG, it requires the largest correction to the measured dynamics and is the one where possible errors in the procedure can have the most visible consequences. The fact that the agreement between the dynamics reconstructed from the coarse-grained simulations and the atomistic-scale representation is good suggests that our procedure is robust.

The "entropic" and "frictional" corrections to the mesoscale dynamics enter the Langevin equation of the coarse-grained system and rescale its dynamics. In the Langevin equation, the energy dissipated is calculated by adopting a bead-spring monomer representation of the polyethylene chain. The correction term that must be included in rescaling of the dynamics of coarse-grained simulations to account for the missing entropic degrees of freedom, starting from a freely rotating chain model, is equivalent to Equation 2.28 [55,56].

The second rescaling addresses the change in the polymer friction coefficient due to the reduction of surface when the chain is coarse grained. This rescaling factor is derived from the solution of the memory functions in the Langevin equations describing the dynamics of the polymer chains in the two levels of representation.

Using the proposed rescaling procedure, dynamical data from mesoscale simulations of polyethylene melts were compared with the ones measured in atomistic simulations and in experiments. The agreement between predicted and known properties was found to be almost exact. Furthermore, the procedure allowed for the prediction of new values of dynamical parameters, that is diffusion coefficients, for systems that were not yet simulated or measured experimentally.

It is possible to take advantage of the artificial acceleration of the dynamics when coarse-grained representations are used to rapidly reach an equilibrated state before starting an atomistic molecular dynamics simulation. In that case, the variable level of CG allowed by our model is used to seamlessly change the resolution in CG.

2.11 A COARSE-GRAINED METHOD FOR PROTEIN DYNAMICS: THE LANGEVIN EQUATION FOR PROTEIN DYNAMICS

We conclude by presenting a coarse-grained approach to describe the fluctuating dynamics of folded proteins in dilute solutions. Many of the concepts presented in this chapter were used in building this coarse-grained model for the dynamics of proteins [11,58,59].

Because we are interested in the motion of a large macromolecule immersed in a liquid of small water molecules and ions, the difference in size and in the characteristic timescale of the dynamics between solute and solvent suggests that it is appropriate to treat the protein and the solvent with different coarse-grained formalisms. The protein is simply treated as a collection of units centered on the position of the alpha-carbons and connected through effective springs along the primary sequence of the protein. The dynamics of each unit is coupled to the remaining others through effective pair potentials. The solvent is treated as a continuum, following the tradition of polymer physics. The solvent affects the dynamics of the protein through the viscosity, the random collision with the protein, and the friction.

In this model, the dynamics follows a modified Rouse–Zimm Langevin equation that we named the Langevin equation for protein dynamics (LE4PD). Interprotein interactions are not important because the solution is diluted. Each coarse-grained macromolecular unit represents one amino acid, and the specific shape of the side-chain is accounted for through the hydrodynamic radius of the residue, its friction coefficient, and the effective interaction with others units. Because the level of CG is contained, and each unit represents a relatively small number of atoms, a numerical evaluation of the potential from atomistic simulations or experiments is appropriate; the resulting error is small and can affect only slightly the large-scale properties of the protein. In the hydrophobic core, the hydrodynamic interaction is screened.

In several points, the LE4PD departs from the traditional model for polymer dynamics. The LE4PD approach uses a harmonically coupled description, but complete with site-specific dissipation, hydrodynamic coupling, and barriers to internal fluctuations, calculated directly from the structural ensemble created by the simulation of the protein in aqueous solvent. The knowledge of the roughness of the free energy landscape, that is the sampled energy barriers, provides information on the long-time dynamics. The predictions for the time correlation functions calculated from the theory exceed in timescale the time correlation function directly calculated from the simulation trajectories. Assuming that in a longer timescale, the protein still samples the configurational space covered by the simulations, then the motion described by our approach in its present form ensures an accurate determination of the dynamics over a wider range of timescales than the simulation itself.

The approach is based on the fundamental picture of proteins as heterogenous polymers that are collapsed into a definite tridimensional structure, which nevertheless retains some amount of flexibility. As opposed to a rigid body, where the global modes are the only degrees of freedom in the system, protein dynamics include both rotational and internal fluctuation modes. Our description accounts for internal dissipation due to fluctuations in the hydrophobic region by accounting for an effective protein internal viscosity and considering the relative exposure of each amino acid to the hydrophobic region. With the correct dissipation, the linear modes of harmonically coupled objects provide a simple but accurate description of the fluctuations of the molecule [58,59]. The theory predicts local dynamics in close agreement with experimentally measured time correlation functions, such as T_1, T_2, and *nuclear Overhauser effect* (NOE) data from NMR experiments [11].

2.12 SUMMARY

In our approach, the CG of macromolecular systems is based on the integral equation theory; pair distribution functions and effective potentials for the coarse-grained units have been calculated for models that have variable levels of resolution. The approach affords the analytical solution of the potential, from which we developed an analysis of the structural, thermodynamic, and dynamical properties of the coarse-grained description as a function of the resolution of the model, or level of CG. Analytical results are always tested against the full numerical solution of the coarse-grained potential and show high accuracy in the region where the analytical approximation applies. Structural and thermodynamic properties between the persistent length resolution and the most extreme level of CG, where the whole molecule is represented as a point particle interacting through a soft long-ranged potential, are properly described by our coarse-grained approach. For resolution smaller than the polymer persistence length, the general properties of our coarse-grained model still hold, but the solution of the integral equation theory has to be performed numerically. We expect that thermodynamic and structural consistency are maintained also on the very local scale, once the proper closure is adopted.

Our theory differs from most alternative CG approaches because it does not require performing high-resolution simulations to numerically parameterize the coarse-grained model. Atomistic simulations can be used only as a test of the consistency of the coarse-grained description. We find that our model reproduces the correct pressure, structural distributions, compressibility, and free energy of the underlying system. Internal energy and entropy are, instead, depending on the degree of resolution of the coarse-grained model.

All the quantities depend on one nontrivial parameter, that is, the total correlation function at $k \to 0$, c_0, which can be directly determined from the experimental isothermal compressibility of the liquid. This parameter is system specific, and depends on the thermodynamic conditions, but does not depend on the resolution of the selected coarse-grained description.

In numerical procedures of optimization, it is common to rely on pair distribution functions obtained either experimentally or from atomistic simulations; in both cases, the function is truncated at large distance. We have shown that the truncation leads to a consequent error in the thermodynamics of the coarse-grained simulation.

CG also produces a speeding up of the dynamics that allows for the fast simulations of molecules within a reduced description. The dynamical properties, however, in the coarse-grained simulations are unrealistically accelerated and need to be properly rescaled to give realistic values of the dynamics. We have shown that both entropic rescaling of the degrees of freedom and rescaling of the effective friction coefficients are important to reconstruct the correct dynamics. Global dynamics and diffusion coefficients are well predicted with our rescaling procedure applied *a posteriori* on data from coarse-grained simulations.

Finally, we briefly described a fine-grained numerical model for the dynamics of a protein in solution. This model is formulated in a normal-mode description, which includes both the rotational relaxation and the local energy barriers. The theory, called "LE4PD," predicts dynamics in agreement with experimental NMR relaxation and does not require direct fitting of the experimental data.

ACKNOWLEDGMENTS

This work is largely that of Jay McCarty, Anthony Clark, Jeremy Copperman, Ivan Lyubimov, and Mohammadhasan Dinpajooh. Without them none of this would have been possible.

My thanks go to my mentor Ken Schweizer who first introduced me to the theory of liquids. His work has been a continuous font of inspiration and his theory, PRISM, is the basis of our coarse-graining method.

In addition, I thank Paula J. Seeger for carefully reading the manuscript.

This work was supported by the National Science Foundation Grant No. CHE-1362500. CPU time was provided by NSF Grant No- ACI-1053575 through Extreme Science and Engineering Discovery Environment (XSEDE) resources.

REFERENCES

1. E. M. Curtis and C. K. Hall, Molecular dynamics simulations of DPPC bilayers using "LIME", a new coarse-grained model. *J. Phys. Chem. B*, **117**, 5019 (2013).
2. J. C. Johnston, N. Kastelowitz, and V. Molinero, Liquid to quasicrystal transition in bilayer water. *J. Chem. Phys.*, **133**, 154516 (2010).
3. S. J. Marrink, H. J. Risselada, S. Yefimov, D. P. Tieleman, and A. H. de Vries, The MARTINI force field: Coarse grained model for biomolecular simulations. *J. Phys. Chem. B*, **111**, 7812 (2007).
4. Z. Wu, Q. Cui, and A. Yethiraj, A new coarse-grained model for water: The importance of electrostatic interactions. *J. Phys. Chem. B*, **114**, 10524 (2010).
5. V. A. Harmandaris and K. Kremer, Dynamics of polystyrene melts through hierarchical multiscale simulations. *Macromolecules*, **42**, 791 (2009).
6. M. Muller and J. de Pablo, Computational approaches for the dynamics of structure formation in self-assembling polymeric materials. *Annu. Rev. Mater. Res.*, **43**, 1 (2013).

7. M. Putz, J. G. Curro, and G. S. Grest, Self-consistent integral equation theory for polyolefins: Comparison to molecular dynamics simulations and x-ray scattering. *J. Chem. Phys.*, **114**, 2847 (2001).

8. A. J. Clark, J. McCarty, and M. G. Guenza, Effective potentials for representing polymer melts as chains of interacting soft particles. *J. Chem. Phys.*, **139**, 124906 (2013).

9. J. McCarty, A. Clark, J. Copperman, and M. G. Guenza, An analytical coarse-graining method which preserves the free energy, structural correlations, and thermodynamic state of polymer melts from the atomistic to the mesoscale. *J. Chem. Phys.*, **140**, 204913 (2014).

10. P. Español, Dissipative particle dynamics, In *Handbook of Materials Modeling*, S. Yip (Ed.). The Netherlands: Springer, 2503 (2005).

11. J. Copperman and M. G. Guenza, Coarse-grained Langevin equation for protein dynamics: Global anisotropy and a mode approach to local complexity. *J. Phys. Chem. B*, **119**, 9195 (2015).

12. I. Y. Lyubimov and M. G. Guenza, Theoretical reconstruction of realistic dynamics of highly coarse-grained cis-1,4-polybutadiene melts. *J. Chem. Phys.*, **138**, 12A546 (2013).

13. R. L. Henderson, A uniqueness theorem for fluid pair correlation functions. *Phys. Lett. A*, **49**, 197 (1974).

14. H. J. Qian, P. Carbone, X. Chen, H. A. Karimi-Varzaneh, C. C. Liew, and F. Mller-Plathe, Temperature-transferable coarse-grained potentials for ethylbenzene, polystyrene, and their mixtures. *Macromolecules*, **41**, 9919 (2008).

15. J.-P. Hansen and I. R. McDonald, *Theory of Simple Liquids*. New York, NY: Academic Press, 2nd Ed. (1990).

16. M. E. Johnson, T. Head-Gordon, and A. A. Louis, Representability problems for coarse-grained water potentials. *J. Chem. Phys.*, **126**, 144509 (2007).

17. J. F. Dama, A. V. Sinitskiy, M. McCullagh, J. Weare, B. Roux, A. R. Dinner, and G. A. Voth, The theory of ultra-coarse-graining. 1. General principles. *J. Chem. Theory Comput.*, **9**, 2466 (2013).

18. P. Ganguly and N. F. A. van der Vegt, Representability and transferability of Kirkwood-Buff iterative Boltzmann inversion models for multicomponent aqueous systems. *J. Chem. Theory Comput.*, **9**, 5247 (2013).

19. M. S. Shell, The relative entropy is fundamental to multiscale and inverse thermodynamic problems. *J. Chem. Phys.*, **129**, 144108 (2008).

20. M. Zgarbova, M. Otyepka, J. Sponer, A. Mladek, P. Banas, T. E. Cheatham III, and P. Jurecka, Refinement of the Cornell et al. nucleic acids force field based on reference quantum chemical calculations of glycosidic torsion profiles. *J. Chem. Theory Comput.*, **7**, 2886 (2011).

21. A. A. Chen and A. E. Garcia, High-resolution reversible folding of hyperstable RNA tetraloops using molecular dynamics simulations. *Proc. Natl. Acad. Sci.*, **110**, 16820 (2013).

22. A. Savelyev and G. A. Papoian, Chemically accurate coarse graining of double-stranded DNA. *Proc. Natl. Acad. Sci.*, **107**, 20340 (2010).

23. A. Y. Grosberg and A. R. Khokhlov, *Statistical Physics of Macromolecules*. AIP Series in Polymers and Complex Materials New York, NY: AIP Press (1994).

24. M. Doi and S. F. Edwards, *The Theory of Polymer Dynamics*. New York, NY: Oxford University Press (1986).

25. B. A. Camley and F. L. H. Brown, Fluctuating hydrodynamics of multicomponent membranes with embedded proteins. *J. Chem. Phys.*, **141**, 075103 (2014).

26. D. M. Hinckley, J. P. Lequieu, and J. J. de Pablo, Coarse-grained modeling of DNA oligomer hybridization: Length, sequence, and salt effects. *J. Chem. Phys.*, **141**, 035102 (2014).

27. C. Clementi, Coarse-grained models of protein folding: Toy models or predictive tools? *Curr. Opin. Struct. Biol.*, **18**, 10 (2008).

28. P. G. Wolynes, Recent successes of the energy landscape theory of protein folding and function. *Q. Rev. Biophys.*, **38**, 405 (2005).

29. V. Tozzini, Coarse-grained models for proteins. *Curr. Opin. Struct. Biol.*, **15**, 144 (2005).

30. R. Galindo-Murillo, D. R. Roe, and T. E. Cheatham III, Convergence and reproducibility in molecular dynamics simulations of the DNA duplex d(GCACGAACGAACGAACGC) *Biochim. Biophys. Acta*, **1850**, 1041 (2015).

31. M. Dinpajooh and M. G. Guenza, Thermodynamic consistency in the structure-based integral equation coarse-grained method. *Polymers*, **117**, 282 (2017).

32. E. J. Sambriski and M. G. Guenza, Theoretical coarse-graining approach to bridge length scales in diblock copolymer liquids. *Phys. Rev. E*, **76**, 051801 (2007).

33. A. J. Clark and M. G. Guenza, Mapping of polymer melts onto soft-colloidal chains. *J. Chem. Phys.*, **132**, 044902 (2010).

34. A. J. Clark, J. McCarty, I. Y. Lyubimov, and M. G. Guenza, Thermodynamic consistency in variable-level coarse-graining of polymeric liquids. *Phys. Rev. Lett.*, **109**, 168301 (2012).

35. G. Yatsenko, E. J. Sambriski, M. A. Nemirovskaya, and M. Guenza, Analytical soft-core potentials for macromolecular fluids and mixtures. *Phys. Rev. Lett.*, **93**, 257803 (2004).

36. G. Yatsenko, E. J. Sambriski, and M. G. Guenza, Coarse-grained description of polymer blends as interacting soft-colloidal particles. *J. Chem. Phys.*, **122**, 054907 (2005).

37. E. J. Sambriski, G. Yatsenko, M. A. Nemirovskaya, and M. G. Guenza, An analytical coarse-grained description for polymer melts. *J. Chem. Phys.*, **125**, 234902 (2006).

38. A. Morriss-Andrews, F. L. H. Brown, and J.-E. Shea, A coarse-grained model for peptide aggregation on a membrane surface. *J. Phys. Chem. B*, **118**, 8420 (2014).

39. T. E. Ouldridge, *Coarse-Grained Modeling of DNA and DNA Self-Assembly*. Dordrecht: Springer (2012).

40. A. A. Louis, Beware of density dependent pair potentials. *J. Phys. Condens. Matter*, **14**, 9187 (2002).
41. Q. Sun, J. Ghosh, and R. Faller, State point dependence and transferability of potentials in systematic structural coarse graining, In *Coarse-Graining of Condensed Phase and Biomolecular Systems*, G. Voth (Ed.). Boca Raton, FL: CRC Press (2008).
42. G. R. Bowman, V. S. Pande, and F. Noè (Eds.), *An Introduction to Markov State Models and Their Application to Long Timescale Molecular Simulation*. New York, NY: Springer (2014).
43. A. Laio and F. L. Gervasio, Metadynamics: A method to simulate rare events and reconstruct the free energy in biophysics, chemistry and material science. *Rep. Prog. Phys.*, **71**, 126601 (2008).
44. A. Laio and M. Parrinello, Escaping free-energy minima. *Proc. Natl. Acad. Sci.*, **99**, 12562 (2002).
45. J. O. B. Tempkin, B. Qi, M. G. Saunders, B. Roux, A. R. Dinner, and J. Weare, Using multiscale preconditioning to accelerate the convergence of iterative molecular calculations. *J. Chem. Phys.*, **140**, 184114 (2014).
46. D. Reith, M. Putz, F. Muller-Plathe, Deriving effective mesoscale potentials from atomistic simulations. *J. Comput. Chem.*, **24**, 1624 (2003).
47. W. Shinoda, R. Devane, and M. L. Klein, Multi-property fitting and parametrization of a coarse grained model fro aqueous surfactants. *Mol. Simul.*, **33**, 27 (2007).
48. D. A. McQuarrie, *Statistical Mechanics*. Sausalito, CA: University Science (2000).
49. K. S. Schweizer and J. G. Curro, RISM theory of polymer liquids: Analytical results for continuum models of melts and alloys. *Chem. Phys.*, **149**, 105 (1990).
50. C. Dalgicdir, O. Sensoy, C. Peter, and M. Sayar, A transferable coarse-grained model for diphenylalanine: How to represent an environment driven conformational transition. *J. Chem. Phys.*, **139**, 234115 (2013).
51. C. N. Likos, A. Lang, M. Watzlawek, and H. Lowen, Criterion for determining clustering versus reentrant melting behavior for bounded interaction potentials. *Phys. Rev. E*, **63**, 031206 (2001).
52. J. McCarty, A. J. Clark, I. Y. Lyubimov, and M. G. Guenza, *Macromolecules*, **45**, 8482 (2012).
53. J. McCarty, I. Y. Lyubimov, and M. G. Guenza, Effective soft-core potentials and mesoscopic simulations of binary polymer mixtures. *J. Phys. Chem. B*, **113**, 11876 (2009).
54. J. F. Rudzinski and W. G. Noid, Coarse-graining entropy, forces, and structures. *J. Chem. Phys.*, **135**, 214101 (2011).
55. I. Y. Lyubimov, J. McCarty, A. Clark, and M. G. Guenza, Analytical rescaling of polymer dynamics from mesoscale simulations. *J. Phys. Chem.*, **133**, 094904 (2010).

56. I. Y. Lyubimov and M. G. Guenza, A first principle approach to rescale the dynamics of simulated coarse-grained macromolecular liquids. *Phys. Rev. E*, **84**, 031801 (2011).

57. S. O. Nielsen, C. F. Lopez, G. Srinivas, and M. L. Klein, Coarse grain models and the computer simulation of soft materials. *J. Phys. Condens. Matter*, **16**, R481 (2004).

58. J. Copperman and M. G. Guenza, Predicting protein dynamics from structural ensembles. *J. Chem. Phys.*, **143**, 243131 (2015).

59. J. Copperman and M. G. Guenza, Mode localization in the cooperative dynamics of protein recognition. *J. Chem. Phys.*, **145**, 015101 (2016).

Microscopic Physics-Based Models of Proteins and Nucleic Acids

UNRES and NARES

Maciej Baranowski, Cezary Czaplewski, Ewa I. Gołaś,
Yi He, Dawid Jagieła, Paweł Krupa, Adam Liwo,
Gia G. Maisuradze, Mariusz Makowski,
Magdalena A. Mozolewska, Andrei Niadzvedtski,
Antti J. Niemi, Shelly Rackovsky, Rafał Ślusarz,
Adam K. Sieradzan, Stanisław Ołdziej,
Tomasz Wirecki, Yanping Yin, Bartłomiej Zaborowski,
and Harold A. Scheraga

CONTENTS

3.1	Introduction	68
3.2	Theory	70
	3.2.1 Designing an Effective Coarse-Grained Energy Function	70
	3.2.2 PMF of a Coarse-Grained System and Its Cluster-Cumulant Expansion	71
3.3	UNRES Model	73
	3.3.1 Effective Energy Function	73
	3.3.2 Expressions of UNRES Energy Terms and Their Parameterization	76
	3.3.2.1 SC–SC Interaction Potentials	76

　　　3.3.2.2　Implementation of New SC–SC Potentials　78
　　　3.3.2.3　Local Potentials　78
　　　3.3.2.4　SC Torsional Potentials　78
　　　3.3.2.5　Backbone-Electrostatic and Correlation Terms　80
　3.3.3　Calibration of the Energy Function　84
　3.3.4　Treatment of Disulfide Bonds　85
　3.3.5　Treatment of D-Amino-Acid Residues　85
　3.3.6　Treatment of the Trans–Cis Isomerization of Proline Residue　86
　3.3.7　Coarse-Grained MD and Its Extensions with UNRES and NARES-2P　86
　3.3.8　UNRES in Blind Prediction of Protein Structures　88
　3.3.9　Investigation of Protein-Folding Pathways and FELs　90
　3.3.10　Protein and Kinks with UNRES　91
　3.3.11　Examples of Biological Applications　93
　　　3.3.11.1　Amyloid Formation by the Aβ40 Peptide　93
　　　3.3.11.2　PICK1–BAR Interactions　94
　　　3.3.11.3　Molecular Chaperones　95
3.4　Coarse-Grained Simulations of Nucleic Acids: The NARES-2P Model　98
　3.4.1　Model and Effective Energy Function　98
　3.4.2　Parameterization　98
　3.4.3　Calibration of the Force Field　100
　3.4.4　Simulations of Folding Pathways of Small DNAs and RNAs　101
　3.4.5　Simulations of Thermodynamics of DNA Folding　101
　3.4.6　Simulations of Premelting Transition (Internal-Loop or Bubble Formation)　103
3.5　A Coarse-Grained Model for Protein–DNA Interactions　103
3.6　Conclusions and Outlook　104
3.7　Program Availability　105
　　　Appendix 3A: Expansion (Factorization) of the Potential of Mean Force into Cluster-Cumulant Functions (Factors) and of the Expansion of the Factors into the Kubo Cluster Cumulants　106
　　　Acknowledgments　109
　　　References　109

3.1　INTRODUCTION

The need to account for experimental observations motivated the development of computer simulations in biophysics, biochemistry, and related sciences. Early work of Pauling et al. (1951) made use of hydrogen bonding in the polypeptide backbone to propose the formation of α and β structures. However, the detailed structures and functions of proteins rely on interactions among the side chains (SCs) of their amino-acid residues. Therefore, early computations considered the role of SC–hydrogen bonding (Laskowski and Scheraga, 1954), hydrophobic interactions (Némethy and Scheraga, 1962), and their combination (Némethy et al., 1963), in determining protein structure and function. These interactions affect the pKs of ionizable SCs and the mechanism of protein–protein interactions, e.g., as to how distant constraints affect folding pathways and native structures of proteins (Némethy and Scheraga, 1965) and how they influence biological processes such as the thrombin-induced conversion of fibrinogen to fibrin (Scheraga, 2004).

This chapter is devoted to highlights of the early development of the physical approach to biology and its evolution from an initial all-atom (Momany et al., 1975) to a coarse-grained (Liwo et al., 2008) treatment to simulate, first, the structure, folding pathways, and function of proteins (Liwo et al., 1999; He et al., 2013b), and subsequently those of nucleic acids (He et al., 2013a).

A natural way to run molecular simulations is the use of an all-atom approach (Momany et al., 1975; Brooks et al., 1983; Weiner et al., 1986; van Gunsteren and Berendsen, 1987). A big advantage of such a treatment is that atomic nuclei are well-defined objects and their spatial arrangement describes the complete geometry of a molecular system. A variety of all-atom force fields for molecular simulations has been constructed and is still being developed. One disadvantage of the all-atom approach, though, is that it is very time consuming and feasible for biological applications only with high-speed supercomputers; another one is that it takes quite an effort to extract the essential dynamics of the system from the atomic motions. Coarse graining, in which several atoms are merged into a single interaction site (Liwo et al., 2001, 2002; Koliński and Skolnick, 2004; Tozzini, 2005; Colombo and Micheletti, 2006; Clementi, 2008; Czaplewski et al., 2010), is a natural way to overcome both obstacles; the reduced number of interaction sites and degrees of freedom result in both reduction of energy cost and in simplification of system motions, which are more transparent for analysis.

Coarse-grained force fields, applicable for computation of biomolecular structures and dynamics, are divided into two main categories: knowledge-based force fields derived from structural data and physics-based force fields, which are based on the relation between all-atom and coarse-grained energy surfaces (Czaplewski et al., 2010). In the knowledge-based coarse-grained force fields, the specific terms are derived from the distributions of, e.g., distances between SC centers or angles between the coarse-grained sites. Conversely, in the physics-based force field, the coarse-grained energy surface is constructed from the all-atom energy surface by averaging out the degrees of freedom not included in the model and, often, also by smoothing and simplifying the resulting potentials. This approach is the basis of the coarse-grained UNited-RESidue (UNRES) force field (Liwo et al., 2001, 2004, 2008) developed in our laboratory to simulate the structure and dynamics of proteins and, recently, the Nucleic Acids united-RESidue 2-point model (NARES-2P) force field (He et al., 2013a) developed to simulate the structure and dynamics of nucleic acids.

This chapter is organized as follows. Section 3.2 summarizes the theoretical principles of the construction of physics-based coarse-grained force fields. Section 3.3 describes the UNRES force field, together with its recent extensions to treat disulfide bonds, cis–trans isomerization of peptide groups, and D-amino-acid residues. It next describes its molecular dynamics (MD) implementation, and its applications in physics-based prediction of protein structures, analysis of protein free-energy landscapes (FELs), and description of protein folding in terms of kinks (or dark solitons) as well as applications to study biological problems. Section 3.4, in the spirit of UNRES, describes our recently developed NARES-2P force field that, despite a very simplified description of polynucleotide chains, reproduces the structure and thermodynamics of the folding of small DNA and RNA molecules. Section 3.5 describes our recently developed potentials for protein–nucleic acid interactions. Sections 3.6 and 3.7 conclude the chapter by describing perspectives for further applications and software availability, respectively.

3.2 THEORY

3.2.1 Designing an Effective Coarse-Grained Energy Function

As mentioned in Section 3.1, the physics-based coarse-grained potentials relate the effective energy surface to all-atom energy surfaces and thermodynamic quantities measured for model systems or both. For a coarse-grained model to work, it is absolutely essential that how this relationship is chosen and how it is implemented. Second, it is important that how care is taken of the dominant interactions leading to the formation of the spatial architecture and dynamics. Without paying enough attention to one or both of these issues, many coarse-grained models that have been developed very carefully do not work as well as expected, for example, the very popular MARTINI force field (Monticelli et al., 2008) can be used to study protein aggregation and protein movement through the lipid phase but it cannot be used to study fold proteins.

In the UNRES and NARES-2P models, both points are taken into account. The effective energy function originates from the potentials of mean force (PMFs) of the system under study, which assures its tight relationship with the respective all-atom energy function; this point is described in the remaining part of this section. Special attention has been devoted to develop a simple but accurate representation of the electrostatic interactions between the polar centers of the biopolymer units (peptide groups or nucleic-acid bases, respectively); this is described in Sections 3.3 and 3.4, which are devoted to UNRES and NARES-2P, respectively.

3.2.2 PMF of a Coarse-Grained System and Its Cluster-Cumulant Expansion

Here, we define the prototype of the UNRES as the PMF, which is obtained by computing the part of the configurational integral of a system corresponding to integrating over the secondary degrees of freedom. If X and Y denote the coarse-grained and secondary degrees of freedom (orthogonal to X), respectively, and $E(X; Y)$ denotes the all-atom energy function, the PMF can be expressed by Equation 3.1 (Liwo et al., 1998, 2001; Izvekov and Voth, 2005a,b; Ayton et al., 2007).

$$F(\mathbf{X}) = -RT \ln \left\{ \int_{\Omega_Y} \exp\left[-\frac{E(X; Y)}{RT} \right] dV_Y \right\} \tag{3.1}$$

where Ω_Y is the space spanned by Y, dV_Y is the volume element of that space, R is the universal gas constant, and T is the absolute temperature. The effective energy function defined by Equation 3.1 is strictly related to the probability of a given coarse-grained conformation to occur and also enables us to compute exactly the thermodynamic quantities and ensemble averages that depend only on the coarse-grained coordinates.

To obtain a usable energy function, the function $F(\mathbf{X})$, given by Equation 3.1, must be simplified. For this purpose, we developed a factor approach (Liwo et al., 2001), in which $F(\mathbf{X})$ is expressed as a sum of cluster-cumulant functions, as in Equation 3.2. Each cluster cumulant corresponds to interactions within a smaller subset of the system, containing interactions of the atoms within a given coarse-grained site or those belonging to two interaction sites (e.g., protein SCs), and then to supplements from the interactions within three and more coarse-grained sites.

$$F(\mathbf{X}) = \sum_i \langle\langle \varepsilon_i \rangle\rangle_f + \sum_{i<j} \langle\langle \varepsilon_i \varepsilon_j \rangle\rangle_f + \sum_{i<j<k} \langle\langle \varepsilon_i \varepsilon_j \varepsilon_k \rangle\rangle_f + \dots + \langle\langle \varepsilon_1 \varepsilon_2 \dots \varepsilon_N \rangle\rangle_f$$

(3.2)

where the quantities in double brackets are the cluster-cumulant functions (Kubo, 1962) (the PMF factors $\langle\langle \varepsilon_1 \varepsilon_2 \dots \varepsilon_m \rangle\rangle_f$ of order m) and the respective εs in double brackets represent the energies of interactions within a site or between two sites involved in a factor. The factors are expressed by Equation 3.3. The coarse-grained variables collected in vector $\mathbf{X}$ were omitted from Equation 3.2 and further equations for the sake of clarity.

$$\langle\langle \varepsilon_{i1}, \varepsilon_{i2}, \dots, \varepsilon_{ik} \rangle\rangle_f = \sum_{l=1}^{k} \sum_{\substack{i_{m1}<i_{m2}<\dots<i_{ml} \\ m_i \in [1\dots k]}} (-1)^{k-l} \langle\langle \varepsilon_{i_{m1}}, \varepsilon_{i_{m2}}, \dots, \varepsilon_{i_{ml}} \rangle\rangle \quad (3.3)$$

Each of the $\langle\langle \varepsilon_1, \varepsilon_2, \dots, \varepsilon_m \rangle\rangle$ terms in the expression for $\langle\langle \varepsilon_1, \varepsilon_2, \dots, \varepsilon_m \rangle\rangle_f$ given by Equation 3.3 is a PMF calculated for a subset of the original system that comprises m groups of interactions, each group containing interactions between the atoms of one coarse-grained site or between the atoms that belong to two coarse-grained sites. The terms $\langle\langle \varepsilon_1, \varepsilon_2, \dots, \varepsilon_m \rangle\rangle$ are expressed by Equation 3.4.

$$\langle\langle \varepsilon_{i_1}, \varepsilon_{i_2}, \dots \varepsilon_{i_k} \rangle\rangle = -\frac{1}{\beta} \ln \left\{ \frac{1}{V_{y_I}} \int_{\Omega_I} \exp\left[-\beta \sum_{l=1}^{k} \varepsilon_{i_l}(\mathbf{X}; z_{i_l}) \right] dV_{y_I} \right\} \quad (3.4)$$

where $\beta = 1/RT$, and V_{y_I} is the volume of the subspace spanned by the variables $y_{i1}, y_{i2}, \dots, y_{ik}$, $dV_{y_I} = dy_{i1}, dy_{i2} \dots, dy_{ik}$ is the volume element and $z_{i_l} = y_{i_l}$ if the respective interactions occur within a single coarse-grained site (good examples are, e.g., the interactions between the atoms of a peptide group), which give rise to the $C^\alpha \dots C^\alpha$ virtual-bond-stretching and trans–cis isomerization potential (Sieradzan et al., 2012b), or $z_{i_l} = \left(y_{\kappa(i_l)}, y_{\lambda(i_l)} \right)$ if the respective interactions occurs between two sites with indices $\kappa(i_l)$ and $\lambda(i_l)$; an example is, e.g., the interactions between two SCs or two peptide groups.

The first-order factors $\langle\langle\varepsilon_i\rangle\rangle_f$ are PMFs of the interactions between isolated sites that can be identified with SC (SC–SC) interactions, SC–peptide group (SC–p) interactions, and peptide group (p–p) interactions as well as the PMF of the local interactions within isolated sites. The second-order factors $\langle\langle\varepsilon_i\varepsilon_j\rangle\rangle_f$ contain PMFs of pairs of component interactions minus the sums of the PMFs of the single-component interactions, i.e., $\langle\langle\varepsilon_i\varepsilon_j\rangle\rangle_f = \langle\langle\varepsilon_i\varepsilon_j\rangle\rangle - \langle\langle\varepsilon_i\rangle\rangle - \langle\langle\varepsilon_j\rangle\rangle$. They can be regarded as correlation terms pertaining to component interactions i and j, reflecting the coupling between the secondary degrees of freedom pertaining to these interactions. Factors of order higher than 1 can be identified with multibody or correlation interactions.

The factors $\langle\langle\varepsilon_1, \varepsilon_2, \ldots, \varepsilon_m\rangle\rangle_f$ can be expanded into Kubo's cluster-cumulant series. Such an approach enabled us to develop analytical expressions for the correlation terms in UNRES (Liwo et al., 2001), which are essential to reproduce the secondary structures of proteins (Koliński and Skolnick, 1994). These terms are described in Section 3.3.2.5.

Detailed description of the cluster-cumulant expansion of the PMF and of the use of the expansion of the cluster-cumulant functions into cluster-cumulant series in the derivation of the UNRES terms that account for the coupling between the backbone-local and backbone-electrostatic interactions was presented in our earlier work (Liwo et al., 2001). The low-order cluster-cumulant terms and their expansion are also illustrated with some examples in the Appendix.

3.3 UNRES MODEL

3.3.1 Effective Energy Function

In the UNRES model (Liwo et al., 1993, 1997a,b, 1998, 2001, 2004, 2007, 2008; Ołdziej et al., 2003, 2004; Czaplewski et al., 2004; Kozłowska et al., 2007, 2010; Sieradzan et al., 2012a,b, 2014, 2015; Krupa et al., 2013), a polypeptide chain is represented by a sequence of α-carbon (C^α) atoms linked by virtual bonds with attached united SCs and united peptide groups (p) located in the middle between the consecutive α-carbons (Figure 3.1). The united peptide groups are the centers of mean-field electrostatic (hydrogen-bonding) interactions, which determine the basic architecture of polypeptide chains. Only the united peptide groups and united SCs serve as interaction sites. The α-carbons serve only to define the geometry and they does not serve as interaction sites in the UNRES

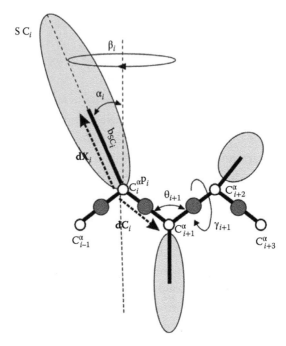

FIGURE 3.1 The UNRES model of polypeptide chains. The interaction sites are peptide-bond centers (p) and side-chain ellipsoids of different sizes (SC) attached to the corresponding α-carbons with different "bond lengths," b_{SC}. The α-carbon atoms are represented by small open circles. The geometry of the chain can be described either by the virtual-bond vectors $\mathbf{dC}_i$ ($C_i^{\alpha} \ldots C_{i+1}^{\alpha}$), $i = 1, 2, \ldots, N - 1$ and $\mathbf{dX}_i$ ($C_i^{\alpha} \ldots SC_i$), $i = 2, 3, \ldots, n - 1$ (represented by thick dashed arrows), where n is the number of residues, or in terms of virtual-bond lengths, backbone virtual-bond angles θ_i, $i = 2, 3, \ldots, n - 1$, backbone virtual-bond-dihedral angles γ_i, $i = 2, 3, \ldots, n - 2$, and the angles α_i and β_i, $i = 2, 3, \ldots, N - 1$ that describe the location of a side chain with respect to the coordinate frame defined by C_{i-1}^{α}, C_i^{α}, and C_{i+1}^{α}.

model (Figure 3.1). The energy of the virtual-bond chain is expressed by Equation 3.5.

$$
\begin{aligned}
U = {} & w_{SCSC} \sum_j \sum_{i<j} U_{SC_i SC_j} + w_{SCp} \sum_j \sum_{i \neq j} U_{SC_i p_j} + w_{pp}^{el} f_2(T) \sum_j \sum_{i<j-1} U_{p_i p_j}^{el} \\
& + w_{pp}^{vdW} \sum_j \sum_{i<j-1} U_{p_i p_j}^{vdW} + w_{tor} f_2(T) \sum_i U_{tor}(\gamma_i) \\
& + w_{tord} f_3(T) \sum_i U_{tord}(\gamma_i, \gamma_{i+1}) + w_{torsc} f_2(T) \sum_i U_{torsc;i,i+1} \\
& + w_b \sum_i U_b(\theta_i, \gamma_{i-1}, \gamma_{i+1}) + w_{rot} \sum_i U_{rot,i} + \sum_{m=2}^{N_{corr}} w_{corr}^{(m)} f_m(T) U_{corr}^{(m)}
\end{aligned}
$$

$$+ w_{turn}^{(3)} f_3(T) U_{turn}^{(3)} + w_{turn}^{(4)} f_4(T) U_{turn}^{(4)} + w_{turn}^{(6)} f_6(T) U_{turn}^{(6)}$$

$$+ w_{bond} U_{bond}(d_i) + w_{SS} \sum_{\substack{disulfide \\ bonds}} U_{SS_i} + n_{SS} E_{SS} \tag{3.5}$$

with

$$f_n(T) = \frac{\ln[\exp(1) + \exp(-1)]}{\ln\{\exp[(T/T_0)^{n-1}] + \exp[-(T/T_0)^{n-1}]\}} \tag{3.6}$$

where $T_0 = 300$ K; the temperature-scaling multipliers $f_n(T)$ were introduced in our previous work (Liwo et al., 2007). The multipliers $f_n(T)$ (Liwo et al., 2007) account for the temperature of those UNRES energy terms that originate from the cumulants of the cluster-cumulant expansion of the PMF (Liwo et al., 2001) and, consequently, scale as $T^{(n-1)}$ for the nth order cumulant.

The terms $U_{SC_i SC_j}$ correspond to the mean free energy of hydrophobic (hydrophilic) interactions between the SCs. These terms implicitly contain the contributions from the interactions of the SCs with the solvent. The terms $U_{SC_i P_j}$ correspond to the excluded-volume potential of the SC–peptide group interactions (Liwo et al., 1993). The terms $U_{P_i P_j}^{el}$ and $U_{P_i P_j}^{vdW}$ represent the energy of average electrostatic and van der Waals interactions between backbone peptide groups, respectively. The terms U_{tor} and U_{tord} are the torsional and double-torsional potentials, respectively, for the rotation about a given virtual bond or two consecutive virtual bonds. The terms U_{sctor} for the SC torsional potentials have been introduced recently (Krupa et al., 2013); they depend on the virtual-bond dihedral angles that involve SC centers and are described in more detail in Section 3.3.2.4. The terms U_b and U_{rot} are the virtual-bond angle-bending and SC-rotamer potentials, respectively. The terms $U_{corr}^{(m)}$ and $U_{turn}^{(m)}$ correspond to the correlations (of order m) between peptide-group electrostatic and backbone local interactions; the terms $U_{turn}^{(m)}$ (the "turn" terms) involve consecutive segments of the chain. The terms $U_{bond}(d_i)$, with d_i being the length of the ith virtual bond (backbone or SC), are Padé rational functions (Kozłowska et al., 2010), which take into account the presence of multiple minima in the virtual-bond-stretching potentials of, e.g., isoleucine or arginine SCs. The virtual-bond lengths are assumed to be fixed in other applications of UNRES. The terms U_{SS_i} (Czaplewski et al., 2004; Chinchio et al., 2007) are the energies of distortion of disulfide bonds from their equilibrium conformation, E_{SS} is the energy of formation of an "un-strained" disulfide bond in the chain (relative to the presence of two free cysteine residues), and n_{SS} is the number of disulfide bonds. The *w*s are the weights of the various energy terms and have been determined by optimization of the potential-energy landscape (Lee et al., 2001; Ołdziej et al., 2004; Liwo et al., 2007); this procedure is described briefly in Section 3.3.3.

3.3.2 Expressions of UNRES Energy Terms and Their Parameterization

The derivation of the UNRES energy terms was accomplished by following the cluster-cumulant approach outlined in Section 3.2.2. For parameterization, model systems that corresponded to the respective terms were constructed, then the corresponding PMFs were calculated and analytical expressions were fitted to them. The derivation and parameterization of the respective UNRES energy-function components are described in the subsequent subsections.

3.3.2.1 SC–SC Interaction Potentials

In the first version of the UNRES force field (Liwo et al., 1997a), the SC–SC interaction potentials were assigned the Gay and Berne (1981) functional forms that take anisotropy of interactions into account. The parameters for SC–SC interactions were determined (Liwo et al., 1997a) by fitting to correlation functions and to SC-contact energies determined from the Protein Data Bank (PDB) (Berman et al., 2000). These potentials are still being used in the current versions of UNRES (Liwo et al., 2007; He et al., 2009).

The results of our work on the development of the new SC–SC interaction potential $\left(U_{SC_iSC_j} \right)$ were presented in the series of papers by Makowski et al. (2007a,b,c, 2008, 2011a,b). This model has been parameterized by fitting analytical formulas that describe the respective free-energy surfaces to PMFs of pairs of interacting SC models in water, calculated by means of MD with the AMBER force field (Pearlman et al., 1995). An important feature of the model is the introduction of two states of charged head groups, one of which corresponds to a smaller and the second one to a larger distance of that head group from the SC center (Makowski et al., 2011b). Consequently, the part of the effective potential corresponding to the interaction of a positively charged group with a negatively charged one is a Boltzmann average over four possible states, each corresponding to one of the two possible distances of the first group from the corresponding SC center and one of the possible distances of the second one from the corresponding SC center. This feature of the energy function enabled us to reproduce the shape and location of the minima and maxima in the PMF curves, especially the minima corresponding to salt-bridge formation. This two-state model of charged SCs is shown in Figure 3.2.

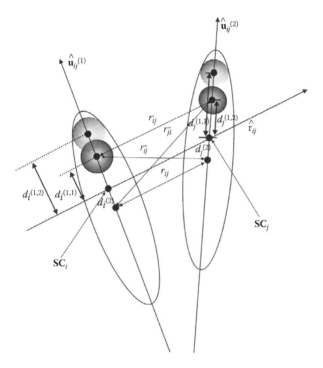

FIGURE 3.2 Illustration of the new model for the interactions of charged and polar side chains. A side chain of this type is assumed to consist of the charged (shaded) and nonpolar (ellipsoidal) parts. The geometric centers of side chains i and j are denoted as SC_i and SC_j, respectively, and represented by small black circles located between the centers of the charged and nonpolar sites. The charged site of each side chain can exist in two possible states; hence two shaded spheres are shown for each charged site. The spheres corresponding to alternative positions of the charged sites (farther away from the centers of side chain i and j, respectively) are bordered by dashed lines and are transparent to indicate that each of them corresponds to the alternative state of a single site and does not represent an additional site. The vector $\hat{\mathbf{u}}_{ij}^{(1)}$ is the unit vector of the long axis of side chain i, $\hat{\mathbf{u}}_{ij}^{(2)}$ is the unit vector of the long axis of side chain j, $\hat{\mathbf{r}}_{ij}$ is the unit vector pointing from the geometric center of the nonpolar site of side chain i to that of side chain j, r_{ij} is the distance between these two centers, r'_{ij} is the distance between the charged/polar centers of the head groups of side chains i and j, r''_{ij} and r''_{ji} are the distances between the charged centers of particle i and the center of particle j, and the charged center of particle j and the center of particle i, respectively (for clarity sake, we show only the distances that involve the polar/charged center in the first possible state), $d_i^{(1,1)}$, $d_i^{(1,2)}$, and $d_j^{(1,1)}$, $d_j^{(1,2)}$ are the distances from the geometrical center of side chain i and j, respectively, to the center of the charge of head group i and j, respectively, and $d_i^{(2)}$ and $d_j^{(2)}$ are the distances from the geometrical center of side chain i and j, respectively, to the nonpolar center of particles i and j, respectively.

3.3.2.2 Implementation of New SC–SC Potentials

The new physics-based potentials were implemented in the UNRES/MD package. Their advantage over the old knowledge-based potential is a much better representation of salt bridges. Calibration of the force field with these new terms is pending.

3.3.2.3 Local Potentials

The local potentials consist of the virtual-bond (U_{bond}), virtual-bond angle (U_b), SC-rotamer (U_{rot}), virtual-bond torsional (U_{tor}), virtual-bond double torsional (U_{tord}), and virtual-bond SC torsional (U_{torsc}). Initially (Liwo et al., 1997a), they were determined as knowledge-based potentials. Subsequently, the local potentials were upgraded to physics-based potentials and determined as the PMF calculated from *ab initio* (U_b [Kozłowska et al., 2007], U_{tor} (Ołdziej et al., 2003), and U_{tord} [Ołdziej et al., 2003]) or semiempirical potential-energy surfaces of model systems [U_{bond} (Kozłowska et al., 2010), U_{rot} (Kozłowska et al., 2010), and U_{torsc} [Krupa et al. 2013]). For U_b, U_{tor}, and U_{tord}, the amino-acid-residue alphabet is reduced to glycine, proline, and alanine, where alanine represents all amino-acid residues except for glycine and proline. For the remaining potentials, which depend on SCs, either the 20 natural amino-acid-residue types were considered or, as in the present U_{torsc} potentials, the alphabet was reduced (Krupa et al., 2013).

Because the local potentials, except the recently introduced U_{torsc}, were described very thoroughly in our previous papers (Ołdziej et al., 2003; Kozłowska et al., 2007, 2010), only the U_{torsc} potentials are described in more detail in Section 3.3.2.4.

3.3.2.4 SC Torsional Potentials

To improve the quality of the structures obtained by UNRES simulation, new SC-backbone torsional potentials (Figure 3.3), U_{torsc}, were introduced and derived by using two methods: a statistical one (i.e., from a statistical analysis of loop regions of 4585 proteins) (Krupa et al., 2013) and a physics-based one (from Austin model 1 [AM1] calculations) (Sieradzan et al., 2014). The implemented potentials were derived for the following torsional angles: $\tau^{(1)}$ (SC–C^α–C^α–C^α), $\tau^{(2)}$ (C^α–C^α–C^α–SC), and $\tau^{(3)}$ (SC–C^α–C^α–SC).

As mentioned above, three types of amino-acid residues are used in the UNRES force field for local potentials: alanine, glycine, and proline. However, for SC correlation potentials, a set of five types of amino-acid

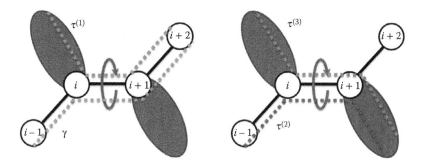

FIGURE 3.3   Illustration of backbone torsional angle γ (lower short-dashed polyline in the left panel) and side-chain backbone torsional angles $\tau^{(1)}$ (upper short-dashed polyline in the left panel), $\tau^{(2)}$ (lower short-dashed polyline in the right panel), and $\tau^{(3)}$ (upper short-dashed polyline in the right panel).

residues, derived in our earlier work on structurally optimized reduced alphabets (Solis and Rackovsky, 2000), was based on a statistical analysis of the properties of amino-acid structures to obtain sufficient statistics. In physics-based potentials (Sieradzan et al., 2015), a full alphabet of 20 amino-acid residues was used. PMFs were derived with the use of Boltzmann summation. An entropy contribution was evaluated with a harmonic approximation from Hessian matrices (Kozłowska et al., 2007; Sieradzan et al. 2015). The new correlation potentials are expressed as one-dimensional Fourier series in the virtual-bond dihedral angles involving SC centroids.

The weights of these new terms were determined by a trial-and-error method, in which multiplexed replica exchange molecular dynamics (MREMD) simulations were run on selected test proteins (five α-helical structures, two α + β structures, and one β-structure). The resulting conformational ensembles were analyzed in detail by using the weighted histogram analysis method (WHAM) (Kumar et al., 1992) and Ward's minimum-variance clustering.

For statistical potentials, analysis showed that the root-mean-square deviations (RMSDs) from the experimental structures dropped by 0.5 Å on average, compared to simulations without the new terms, and the deviations of individual residues in the loop region of the computed structures from their counterparts in the experimental structures (after optimum superposition of the calculated and experimental structure) decreased by up to 8 Å. Consequently, the new U_{torsc} terms improve the description of local structure, especially in loop regions.

Test simulations carried out with the physics-based U_{tors} potentials and preoptimized energy-term weights resulted in a decrease of RMSD of average simulated conformations (from the most probable cluster centroid) by up to 4 Å, and 0.4 Å on average. This is a significant improvement of the resolution of the UNRES force field. However, further optimization of the force field is required for further improvement of the calculated structures and thermodynamic properties.

3.3.2.5 Backbone-Electrostatic and Correlation Terms

As mentioned in Section 3.3.1, the energy components related to the mean-field interactions of the backbone are an essential component of UNRES, because they correspond to the hydrogen-bonding interactions of the backbone that, together with local interactions, determine the architecture of a protein. To derive the analytical expressions, we assumed (Liwo et al., 1993) that backbone-electrostatic interactions can be represented by the interactions of peptide-group dipoles, as shown in Figure 3.4a. The dipole moment of each of the two interacting peptide groups has a component parallel ($p_{\|}$) and perpendicular ($p_{\perp}$) to the virtual-bond axis; the rotation of the dipoles about the $C^{\alpha} \dots C^{\alpha}$ virtual-bond axes are averaged. The first-order factors in the cluster-cumulant expansion of the PMF are PMFs of two isolated peptide groups. Because the perpendicular (rotatable component) of a peptide-group dipole moment is dominant, the PMF can be approximated by a sum of the variance (the first nonvanishing cumulant) of the energy of interaction of the perpendicular dipole-moment components and the energy of the parallel dipole-moment components (which does not average). The respective expression is given by Equation 3.7 (Liwo et al., 1993).

$$
U^{\text{el}}_{p_i p_j} = \frac{A_{ij}}{r_{ij}^3} \left(\omega_{ij}^{(12)} - 3\omega_{ij}^{(1)}\omega_{ij}^{(2)} \right)
$$

$$
- \frac{B_{ij}}{r_{ij}^6} \left[4 + \left(\omega_{ij}^{(12)} - 3\omega_{ij}^{(1)}\omega_{ij}^{(2)} \right)^2 - 3\left(\omega_{ij}^{(1)2} + \omega_{ij}^{(2)2} \right) \right] \qquad (3.7)
$$

where $\omega_{ij}^{(1)}$, $\omega_{ij}^{(2)}$, and $\omega_{ij}^{(12)}$ are the direction cosines of the $C_i^{\alpha} \dots C_{i+1}^{\alpha}$ virtual-bond and the vector from peptide group p_i to peptide group p_j, between the vector from peptide group p_i to peptide group p_j and the $C_j^{\alpha} \dots C_{j+1}^{\alpha}$ virtual-bond axes, and between the $C_i^{\alpha} \dots C_{i+1}^{\alpha}$ and the $C_j^{\alpha} \dots C_{j+1}^{\alpha}$ virtual-bond axes, respectively (see Figure 3.4a), r_{ij} is the distance between the interacting peptide groups, and A_{ij} and B_{ij} are constants.

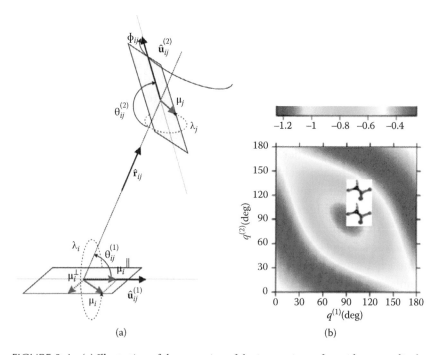

(a)　　　　　　　　　　　　(b)

FIGURE 3.4   (a) Illustration of the averaging of the interactions of peptide-group dipoles to derive the $U^{el}_{p_i p_j}$ potentials in Equation 3.5. The dipole moment of each peptide group is decomposed into the component parallel ($\mu^{\parallel}$) and perpendicular ($\mu^{\perp}$) to the $C^\alpha \ldots C^\alpha$ virtual-bond axis. The energy of the interaction involving the perpendicular components is averaged over the rotation angles λ to give the PMF of peptide-group interaction. $\hat{\mathbf{u}}^{(1)}_{ij}$ and $\hat{\mathbf{u}}^{(1)}_{ij}$ are the unit vector of the $C^\alpha_i \ldots C^\alpha_{i+1}$ and C^α_j and C^α_{j+1} virtual-bond axes, respectively, $\hat{\mathbf{r}}_{ij}$ is the unit vector from peptide group p_i to peptide group p_j, $\theta^{(1)}_{ij}$ and $\theta^{(2)}_{ij}$ are the angles between the respective virtual-bond axes and the vector linking peptide-group center, ϕ_{ij} is the angle of the precession of the C^α_j and C^α_{j+1} virtual-bond axis about the axis linking peptide-group centers, and λ_i and λ_j are the angles for the rotation of the peptide-group plates about the respective virtual-bond axes. (b) Contour plot of the $U^{el}_{p_i p_j}$ UNRES energy (Equation 3.7) in the $\theta^{(1)}$ and $\theta^{(2)}$ angles for $\phi = 0$. It can be seen that the energy minimum ($\theta_1 = 90°, \theta_2 = 90°$) corresponds to the parallel orientation of the $C^\alpha \ldots C^\alpha$ axes, which corresponds to optimal hydrogen-bonding orientation.

The potential expressed by Equation 3.7 excellently reproduces the features of the PMF of two interacting peptide groups out of the context of a polypeptide chain (which can be modeled by two acetamide molecules) and the directional properties of peptide-group hydrogen bonds (Figure 3.4b) (Liwo et al., 2004). These terms can reproduce the geometry of the α-helices but cannot reproduce the geometry of the β-sheets, because the formation of the latter requires coupling with local interactions (see e.g., Equations 3.8 and 3.9; Liwo et al., 2001).

The other backbone-electrostatic components of the UNRES effective energy function are higher order factors in the cluster-cumulant expansion for the PMF and correspond to the coupling of backbone-local and backbone-electrostatic interactions (Liwo et al., 2001, 2004). The most essential lowest order terms are the third-order terms that correspond to the coupling of the interaction between two peptide-group dipoles and the local interactions within residues, each adjacent to one of those peptide groups. These third-order terms account for the fact that peptide-group dipoles cannot rotate freely about the $C^\alpha \dots C^\alpha$ virtual-bond axes but their rotation is restricted by local interactions. Because each peptide group is adjacent to two residues, there are four such terms for each pair of interacting peptide groups. Because, given restricted rotation, the dipole–dipole interactions do not average out to zero, the lowest nonvanishing cumulants in the expansion of the abovementioned third-order factors are the first cumulants. The sum of the first cumulants of all third-order terms containing the interactions between two given peptide groups can be represented as the interactions of two fictitious dipoles whose lengths and orientation depend on the virtual-bond dihedral angles whose axes are the $C^\alpha \dots C^\alpha$ virtual bonds containing two peptide groups (Liwo et al., 2001, 2004; Liwo, 2013), as expressed by Equation 3.8 (Liwo et al., 2001, 2004) and illustrated in Figure 3.5. These third-order terms are essential in reproducing β-sheet structures.

$$U_{\text{loc}-\text{el};ij}^{(3)} = B_{ij}^{1/2} \frac{\mu_i \mu_j - 3 \left(\mu_i \circ \hat{r}_{ij} \right) \left(\mu_j \circ \hat{r}_{ij} \right)}{r_{ij}^3} \tag{3.8}$$

$$\mu_i = \mathbf{R}(\gamma_i) \begin{bmatrix} -b_{21,i} \\ b_{22,i} \end{bmatrix} + \begin{bmatrix} b_{11,i+1} \\ b_{12,i+1} \end{bmatrix} \tag{3.9}$$

where μ_i and μ_j are the fictitious dipole moments, $\mathbf{R}(\gamma)$ is the matrix of anticlockwise rotation about the respective $C^\alpha \dots C^\alpha$ virtual-bond axis by angle γ, $\hat{r}_{ij}$ is the unit vector pointing from peptide group i to peptide group j, and the bs are the coefficients of the Fourier expansion of the local-energy surface in the angles of rotation of the peptide groups about the $C^\alpha \dots C^\alpha$ virtual-bond axes (Equation 3.10).

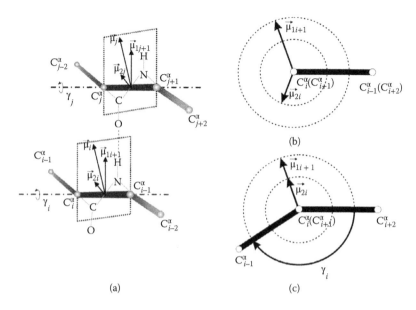

(a)   (b)   (c)

FIGURE 3.5   Illustration of the physical meaning of the third-order correlation terms in the UNRES coarse-grained force field that couple the long-range electrostatic interactions between two peptide groups p_i and p_j located between the C_i^α and C_{i+1}^α (C_j^α and C_{j+1}^α) atoms, respectively, with local conformational states of a residue adjacent to each peptide group ($U_{loc-el}^{(3)}$ of [Liwo et al., 2001]). There are four such terms for each pair of interacting peptide groups. These terms have been derived by the Kubo cluster-cumulant expansion of the potentials of mean force of polypeptide chains (Liwo et al., 2001) and are essential for reproducing regular α-helical and β-sheet structures in the UNRES model (Liwo et al., 2001). (a) Detailed view of the peptide groups. The atom symbols and bonds of the peptide groups are displayed as light-gray letters and lines. The light-gray rectangles mark the planes perpendicular to the $C_i^\alpha \dots C_{i+1}^\alpha$ and $C_j^\alpha \dots C_{j+1}^\alpha$ virtual bonds, respectively. With the use of Kubo cluster-cumulant expansion, the sum of the four terms corresponding to the pair of interacting peptide groups can be represented as the energy of interaction of two fictitious dipoles, μ_i and μ_j, each one ascribed to the respective peptide group; it should be noted that these dipoles are different from the dipoles of the peptide groups calculated from the charge distribution in all-atom representation. Each of these dipoles has a component ascribed to the $C_i^\alpha \dots C_{i+1}^\alpha \dots C_{i+2}^\alpha$ ($\mu_{1,i+1}$) or $C_j^\alpha \dots C_{j+1}^\alpha \dots C_{j+2}^\alpha$ ($\mu_{1,j+1}$), respectively, and a component ascribed to the $C_{i-1}^\alpha \dots C_i^\alpha \dots C_{i+1}^\alpha$ (μ_{2i}) or $C_{j-1}^\alpha \dots C_j^\alpha \dots C_{j+1}^\alpha$ (μ_{2j}), respectively. Thus, the total vectors of the fictitious dipole moments depend on the angles of rotation (γ_i and γ_j, respectively) about the shared $C_i^\alpha \dots C_{i+1}^\alpha$ or $C_j^\alpha \dots C_{j+1}^\alpha$ axis. (b) Side view of the three-virtual-bond arrangement across the $C_i^\alpha \dots C_{i+1}^\alpha$ virtual-bond axis, the C_i^α atom pointing to the front of the picture in the reference configuration ($\gamma_i = 0°$), in which actual relations between the magnitudes and directions of the two dipole-moment components, as determined by fitting the analytical cumulant-based expression to the PMF of nonproline and nonglycine-type residues (Liwo et al., 2004) are shown. (c) After rotation about the $C_i^\alpha \dots C_{i+1}^\alpha$ by angle $\gamma \approx -150°$, the two dipole-moment components fully add to produce the largest total fictitious dipole moment. This situation is encountered in β-sheets. It should be noted that this simple analysis also predicts the left-handed twist of β-strands (which implies right-handed twist of β-sheets).

$$E_{loc}\left(\lambda^{(1)}, \lambda^{(2)}\right) = \mathbf{u}^T\left(\lambda_i^{(1)}\right)\mathbf{b}_{1i} + \mathbf{u}^T\left(\lambda_i^{(2)}\right)\mathbf{b}_{2i} + \mathbf{u}^T\left(\lambda_i^{(1)}\right)\mathbf{Cu}\left(\lambda_i^{(1)}\right)$$
$$+ \mathbf{u}^T\left(\lambda_i^{(2)}\right)\mathbf{Du}\left(\lambda_i^{(2)}\right) + \mathbf{u}^T\left(\lambda_i^{(1)}\right)\mathbf{Eu}\left(\lambda_i^{(2)}\right) + \ldots$$

$$(3.10)$$

with

$$\mathbf{u}(\lambda) = \begin{bmatrix} \cos\lambda \\ \sin\lambda \end{bmatrix} \tag{3.11}$$

3.3.3 Calibration of the Energy Function

Parameterization of the UNRES potential energy function requires the determination of the relative energy-term weights in Equation 3.5, which allows for the reproduction of native structures and the folding thermodynamics of selected proteins. The Z-score and energy gap between the native structure and the lowest energy nonnative structure are maximized and are used as criteria for the assessment of force field parameters. The Z-score is defined as the difference between the mean energies of the native-like structures and the nonnative structures, divided by the standard deviation of the energy of the nonnative structures. Initially, the parameters were optimized by using an iterative procedure in which a set of decoy structures were generated using the conformational space annealing (CSA) method for searching the conformational space, and the energy gap and Z-score were maximized in each iteration (Lee et al., 2001). This approach worked well for small proteins with simple topologies, but not for more complicated ones (especially those containing β and α–β folds) for which the native-like structures are far from the low-energy conformations found with initial-guess parameters. Therefore, a hierarchical procedure was developed (Liwo et al., 2002) for optimizing a potential function based not only on an energy ranking of native-like structures with respect to nonnative structures, but also on the energy ranking of all structures, depending on their native-likeness.

In this hierarchical optimization method (Liwo et al., 2002), the conformational space is divided into levels containing conformations with certain native-like elements. Secondary-structure elements are defined as elementary fragments, and higher hierarchy levels are constructed by assembling them gradually. If some of the elementary fragments or their aggregates persistently fail to appear in probable conformations, or if the optimization fails to progress, the hierarchy is revised by removing weakly stable structural elements from lower levels and moving them to higher levels. To test this hierarchical procedure on proteins, different folds were selected to treat all the structural classes. Therefore, the UNRES potential energy function was optimized using two [1E0G(α + β) and 1E0L(β)], three [1E0G, 1E0L, and 1GAB(α)], and finally four training proteins [1E0G, 1E0L, 1GAB, and 1IGD (α + β)] simultaneously. These training sets and the resulting force fields are referred to as 2P, 3P, and 4P, respectively (Ołdziej et al., 2004). Later (Liwo et al., 2007), the approach was extended to the calibration of the force field for UNRES/MD simulations, and an extensive search of the energy-term-weight space was performed to produce a transferable force field (He et al., 2009).

3.3.4 Treatment of Disulfide Bonds

Dynamic formation and breaking of disulfide bonds are a part of the folding of many proteins such as, e.g., bovine pancreatic trypsin inhibitor and ribonuclease. The earlier version of UNRES designed for global minimization of the energy function handled the formation of disulfide bonds by switching between the nonbonded cysteine–cysteine potential and the harmonic potential with extra energy added (E_{SS} in Equation 3.5), in a Monte Carlo with minimization move (Czaplewski et al., 2004). For MD with UNRES, we designed a bimodal potential (Chinchio et al., 2007), one minimum of which corresponds to two bonded half-cysteines and the second one to nonbonded cysteine SCs. This approach proved successful in simulating the folding pathways of small disulfide-bonded proteins (Chinchio et al., 2007).

3.3.5 Treatment of D-Amino-Acid Residues

D-Amino-acid residues occur in a number of natural peptides and proteins, mostly from bacteria or fungi (such as, e.g., gramicidin S, gramicidin D, phalloidin, etc.). The D-amino-acid residues are much more abundant in designed peptides and proteins because of their importance in drug design; for example, substituting a natural L-amino-acid residue to a D-amino-acid residue counterpart can make a ligand an effective enzyme inhibitor or hormone antagonist. Therefore, recently (Sieradzan et al., 2012b, 2014), we have introduced D-amino-acid residues into UNRES. This extension affected all potentials that are dependent explicitly or implicitly on amino-acid-residue chirality, i.e., the local potentials U_b, U_{rot}, U_{tor}, U_{tord}, and U_{torsc} and the correlation potentials $U_{corr}^{(m)}$ and $U_{turn}^{(m)}$ (cf. Equation 3.5). The extension of the SC-rotamer (U_{rot}) and correlation ($U_{corr}^{(m)}$ and $U_{turn}^{(m)}$) potentials, which can be decomposed into contributions dependent on the chirality of single residues, required only a change of the sign of some of the coefficients (Sieradzan et al., 2012b, 2014). Conversely, for the torsional, double-torsional, and virtual-bond-angle terms that are not reducible to single-residue components; symmetry relations could be utilized only for the potentials corresponding to all-D-amino-acid residues (e.g., that for the D-Ala–D-Ala fragment). The other torsional, double-torsional, and virtual-bond-angle potentials were determined from energy maps of terminally blocked alanine, proline, and glycine; the energy maps of D-residues were obtained by applying the inversion operation to the energy maps of L-amino-acid residues (Sieradzan et al., 2012b, 2014).

The extended UNRES force field to treat D-residues reproduced the change of the free energies of helix formation in model peptides reasonably well (Sieradzan et al., 2014) and was also implemented in the folding of BBAT1, a tetrameric protein with D-amino-acid residues (Sieradzan et al., 2012b).

3.3.6 Treatment of the Trans–Cis Isomerization of Proline Residue

As for disulfide bond formation, proline isomerization can also determine the essential folding steps (e.g., in the folding of ribonuclease A). Therefore, we introduced peptide-bond isomerization in UNRES (Sieradzan et al., 2012b). The respective potentials are bimodal in the $C^\alpha \ldots C^\alpha$ virtual-bond length and depend on the types of residues forming a peptide bond, i.e., nonproline or proline. The potentials were determined, as PMFs, from the energy maps of N-methylacetamide and N-pyrrolidylamide. Comparison of the isomerization rate constants calculated with these coarse-grained potentials with the experimental values and with the values estimated from the energy maps of the two compounds mentioned above revealed a six-order-of-magnitude speed–up of the process with respect to the all-atom approach.

3.3.7 Coarse-Grained MD and Its Extensions with UNRES and NARES-2P

The UNRES force field is a very useful tool for *de novo* protein folding simulations. MD simulations are recommended for this purpose. Due to the fact that part of the interacting sites in the UNRES model (the united peptide group p located between two consecutive C^α atoms) are different from the geometrical sites (C^α atoms) used to describe the polypeptide conformations, the Lagrange formulation of the equations of motion is used for MD simulations (Kilmister, 1967). Lagrange equations are frequently used instead of Newton equations to describe the time evolution of a classical system characterized by a set of generalized coordinates (Khalili et al., 2005a). A modified velocity Verlet integrator is used for calculating the progress of each MD step, because it provides velocities and positions at a specific time, which are used to compute the kinetic energy (necessary to couple the system to a thermal bath) and friction forces (Khalili et al., 2005b; Liwo et al., 2005). Other advantages of the velocity Verlet integrator are that it is symplectic and time reversible (Ruth, 1983; Sanz-Serna and Calvo, 1994; Benettin and Giorgilli, 1994). Thanks to the coarse-grained representation of the polypeptide chain in UNRES, the actual value of one time step is approximately 48.9 ps (Khalili et al., 2005b). UNRES enables the time step value to be increased even more by use of the reversible reference system propagator algorithm (RESPA) (Tuckerman et al., 1992).

Canonical MD simulations require the use of a thermostat algorithm. UNRES offers the use of several thermostats, e.g., Langevin stochastic dynamics, weak coupling Bernedsen-bath (Khalili et al., 2005b), and extended phase Nose–Hoover and Nose–Poincaré thermostats (Kleinerman et al., 2008). Although the Berendsen thermostat does not generate a correct canonical ensemble, it is still the best choice due to the correct folding/computational time efficiency.

The UNRES force field provides a number of extensions of MD algorithms. The most efficient and widely used is replica exchange molecular dynamics (REMD) (Nanias et al., 2006) and MREMD (Czaplewski et al., 2009). In REMD, several copies (replicas) are simulated at the same time. Each one of the replicas is simulated at a different temperature. Periodically, exchanges of temperatures between neighboring replicas are attempted and accepted with a well-defined acceptance probability which is consistent with detailed balance. It enables energy barriers to be overcome and searches of rugged conformational energy landscapes to be carried out more effectively. The MREMD algorithm is even more efficient because it adds additional layers (e.g., for each temperature, there are multiple trajectories) of replicas between which the temperature exchange can occur (Czaplewski et al., 2009). Multiplexing extends the multiple temperature aspect of REMD with the large number of replicas to enhance sampling considerably without increasing the frequency of temperature swaps necessary with increasing number of temperatures.

Several other extensions of the MD algorithm were implemented in the UNRES force field although they proved to be not as efficient as the REMD and MREMD algorithms (Nanias et al., 2006). The multicanonical algorithm (MUCA) was combined with REMD in three different ways. The replica exchange multicanonical algorithm (REMUCA) uses short REMD runs to obtain starting weights for the multicanonical simulation. The multicanonical replica exchange method (MUCAREM) differs from REMD; instead of focusing on replicas with different temperatures, it uses replicas associated with different energy ranges over which multicanonical simulations are carried out. The replica exchange multicanonical with replica exchange method (REMUCAREM) is a combination of the abovementioned two methods. It obtains starting weights from short REMD simulations and then proceeds in the same manner as in MUCAREM (Nanias et al., 2006). The last of the MD extensions is the serial replica exchange method (SREM) (Shen et al., 2008). The REMD algorithm requires synchronization between different replicas on parallel computers. Therefore, the slowest processor determines the overall performance of the simulation. The SREM allows running simulations asynchronously on distributed computers, thereby increasing the performance of the REMD, although it cannot reproduce proper REMD results in the current temperature-dependent version of the UNRES (Shen et al., 2008).

MD with its most efficient extensions, REMD and MREMD, was also implemented for DNA and RNA folding simulations using the NARES-2P model (He et al., 2013a).

3.3.8 UNRES in Blind Prediction of Protein Structures

The community-wide experiment on the critical assessment of techniques for protein structure prediction (CASP) provides an independent mechanism for the assessment of methods of protein structure modeling. Our new physics-based procedure resulted in some good predictions in the CASP7–CASP9 exercises (Liwo et al., 2011). This section presents the performance of our physics-based protein structure prediction method in the CASP10 exercise under the Cornell–Gdansk group name. We demonstrate that use of physics-based methodology can make a difference in predicting the correct global fold.

As the Cornell–Gdansk group, we submitted predictions for 21 targets out of 53 targets available to human predictor groups (He et al., 2013b). As in previous CASP exercises (Liwo et al., 1999, 2011; Lee et al., 2000; Pillardy et al., 2001; Ołdziej et al., 2005), in order to use supercomputer resources effectively, we considered primarily the targets that had below 20% sequence similarity with proteins from the PDB, as assessed by the position-specific iterative basic local alignment search tool (PSI-BLAST) server (Altschul et al., 1997). The best predictions made with UNRES are described below.

Target T0663, classified as a new fold, consists of two α + β domains homologous to those of known proteins, but the domain packing is different from any known protein structure. UNRES predicted the correct symmetry of packing on top of all the participate groups, in which the domains are rotated with respect to each other by 180° in the experimental structure, as shown in Figure 3.6. However, judging by the global-distance-test score, comparative-modeling methods seem to have given better results compared to UNRES (Figure 3.6d). The reason for this is that only a combination of global distance test total score (GDT_TS) with the chirality score (defined as the fraction of tetrahedra of vertices in the C^α atoms of the structure under consideration that have the same chirality as the corresponding tetrahedra of the reference structure) (Taylor et al., 2013) enabled the assessors to distinguish the UNRES models. The chirality scores for our models 1 and 4 are 0.71 and 0.63, compared to 0.58 (model 5 of group 27), 0.52 (model 1 of group 190), and 0.52 (model 1 of

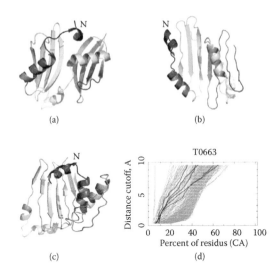

FIGURE 3.6   (a) The experimental 4EXR structure of target T0663. (b) Our CASP model 1. (c) Our CASP model 4. (d) GDT_TS plots of all models of T0663 from all groups with plots corresponding to the Cornell-Gdansk group models shown as black lines, and models from the other groups shown as orange lines. The N termini in a–c are marked with "N". The values of GDT_TS are 23.19, 31.98, and 42.80 for model 1 of the whole protein and its domains D1 and D2, respectively, and 22.04, 31.98, and 40.15 for model 4 of the whole protein and its domains D1 and D2, respectively. The respective GDT_TS values of the models with the highest GDT_TS submitted to CASP are 42.93 (model 1 from group 27), 68.61 (model 3 from group 27), and 98.20 (model 4 from group 27). (The GDT_TS plots have been reproduced with permission from the CASP10 website [www.predictioncenter.org/casp10/results.cgi]. The drawings of the structures were produced from PYMOL [www.pymol.org]).

group 388) of the three groups that scored best by means of the GDT_TS measure. We also obtained very good results, in terms of the GDT_TS measure, for T0668 and T0684 D2 (the second domain of target T0684) (He et al., 2013b).

3.3.9 Investigation of Protein-Folding Pathways and FELs

From theoretical and conceptual points of view, it is well known that a study of FELs holds the key to understand how proteins fold and function (Wales, 2003). However, the selection of a correct model for protein folding kinetics and the coordinates along which the intrinsic folding pathways can be identified in order to interpret experimental data still remains challenging. Some progress to treat this problem has been made recently based on the analysis of MD simulations of protein folding, particularly in the use of principal component analysis (PCA) (Maisuradze et al., 2009a) (i.e., projected FELs) and network analysis (Yin et al, 2012) (i.e., unprojected FELs). For projected FELs, this involves not only proper choices for the reaction coordinates, but also dimension of FELs (Maisuradze et al., 2009b, 2010a,b) and length of MD simulation (Maisuradze et al., 2009a). It appears that the low-dimensional FELs are not always sufficient for the description of folding processes (Maisuradze et al., 2009b, 2010a) (Figure 3.7); although PCA is very efficient for characterizing the general folding and nonfolding features of proteins, one should be careful in interpreting the results of this analysis, because of its dependence on the width of the sampling window (Maisuradze et al., 2009a). In order to avoid these problems, criteria for the determination of the minimal dimensionality of an FEL (Maisuradze et al., 2009b, 2010a) and the minimal simulation time for sampling (Maisuradze et al., 2009a), which would be sufficient for a correct description of protein folding dynamics, were introduced. The advantage of unprojected FELs (network analysis) over projected FELs along the order parameters is shown by identifying the large spectrum of folding pathways, hidden in the FELs along the order parameters (Yin et al, 2012). Another view of the folding thermodynamics and pathways is provided by analysis of sections of the free-energy profiles (FEPs) along the virtual-bond angle θ and virtual-bond dihedral angle γ of the backbone (Figure 3.7) (Maisuradze et al., 2010a, 2013). This analysis provides a more detailed insight into the conformational changes of chain segments, and it reveals the key residues involved in the transition between the basins of the FEL.

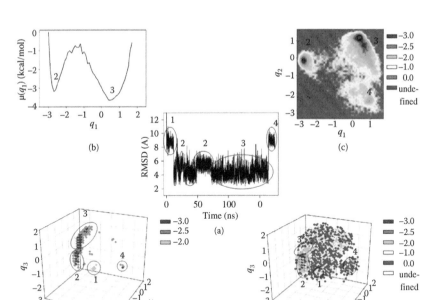

FIGURE 3.7 RMSD as a function of time (a). One-dimensional (b), two-dimensional (c), and three-dimensional (d, e) FELs ($\mu(q_1, \cdots, q_n) = -k_B T \ln P(q_1, \cdots, q_n)$) along the Principal Components (PCs) (q_i, with $i = 1, \ldots, n$), where $P(q_1, \cdots, q_n)$, T, and k_B are the probability density function, the absolute temperature, and the Boltzmann constant, respectively) of the B-domain of staphylococcal protein A (1BDD). The numbers in all panels correspond to the conformational states. Panel (d) shows all points in the three-dimensional FEL space with $\mu \leq 0$ kcal/mol, and the folding-unfolding pathways are not clearly illustrated in this plot because of strong overlapping of points corresponding to diverse energies. Therefore, the same FEL with only the lowest free energy points is plotted in panel (e).

3.3.10 Protein and Kinks with UNRES

Another means of analysis of protein structure and folding that has been developed recently (Krokhotin et al., 2012, 2014) involves the application of the discrete nonlinear Schrodinger (DNLS) equation (Chernodub et al., 2010; Molkenthin et al., 2011). It has been found that crystallographic protein structures in the PDB have a modular build (Murzin et al., 1995; Orengo et al., 1997). Furthermore, despite a rapid increase in the number of resolved structures, novel topologies are now becoming rare. In fact, it has been proposed that probably most of the modular building blocks of folded proteins have already been observed (Skolnick et al., 2009). This apparent convergence in the protein architecture can be interpreted as a manifestation that protein folding is a self-organization process with a structural hierarchy.

A detailed investigation of the loop profiles in crystallographic protein structures has revealed that the modular structure of folded proteins can be understood in terms of the kink (heteroclinic standing wave) solution of a generalized DNLS equation (Chernodub et al., 2010; Molkenthin et al., 2011). The DNLS equation is one of the most fundamental lattice equations. It already plays a prominent role in theories of optical waveguides, Bose–Einstein condensates, string theory, and so forth (Faddeev and Takhtajan, 1987; Ablowitz et al., 2003). In the case of proteins, the DNLS equation describes the minimum energy configuration of an effective free energy that approximates the thermodynamical Helmholtz free energy. This effective free energy can be derived from the Wilsonian principle of universality; it emerges as the long wavelength limit of the microscopic and atomic level free energy and describes the shape of the protein in terms of the backbone C^α trace (Niemi, 2003; Danielsson et al., 2010). The kink is an example of the kind of nonperturbative structures, which arise generically when nonlinear interactions merge elementary constituents such as atoms into a localized collective excitation that is stable against small perturbations. In particular, the kink does not decay, unwrap, or disentangle through local disturbances. It has already been established that over 92% of super-secondary structures in the PDB can be modeled in terms of different parameterizations of the unique kink profile of the DNLS equation (Krokhotin et al., 2012). The kink provides a systematic method to understand the apparent structural self-organization that drives the protein folding process. Moreover, the concept of a kink is a very effective tool with which to leap over the various hurdles and obstacles in the energy landscape that cause the numerous bottlenecks in classical all-atom MD approaches.

In a detailed numerical investigation (Krokhotin et al., 2014), we have employed the DNLS equation to identify the three individual kink profiles in the experimental structure of protein A using the UNRES energy function (Liwo et al., 2008). These kink profiles correspond to the two loops in the native structure; the first loop is a single DNLS kink whereas the second loop is a kink–kink pair. We have found that all three kinks emerge during a coarse-grained UNRES simulation. Moreover, we have found that the formation of each is accompanied by a local free energy increase that is about 7 kcal/mol of UNRES energy. During the course of our UNRES simulations, we found that the formation of a kink is always initiated by an abrupt change in the orientation of a pair of consecutive side chains in the loop region. Conceptually, the process resembles the formation of a Bloch wall along a magnetic (Heisenberg) spin chain (Landau and Lifshitz, 1960), where the C^α backbone corresponds to the underlying spin-chain lattice, and the amino-acid SCs are interpreted as the ensuing spin variables. Our analysis proposes a plausible mechanism for the kink formation along protein backbones and reveals the relevance of kinks for structure formation, in a manner that involves a coherent interplay between the backbone and the SCs.

3.3.11 Examples of Biological Applications

In the following subsections, examples of applications of UNRES to study biological problems are described.

3.3.11.1 Amyloid Formation by the Aβ40 Peptide

UNRES has been used, together with various simulation techniques, in several biological applications to gain an understanding, for example, of amyloid formation in several pathological conditions such as Alzheimer's, Parkinson's, and Huntington's diseases. Each of these diseases involves a relatively small polypeptide, which differs in each of these diseases and interferes with a naturally functional biological process. In Alzheimer's disease, the aggregation of a 40- or 42-residue polypeptide, known as a β-amyloid peptide (Aβ), forms elongated fibrils that interfere with neuronal function in the brain. The aggregation of Aβ has been studied by solid state nuclear magnetic resonance (NMR) spectroscopy (Tycko, 2006) and by MD (Rojas et al., 2010). The structure of the fibrils appears to be a stacked array of hairpin-like dimers (Rojas et al., 2010) involving hydrogen-bonding and hydrophobic interactions (Rojas et al., 2010). In the initial formation of the fibril, a partial α-helix appears in part of a monomer but later undergoes an α-to-β transition (Rojas et al., 2011). Fibril elongation follows a mechanism in which monomers attach in two distinct stages, docking and then locking, with hydrogen-bonding and hydrophobic interactions contributing to the stability of the fibril (Rojas et al., 2010). Simulated two-dimensional ultraviolet spectroscopy supports this mechanism (Lam et al., 2013).

3.3.11.2 PICK1–BAR Interactions

Protein interacting with C kinase 1 (PICK1) is a multidomain mammalian membrane protein (Staudinger et al., 1995). Its monomeric form contains one postsynaptic density 95/discs large/zonula occludens-1 (PDZ) (Sheng and Sala, 2001; Hung and Sheng, 2002) and one Bin/Ampiphysin/ Rvs (BAR) (Takei et al., 1999) domain. Although PDZ and BAR domains are common protein functional domains, PICK1 is the only protein that contains both PDZ and BAR domains and interacts with over 40 proteins in the cell, including receptors, transporters, and ionic channels (Staudinger et al., 1997; Torres et al., 1998, 2001; Dev et al., 1999; Takeya et al., 2000; Boudin et al., 2000; El Far et al., 2000; Cowan et al., 2000; Lin et al., 2001a,b; Penzes et al., 2001; Jaulin-Bastard et al., 2001; Perroy et al., 2002; Duggan et al., 2002; Hruska-Hageman et al., 2002; Williams et al., 2003; Enz and Croci, 2003; Leonard et al., 2003; Hirbec et al., 2003; Meyer et al., 2004; Excoffon et al., 2004; Reymond et al., 2005). Given the lack of detailed structural information about the complete PICK1 and its putative dimer, it becomes necessary to evaluate the proposed complex with computational modeling. Because PICK1 is the only protein that contains both a PDZ domain and a BAR domain, making it impossible to employ homology modeling to determine their modes of interaction, the structural and energetic feasibility of the putative auto-inhibited form of PICK1, and the manner in which the PICK1-PDZ domain finds its way to the binding sites on the crescent-shaped PICK1-BAR dimer. Therefore, UNRES was used to facilitate the long computational explorations of the large PDZ/BAR complex system. We identify interactions between the component PDZ and BAR domains of PICK1 by calculating possible binding sites for the PDZ domain of PICK1 (PICK1-PDZ) to the homology modeled, crescent-shaped dimer of the PICK1-BAR domain using MREMD and canonical MD simulations with the coarse-grained UNRES force field. The MREMD results show that the preferred binding site for the single PDZ domain is the concave cavity of the BAR dimer. A second possible binding site is near the N-terminus of the BAR domain that is linked directly to the PDZ domain. Subsequent short canonical MD simulations used to determine how the PICK1-PDZ domain moves to the preferred binding site on the BAR domain of PICK1 revealed that initial hydrophobic interactions drive the progress of the simulated binding. Thus, the concave face of the BAR dimer accommodates the PDZ domain first by weak hydrophobic interactions and then the PDZ domain slides to the center of the concave face, where more favorable hydrophobic interactions take over as shown in Figure 3.8.

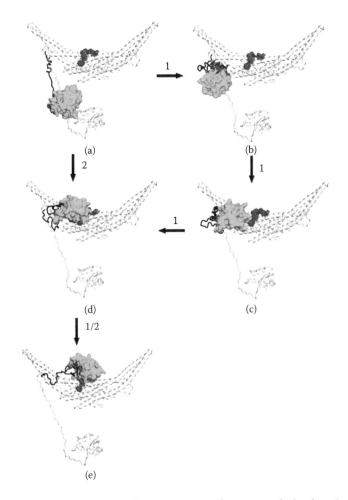

FIGURE 3.8 Snapshots corresponding to two types of trajectories for binding the PICK1-PDZ domain to the dual-BAR domain, starting from the initial structure (a) in which the PICK1-PDZ domain faces the convex surface of the PICK1-BAR dimer. The initial structure of the PICK1 dual-BAR construct is colored light gray and shown in cartoon representation as a background to compare with the snapshots from the folding trajectory. The PDZ domain in all five sites is shown as a surface view and is colored green. The 40-residue linker region is colored black. The interaction surfaces (i.e., the interface in the final complex) of both the PICK1-PDZ and PICK1-BAR domains are shown as a surface view and colored red. These structures were drawn with PYMOL (www.pymol.org). The two binding pathways are marked by arrows with 1 or 2 (or both if the conformations are similar in the two pathways). (b and c) Illustrate the initial stages of binding pathway 1; (a) and (d) illustrate the initial stage of binding pathway 2. The final bound structure is shown in (e).

3.3.11.3 Molecular Chaperones

The heat shock proteins (Hsps) are a family of chaperones that are ubiquitous in the cells of all organisms (Srivastava and Zihai, 2003; Ellis, 2013). Members are distinguished by their mass (kDa); Hsp70 and Hsp90 focus on unfolded and/or damaged proteins, and these proteins identify and bind to short hydrophobic stretches on their clients (Calderwood et al., 2009). It is believed that Hsp70 operates by grasping and unfolding denatured sections, which enables their correct refolding upon release (Broadley and Hartl, 2009). Hsp70 consists of two domains, the substrate binding domain (SBD) and the nucleotide binding domain (NBD), with the former binding the denatured substrate, and the latter acting as an adenosine triphosphatase (ATPase). Understanding the communication between these two domains and their dynamics continues to be of significant research interest and is an experimental challenge. Modeling the bacterial Hsp70 DnaK (PDB code 2KHO) by UNRES, however, has provided substantial insight into the mechanics of this chaperone. Canonical UNRES/MD simulations revealed three binding modes of the SBD to the NBD (Gołaś et al., 2012, Figure 3.9). One of these modes entails binding of the closed SBD to the NBD, and the remaining two entail the opening of the α- and β-subdomain constituents of the SBD prior to its binding to the NBD. Of the open-binding modes, one is mirrored in the experimentally determined open structure of the adenosine triphosphate (ATP)-bound (open) DnaK structure (Kityk et al., 2013), which was solved after our results were published. The frequency of the open binding event was further linked to an "ATP-bound" state of the NBD (Craig and Kampinga, 2010). Thus, coarse-grained modeling enables a significant advance in the understanding of the dynamics of the Hsp70 molecular chaperone.

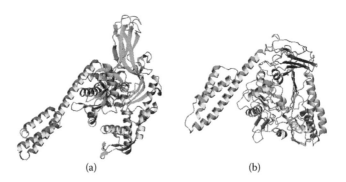

(a)   (b)

FIGURE 3.9 (a) The recently solved experimental structure of the open conformation of the DnaK chaperone (PDB: 1bq9) (From Kityk, R. et al., *Mol. Cell.*, 48, 863–874, 2013.). (b) The open conformation of DnaK obtained by canonical MD simulations with the UNRES force fields, starting from the closed conformation of DnaK (PDB: 2KHO) before the experimental structure was determined (From Gołaś, E. et al., *J. Chem. Theory Comput.*, 8, 1750–1764, 2012.).

Hsp70s, together with Hsp40 chaperones (the J-proteins), also participate in the transfer of Fe–S clusters from the Iron–Sulfur binding (Isu1)
protein to the target proteins. One of the key steps of the iron–sulfur cluster transfer is the interaction of Jac1 with Isu1 (Lill, 2009). Until now, only
an experimental study of the interaction between these two proteins was
carried out, which demonstrated that only the L^{105}, L^{109}, and Y^{163} residues
from Jac1 are crucial in this process. However, the study did not determine
which residues in Isu1 are crucial to bind Jac1 (Ciesielski et al., 2012).

By using homology modeling, molecular docking, and coarse-grained
and all-atom MD simulations, we studied the binding of Isu1 to Jac1. We
found that almost all trajectories, starting from different initial structures,
converged to structures similar to one of the predicted groups, in which
the β-sheet part of Isu1 (Figure 3.10) interacts with the side of Jac1. Further studies revealed that replacing Tyr^{163} from Jac1 with alanine has an
impact to lower the stability of the complex. Replacement of other residues
(Leu^{105}, Leu^{109}) by alanine confirmed their important role in binding the
two molecules, which is in complete agreement with experiment (Ciesielski
et al., 2012). All these findings suggest that the predicted structure of the
Isu1–Jac1 complex is a reasonable one.

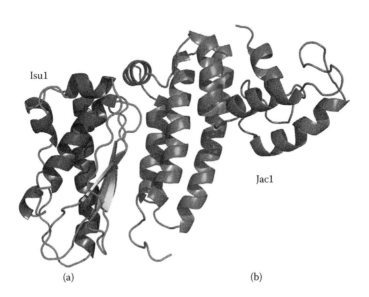

(a) (b)

FIGURE 3.10   The structure of the Isu1 (a)–Jac1 (b) complex simulated by coarse-grained
molecular dynamics with UNRES (Mozolewska, M. et al., 2015).

3.4 COARSE-GRAINED SIMULATIONS OF NUCLEIC ACIDS: THE NARES-2P MODEL

In this section, we describe our simplified model of nucleic acids that was constructed following the philosophy of UNRES.

3.4.1 Model and Effective Energy Function

As in UNRES, in the NARES-2P model (He et al., 2013a), there are two interaction sites per nucleotide. One interaction site is the phosphate group and the second one is the nucleic-acid base merged with the sugar ring; these sites serve as the polar units interacting by mean-field dipole–dipole interactions. The energy of the virtual-bond chain in the present NARES-2P model is expressed by Equation 3.12.

$$
U = w_{BB}^{GB} \sum_{i} \sum_{j<i} U_{B_iB_j}^{GB} + w_{BB}^{dip} f_2(T) \sum_{i} \sum_{j<i} U_{B_iB_j}^{dip} + w_{PP} \sum_{i} \sum_{j<i} U_{P_iP_j}
$$

$$
+ w_{PB} \sum_{i} \sum_{j} U_{P_iB_j} + w_{bond} \sum_{i} U_{bond}(d_i) + w_{ang} \sum_{i} U_{ang}(\theta_i)
$$

$$
+ w_{tor} f_2(T) \sum_{i} U_{tor}(\gamma_i) + w_{rot} \sum_{i} U_{rot}(\alpha_i, \beta_i) + U_{restr} \qquad (3.12)
$$

where $U_{B_iB_j}^{GB}$ denotes the nonbonded interactions between the coarse-grained sugar-base sites, which are described by the Gay and Berne (1981) anisotropic potential, and $U_{B_iB_j}^{dip}$ denotes the mean-field interactions between nucleic-base dipoles. $U_{P_iP_j}$ and $U_{P_iB_j}$ denote the interaction energies between phosphate–phosphate and phosphate bases. $U_{bond}(d_i)$, $U_{ang}(\theta_i)$, $U_{tor}(\gamma_i)$, and $U_{rot}(\alpha_i, \beta_i)$ denote the energies of virtual-bond stretching, of virtual-bond angle bending, of rotation about backbone virtual bonds, and of sugar base rotamers, respectively. The $U_{tor}(\gamma_i)$ terms provide correct right-handed chirality of the strands. The factors $f_n(T)$ account for the temperature dependence of terms of order higher than 1 of the generalized-cumulant expansion (Liwo et al., 2001). The term U_{restr} restricts the distance between the 5' end of one chain and the 3' end of the second chain to distances less than d_{max} (selected to correspond to the desired monomer concentration). The temperature factors $f_n(T)$ are defined by Equation 3.6.

3.4.2 Parameterization

U_{bond}, U_{ang}, U_{tor}, and U_{rot} were determined as statistical potentials derived from the statistics collected from the DNA and RNA molecules. U_{bond},

U_{ang}, and U_{tor} were determined by Boltzmann inversion of the distribution functions of virtual-bond length, virtual-bond angles, and virtual-bond dihedral angles, respectively, as given by Equation 3.13, followed by least-squares fitting, while U_{rot} was determined by minimization of the negative of the maximum likelihood, as given by Equation 3.14. The procedure is described in detail in our earlier work on the UNRES force field (Liwo et al., 1997b).

$$W(\bar{x}) = -RT \ln \frac{N([x])}{N_0([x])} \tag{3.13}$$

where $W(\bar{x})$ is the statistical potential computed at the center of a bin, $N([x])$ is the number of points in a given bin, $N_0([x])$ is the number of points in that bin expected in the absence of the interactions under consideration, R is the universal gas constant, and T is the absolute temperature; we took $T = 298$ K.

$$\ell(\mathbf{p}) = \sum_{\substack{data \\ points}} P_i^{exp} \ln P_i^{calc} = \sum_{\substack{data \\ points}} \frac{1}{N} \ln P_i^{calc}$$

$$= \sum_{\substack{data \\ points}} \frac{1}{N} \ln \frac{\exp\left[-\frac{U_{rot}(\mathbf{p};\alpha_i,\beta_i)}{RT}\right] \sin\alpha_i}{\int\limits_{-\pi}^{\pi} \int\limits_{0}^{2\pi} \exp\left[-\frac{U_{rot}(\mathbf{p};\alpha,\beta)}{RT}\right] \sin\alpha\, d\beta\, d\alpha}$$

$$= -\frac{1}{N} \sum_{\substack{data \\ points}} \left[\frac{1}{RT} \sum_{\substack{data \\ points}} U_{rot}(\mathbf{p};\alpha_i,\beta_i) \right.$$

$$\left. + \ln \int\limits_{-\pi}^{\pi} \int\limits_{0}^{2\pi} \exp\left[-\frac{U_{rot}(\mathbf{p};\alpha,\beta)}{RT}\right] \sin\alpha\, d\beta\, d\alpha \right] + A$$

$$A = \frac{1}{N} \sum_{\substack{data \\ points}} \sin\alpha_i \tag{3.14}$$

where $\mathbf{p}$ denotes the vector of parameters of U_{rot}, α and β are the angular coordinates of a SC center in spherical coordinate system, and N is the number of data points; the constant factor A can be omitted because it does not depend on force-field parameters. Because each data point is weighted equally, the corresponding "experimental" probability of occurrence is equal to $1/N$. The calculated probability is the Boltzmann probability calculated for the potential-energy function that is to be parameterized and depends on the parameters of the energy function. Minimizing the expression given by Equation 3.14 makes the calculated probability reflect the experimental density of points in the (α, β) space.

The base–base interaction terms were parameterized by fitting the PMF computed by numerical integration of the AMBER (Weiner et al., 1986) energy surfaces of all pairs of nucleic bases. The Boltzmann factor was computed from the AMBER force field and numerically integrated over the χ_1 and χ_2 angles defined in our earlier paper (Liwo et al., 1993). In summary, the PMF at each grid point were calculated from Equation 3.15.

$$W_{B_iB_j}\left(\theta_{1k}, \theta_{2k}, \varphi_k\right)$$

$$= -RT \ln \sum_{l=0}^{36} \sum_{m=0}^{36} \exp\left[-\frac{E_{B,B}^{AMBER}\left(\theta_{1k}, \theta_{2k}, \chi_{1l}, \chi_{2m}\right)}{RT}\right] (\Delta\chi)^2$$

$$\chi_{1l} = l\Delta\chi, \chi_{1m} = m\Delta\chi \tag{3.15}$$

where $W_{B_iB_j}\left(\theta_{1k}, \theta_{2k}, \varphi_k\right)$ is the computed PMF, $E_{B,B}^{AMBER}\left(\theta_{1k}, \theta_{2k}, \chi_{1l}, \chi_{2m}\right)$ is the energy computed with the AMBER force field (Weiner et al., 1986), using the dielectric constant $D = 4$, which corresponds to the interior of a biomolecule, and $\Delta\chi = 10°$. The parameters of the total $\left(U_{B_iB_j}^{dip} + U_{B_iB_j}^{GB}\right)$ potentials were obtained by nonlinear least-square fitting of the sum of the analytical forms of these expressions.

To parameterize the phosphate–phosphate/phosphate-base interaction terms, umbrella-sampling MD simulations with the AMBER force field for two dimethylphosphate ions in a TIP3P water box, including chloride counter ions, were performed. Then, the PMF was calculated from raw simulation data as a function of the distance between the phosphorus atoms by using the weighted-histogram-analysis method (Kumar et al., 1992). For the phosphate–base interaction, we use a simple arbitrary excluded volume potential at present, the purpose of which is to prevent the collapse of base centers onto the phosphate centers.

3.4.3 Calibration of the Force Field

The NARES-2P force field reported in our first paper (He et al., 2013a) was not specifically tuned to reproduce the structure and thermodynamics of DNA and RNA molecules. We are currently working on the calibration of the force field by a combination of maximum-likelihood fitting to the double-helix structures and heat-capacity curves of model DNA and RNA molecules.

3.4.4 Simulations of Folding Pathways of Small DNAs and RNAs

The NARES-2P model was built into the UNRES/MD platform (Khalili et al., 2005a,b; Liwo et al., 2005), which enables canonical and replica-exchange simulations of nucleic acids to be carried out. To check if the model is able to reproduce the double-helix structure, MREMD simulations of the following two DNA systems were carried out, the experimental structures of which have been determined in aqueous solution by NMR spectroscopy: the Dickerson–Drew dodecamer (DNA; 2 × 12 residues; PDB: 9BNA) and 2JYK (DNA; 2 × 21 residues), using the simulation protocol developed earlier (Liwo et al., 2007). For test purposes, the MREMD simulations were also carried out for two RNA molecules, 2KPC (RNA; 17 residues) and 2KX8 (RNA; 44 residues); the molecules are identified by their PDB structure names. We also ran additional MREMD simulations to assess the reproduction of folding thermodynamics and internal-loop (bubble) formation. All of these systems have a simple double-helix topology without long loops, bulges, multiplexes, or defects. The structures obtained from MREMD simulations with NARES-2P are compared with the respective experimental structures of the DNA and RNA test systems in Figure 3.11.

As can be seen in Figure 3.11, the calculated structures have the correct double-right-handed-helix topology. It can also be noted that, despite the early stage of development of the model, the resolution is of the order of that of the all-atom simulation. The NARES-2P force field is now being refined for better resolution. The force field is also ergodic; canonical simulations lead to a right-handed double-helix structure.

3.4.5 Simulations of Thermodynamics of DNA Folding

To assess how well NARES-2P reproduces the thermodynamic properties of DNA hybridization, we ran calculations for small DNA molecules, for which the heat-capacity curves were measured by differential-scanning calorimetry (DSC), from which the melting temperatures and standard enthalpies and entropies were determined (Hughesman et al., 2011a,b). The calculated melting temperatures agreed well with the experimental values except for two systems, while the calculated enthalpies and entropies of melting were generally smaller than the experimental values; this difference was also manifested in the fact that the calculated heat capacity curves were flatter compared to the experimental ones (He et al., 2013a). This feature of the present NARES-2P force field suggests that correlation terms must be introduced to make the melting transition sharper.

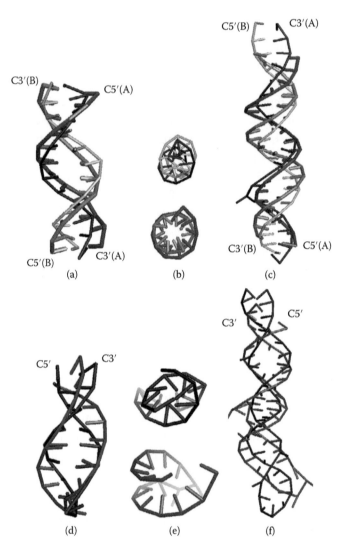

FIGURE 3.11 Calculated ensemble-averaged structures at $T = 300$ K obtained in NARES-2P MREMD simulations of the molecules studied (blue sticks) compared to the respective experimental structures (light and dark brown sticks for DNA and dark brown sticks for RNA, respectively). Sticks correspond to the S...S and S...B virtual bonds. Calculated structures of 9BNA, 2JYK, 2KPC, and 2KX8 are superposed on the corresponding experimental structures and shown as side views in panels a, c, d, and f, respectively. Top views of 9BNA and 2KPC (from the C5′ ends) are shown in panels b and e, respectively, with calculated structures above and the experimental structures below. RMSD values over the S centers of the calculated structures averaged over all native-like clusters are 4.7 Å, 10.7 Å, 5.10 Å, and 9.8 Å for 9BNA, 2JYK, 2KPC, and 2KX8, respectively, with respect to each experimental structure. The lowest RMSD values obtained in the respective MREMD runs are 2.6 Å, 4.2 Å, 3.2 Å, and 6.0 Å for 9BNA, 2JYK, 2KPC, and 2KX8, respectively.

3.4.6 Simulations of Premelting Transition (Internal-Loop or Bubble Formation)

One of the important features of the melting of DNA duplexes that contain substantial amounts of A–T-rich fragments is the premelting transition, which causes the disruption of the A–T pairs (which are weaker bonded than the C–G pairs) at temperatures lower than that of duplex dissociation. This transition is manifested as the difference between the fraction of paired bases and the fraction of paired chains (Zeng et al, 2004) or the disruption of the C–G pairs that are about 10 nucleotides away from an A–T-rich fragment (Cuesta-Lopez et al., 2011). By comparing the fractions of paired bases and paired chains calculated with NARES-2P for the sequences studied experimentally (Zeng et al, 2004; Cuesta-Lopez et al., 2011), we demonstrated (He et al., 2013a) that this force field reproduces qualitatively the premelting transition.

3.5 A COARSE-GRAINED MODEL FOR PROTEIN–DNA INTERACTIONS

UNRES and NARES-2P can be combined to describe protein–DNA interactions. The complete coarse-grained energy function to describe protein–DNA systems is given by Equation 3.16 (Yin et al, 2015).

$$U = U_\mathrm{P} + U_\mathrm{N} + U_\mathrm{PN} \tag{3.16}$$

where U_P is the effective energy function of the protein, U_N is the effective energy function of the nucleic acid, and U_PN is the effective energy function of the combined protein–nucleic acid system expressed by Equation 3.17.

$$U_\mathrm{PN} = w_\mathrm{p-P} \sum_i \sum_j U_\mathrm{p-P} + w_\mathrm{p-B} \sum_i \sum_j U_\mathrm{p-B} + w_\mathrm{SC-P} \sum_i \sum_j U_\mathrm{SC-P}$$
$$+ w_\mathrm{SC-B} \sum_i \sum_j U_\mathrm{SC-B} \tag{3.17}$$

where w is the weight for each corresponding potential term, $U_\mathrm{p-P}$ is the peptide group–phosphate group interaction potential, $U_\mathrm{p-B}$ is the peptide group–base interaction potential, $U_\mathrm{SC-P}$ is the SC–phosphate group interaction potential, and $U_\mathrm{SC-B}$ is the SC–base interaction potential.

The force field was parameterized as were the separate UNRES and NARES-2P terms (Yin et al, 2015). In order to demonstrate the reliability of the PMFs from MD simulation, the calculated SC–base equilibrium

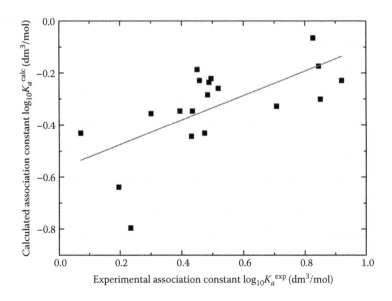

FIGURE 3.12 Correlation of the decimal logarithm of calculated association constant $\log K_a^{calc}$ and the decimal logarithm of experimental association constant $\log K_a^{exp}$.

association constants, K_a^{calc}, were calculated from the PMFs and then compared to the experimental association constants K_a^{exp} between amino acids and nucleotides produced by Reuben and Polk (1980), Kamiichi et al. (1987), and Ishida et al. (1989), as shown in Figure 3.12.

It can be seen from Figure 3.12 that the values of K_a^{calc} and K_a^{exp} are reasonably correlated, with a correlation coefficient of 0.65 (Yin et al, 2015). It should be noted that the experimental data pertain to the association constant of an amino acid and a nucleotide, but the theoretical association constants calculated here are between an amino-acid SC and a nucleoside, not the whole molecule. This explains why the experimental association constants are larger in value than the calculated constants. This result suggests that the analytical potential expressions presented here are a good representation of the UNRES + NARES-2P force field for protein–DNA systems.

3.6 CONCLUSIONS AND OUTLOOK

Physics-based coarse-grained models of proteins, such as UNRES, have facilitated the extension of the time scale in MD simulations. As a result, it has been possible to treat larger molecules and larger molecular complexes in real time. The philosophy of UNRES has been applied to develop a coarse-grained model of nucleic acids, namely NARES-2P. With the further development of these two coarse-grained models in the near future, we may envisage the possibility to treat biological protein–protein and protein–nucleic acid interactions, with the hope of ultimately being able to cure related diseases. All of this depends on bringing the coarse-grained models of proteins and nucleic acids to a higher degree of sensitivity and accuracy.

3.7 PROGRAM AVAILABILITY

The UNRES software is provided free of charge to academic users. Its source code and some precompiled binaries are available on the project website http://www.unres.pl.

Before downloading the UNRES program package one must agree on the license terms stating that no part of it will be sold or used in any other way for commercial purposes. This includes, but is not limited to, its incorporation into commercial software packages without written consent from the authors. For permission, one must contact Prof. H. A. Scheraga, Cornell University. The contact information is available on the website http://www.unres.pl.

Reports or publications using this software package must contain an acknowledgment to the authors and the National Institutes of Health (NIH) resource in the form commonly used in academic research. For further information about availability of UNRES software, consult the UNRES project website: http://www.unres.pl.

APPENDIX 3A: EXPANSION (FACTORIZATION) OF THE POTENTIAL OF MEAN FORCE INTO CLUSTER-CUMULANT FUNCTIONS (FACTORS) AND OF THE EXPANSION OF THE FACTORS INTO THE KUBO CLUSTER CUMULANTS

Let us consider a system composed of three interaction sites such as, e.g., the terminally blocked diglycine. The energy of this system can be expressed by Equation 3A.1.

$$
\begin{aligned}
E\left(X_1, y_1; X_2, y_2; X_3, y_3\right) &= E_{12}\left(X_1, y_1; X_2, y_2\right) + E_{13}\left(X_1, y_1; X_3, y_3\right) \\
&\quad + E_{23}\left(X_2, y_2; X_3, y_3\right) + E_1\left(X_1, y_1\right) + E_2\left(X_2, y_2\right) \\
&\quad + E_3\left(X_3, y_3\right) \\
&= E_{12}\left(X_1, X_2; y_1, y_2\right) + E_{13}\left(X_1, X_3; y_1, y_3\right) \\
&\quad + E_{23}\left(X_2, X_3; y_2, y_3\right) + E_1\left(X_1, y_1\right) + E_2\left(X_2, y_2\right) \\
&\quad + E_3\left(X_3, y_3\right) \\
&= \varepsilon_1\left(X; z_1\right) + \varepsilon_2\left(X; z_2\right) + \varepsilon_3\left(X; z_3\right) + \varepsilon_4\left(X; z_4\right) \\
&\quad + \varepsilon_5\left(X; z_5\right) + \varepsilon_6\left(X; z_6\right) \quad\quad (3A.1)
\end{aligned}
$$

with

$$
\begin{aligned}
\varepsilon_1\left(X; z_1\right) &= E_{12}\left(X_1, X_2; y_1, y_2\right) \\
\varepsilon_2\left(X; z_2\right) &= E_{13}\left(X_1, X_3; y_1, y_3\right) \\
\varepsilon_3\left(X; z_3\right) &= E_{23}\left(X_2, X_3; y_2, y_3\right) \\
\varepsilon_4\left(X; z_4\right) &= E_1\left(X_1, y_1\right) \\
\varepsilon_5\left(X; z_5\right) &= E_2\left(X_2, y_2\right) \\
\varepsilon_6\left(X; z_6\right) &= E_3\left(X_3, y_3\right) \quad\quad (3A.2)
\end{aligned}
$$

and $X = \left(X_1, X_2, X_3\right)$, $z_1 = \left(y_1, y_2\right)$, $z_2 = \left(y_1, y_3\right)$, $z_3 = \left(y_2, y_3\right)$, $z_4 = y_1$, $z_5 = y_2$, $z_6 = y_3$. The variables X_1, X_2, and X_3 are the coarse-grained variables (such as, e.g., the positions of the centers and the orientation of the axes of the peptide groups) while y_1, y_2, and y_3 are the secondary variables that are integrated out in the coarse-grained model (such as, e.g., the angles of rotation of the peptide groups about the $C^\alpha \dots C^\alpha$ axes) of sites 1, 2, and 3. The quantities $\varepsilon_1, \dots, \varepsilon_6$ are *component interaction energies* (Liwo et al., 2001). Even though a particular component-interaction energy depends only on part of coarse-grained variables, for uniform notation, all these variables are collected in the vector $\mathbf{X}$ on the left-hand side of Equation 3A.2.

The PMF of this system is obtained by integrating out the secondary variables, as given by Equation 3A.3 (cf. Equation 3.1).

$$F(\mathbf{X}) = -RT \ln \frac{1}{V_y} \int_{\Omega_y} \exp\left[-\frac{E\left(X_1, y_1; X_2, y_2; X_3, y_3\right)}{RT}\right] dy_1 dy_2 dy_3 \quad (3A.3)$$

Now, we rewrite the potential of mean force (PMF) of this three-site system as a linear combination of potentials of mean force corresponding to single interactions, pairs of interactions, ending up at the triplets of interactions as in the original PMF. For simplicity, let us assume that all sites are rigid, i.e., ε_4–ε_6 are constant. Then Equation 3A.3 becomes Equation 3A.4.

$$F(\mathbf{X}) = \langle\langle\varepsilon_1\varepsilon_2\varepsilon_3\rangle\rangle =$$

$$\overbrace{}^{\langle\langle\varepsilon_1\varepsilon_2\varepsilon_3\rangle\rangle_f}$$

$$\begin{bmatrix} \langle\langle\varepsilon_1\varepsilon_2\varepsilon_3\rangle\rangle - \left(\langle\langle\varepsilon_1\varepsilon_2\rangle\rangle - \langle\langle\varepsilon_1\rangle\rangle - \langle\langle\varepsilon_2\rangle\rangle\right) \\ - \left(\langle\langle\varepsilon_1\varepsilon_3\rangle\rangle - \langle\langle\varepsilon_1\rangle\rangle - \langle\langle\varepsilon_3\rangle\rangle\right) - \left(\langle\langle\varepsilon_2\varepsilon_3\rangle\rangle - \langle\langle\varepsilon_2\rangle\rangle - \langle\langle\varepsilon_3\rangle\rangle\right) \\ - \langle\langle\varepsilon_1\rangle\rangle - \langle\langle\varepsilon_2\rangle\rangle - \langle\langle\varepsilon_3\rangle\rangle \end{bmatrix}$$

$$+ \overbrace{\left[\langle\langle\varepsilon_1\varepsilon_2\rangle\rangle - \langle\langle\varepsilon_1\rangle\rangle - \langle\langle\varepsilon_2\rangle\rangle\right]}^{\langle\langle\varepsilon_1\varepsilon_2\rangle\rangle_f} + \overbrace{\left[\langle\langle\varepsilon_1\varepsilon_3\rangle\rangle - \langle\langle\varepsilon_1\rangle\rangle - \langle\langle\varepsilon_3\rangle\rangle\right]}^{\langle\langle\varepsilon_1\varepsilon_3\rangle\rangle_f}$$

$$+ \overbrace{\left[\langle\langle\varepsilon_2\varepsilon_3\rangle\rangle - \langle\langle\varepsilon_2\rangle\rangle - \langle\langle\varepsilon_3\rangle\rangle\right]}^{\langle\langle\varepsilon_2\varepsilon_3\rangle\rangle_f} + \overbrace{\langle\langle\varepsilon_1\rangle\rangle}^{\langle\langle\varepsilon_1\rangle\rangle_f} + \overbrace{\langle\langle\varepsilon_2\rangle\rangle}^{\langle\langle\varepsilon_1\rangle\rangle_f} + \overbrace{\langle\langle\varepsilon_3\rangle\rangle}^{\langle\langle\varepsilon_3\rangle\rangle_f} \quad (3A.4)$$

where the double bracket denotes the PMF computed with the use of a subset of component energies, as in Equation 3.4, and the coarse-grained variables were omitted from the right-hand-side expressions for clarity. It is clearly seen that the factors, denoted by double brackets with the "f" superscript (Equation 3.2 of the main text) are obtained by recursive subtractions of lower-order factors from the PMFs corresponding to subsets of component energies contained in a given factor. At the end, the factors corresponding to integrals of single component energies are equal to the respective potentials of mean force. Eventually, we arrive at the general expression given by Equation 3.3 in which the signs alternate as the order of the PMF decreases. From Equation 3A.4, this is obvious for the second-order factor and we order the expression for the third-order factor, getting

Equation 3A.5, which again conforms to Equation 3.3.

$$\langle\langle\varepsilon_1\varepsilon_2\varepsilon_3\rangle\rangle_f = \langle\langle\varepsilon_1\varepsilon_2\varepsilon_3\rangle\rangle - \langle\langle\varepsilon_1\varepsilon_2\rangle\rangle - \langle\langle\varepsilon_1\varepsilon_3\rangle\rangle - \langle\langle\varepsilon_2\varepsilon_3\rangle\rangle + \langle\langle\varepsilon_1\rangle\rangle$$
$$+ \langle\langle\varepsilon_2\rangle\rangle + \langle\langle\varepsilon_3\rangle\rangle \tag{3A.5}$$

It seems that the decomposition given by Equation 3A.4 (or Equation 3.2) makes the problem of PMF determination even more complicated. It would be so, if we wanted to keep all the terms up to order n, where n is the number of component interactions. However, the expansion can be cut at order $k < n$, where k is selected so that the simplified PMF still has the essential features of the full PMF. In most of the coarse-grained force fields, the expansions are cut at order 1 for long-range interactions (i.e., only the PMFs of isolated pairs of sites) or 2 for local interactions (which result in the appearance of torsional terms [Liwo et al., 2001; Ołdziej et al., 2003]). Such an approach ignores most of the multibody terms. As seen from Equation 3A.4 and Equations 3.2 and 3.3, the higher order factors such as $\langle\langle\varepsilon_1, \varepsilon_2, \varepsilon_3\rangle\rangle_f$ contain only the genuine multibody contributions, which cannot be expressed in terms of PMFs involving a smaller number of component energies. This approach prevents us from double or multiple counting of same contributions to the effective energy, which is rather unavoidable when statistically derived or heurestic potentials are used. Moreover, approximate analytical expressions for the multibody terms can be obtained by using Kubo's cluster-cumulant expansion in $\beta = 1/RT$ (Kubo, 1963), which is very important because, usually, these terms have no analogs in all-atom force fields. As an example, the first terms of the expansion for $\langle\langle\varepsilon_i\rangle\rangle_f$, $\langle\langle\varepsilon_i\varepsilon_j\rangle\rangle_f$, and $\langle\langle\varepsilon_i\varepsilon_j\varepsilon_k\rangle\rangle_f$ are given by Equation 3A.6:

$$\langle\langle\varepsilon_i\rangle\rangle_f = \langle\varepsilon_i\rangle + \dots$$
$$\langle\langle\varepsilon_i\varepsilon_j\rangle\rangle_f \approx -\beta\langle\varepsilon_i\varepsilon_j\rangle_c = -\beta\left[\langle\varepsilon_i\varepsilon_j\rangle - \langle\varepsilon_i\rangle\langle\varepsilon_j\rangle\right] + \dots$$
$$\langle\langle\varepsilon_i\varepsilon_j\varepsilon_k\rangle\rangle_f \approx \beta^2\langle\varepsilon_i\varepsilon_j\varepsilon_k\rangle_c = \beta^2\left[\langle\varepsilon_i\varepsilon_j\varepsilon_k\rangle - (\langle\varepsilon_i\varepsilon_j\rangle\langle\varepsilon_k\rangle + \langle\varepsilon_i\varepsilon_k\rangle\langle\varepsilon_j\rangle\right.$$
$$\left. + \langle\varepsilon_j\varepsilon_k\rangle\langle\varepsilon_i\rangle) + 2\langle\varepsilon_i\rangle\langle\varepsilon_j\rangle\langle\varepsilon_k\rangle\right] + \dots \tag{3A.6}$$

It can be seen that the higher the order of the factor, the higher the power of β at the first term in its expansion. UNRES and NARES-2P reflect this fact by the introduction of temperature dependence of the multibody terms.

ACKNOWLEDGMENTS

This work was supported by grant GM-14312 from the US National Institutes of Health (to HAS), by grant MCB10-19767 (to HAS) from the US National Science Foundation, by grants DEC-2012/06/A/ST4/00376 (to AL), DEC-2013/10/E/ST4/00755 (to MM) and DEC-2015/17/D/ST4/00509 (to AKS) from the National Science Center of Poland, by grant 530-8370-D498-14 from the Polish Ministry of Science and Higher Education (to CC), and by grant Mistrz7.1./2013 (to AL) from the Foundation for Polish Science. This research was supported by an allocation of advanced computing resources provided by the National Science Foundation (http://www.nics.tennessee.edu/), and by the National Science Foundation through TeraGrid resources provided by the Pittsburgh Supercomputing Center. Computational resources were also provided by (a) the supercomputer resources at the Informatics Center of the Metropolitan Academic Network (IC MAN) in Gdansk, (b) the 952-processor Beowulf cluster at the Baker Laboratory of Chemistry, Cornell University, (c) our 692-processor Beowulf cluster at the Faculty of Chemistry, University of Gdansk, and (d) the Interdisciplinary Center of Mathematical and Computer Modeling (ICM) of the University of Warsaw, Warsaw, Poland.

REFERENCES

Ablowitz, M.J., Prinari, B. and Trubatch, A.D. 2003. *Discrete and Continuous Nonlinear Schrödinger Systems*. Cambridge: Cambridge University Press.

Altschul, S.F., Madden, T.L. and Schaffer, A.A. 1997. Gapped BLAST and PSI-BLAST: A new generation of protein database search programs. *Nucleic Acids Res.* 25(17): 3389–3402.

Ayton, G.S., Noid, W.G. and Voth, G.A. 2007. Multiscale modeling of biomolecular systems: In serial and in parallel. *Curr. Opin. Struct. Biol.* 17: 192–198.

Benettin, G., and Giorgilli, A. 1994. On the Hamiltonian interpolation of near-to-the identity symplectic mappings with application to symplectic integration algorithms. *J. Stat. Phys.* 74(5–6): 1117–1143.

Berman, H.M., Westbrook, J., Feng, Z. et al. 2000. The protein data bank. *Nucleic Acids Res.* 28: 235–242.

Boudin, H., Doan, A. and Xia, J. 2000. Presynaptic clustering of mGluR7a requires the PICK1 PDZ domain binding site. *Neuron* 28(2): 485–497.

Broadley, S., and Hartl, F. 2009. The role of molecular chaperones in human misfolding diseases. *FEBS Lett.* 583: 2647–2653.

Brooks, B.R., Bruccoleri, R.E., Olafson, B.D. et al. 1983. CHARMM: A program for macromolecular energy, minimization, and dynamics calculations. *J. Comput. Chem.* 4: 187–217.

Calderwood, S., Murshid, A. and Prince, T. 2009. The shock of aging: Molecular chaperones and the heat shock response in longevity and aging—A mini-review. *Gerontology* **55**: 550–558.

Chernodub, M., Hu, S. and Niemi, A.J. 2010. Topological solitons and folded proteins *Phys. Rev. E* **82**: 011916.

Chinchio, M., Czaplewski, C., Liwo, A. et al. 2007. Dynamic formation and breaking of disulfide bonds in molecular dynamics simulations with the UNRES force field. *J. Chem. Theory Comput.* **3**: 1236–1248.

Ciesielski, S., Schilke, B., Osipiuk, J. et al. 2012. Interaction of J-protein co-chaperone Jac1 with Fe-S scaffold Isu is indispensable in vivo and conserved in evolution. *J. Mol. Biol.* **417**: 1–12.

Clementi, C. 2008. Coarse-grained models of protein folding: Toy models or predictive tools? *Curr. Opin. Struct. Biol.* **18**: 10–15.

Colombo, G., and Micheletti, C. 2006. Protein folding simulations: Combining coarse-grained models and all-atom molecular dynamics. *Theor. Chem. Acc.* **116**: 75–86.

Cowan, C.A., Yokoyama, N., Bianchi, L.M. et al. 2000. EphB2 guides axons at the midline and is necessary for normal vestibular function. *Neuron* **26**(2): 417–430.

Craig, E.A., and Kampinga, H.H. 2010. The HSP70 chaperone machinery: J proteins as drivers of functional specificity. *Nat. Rev.* **11**: 579–592.

Cuesta-Lopez, S., Menoni, H., Angelov, D. et al. 2011. Guanine radical chemistry reveals the effect of thermal fluctuations in gene promoter regions. *Nucleic Acid Res.* **39**: 5276–5283.

Czaplewski, C., Kalinowski, S., Liwo, A. et al. 2009. Application of multiplexed replica exchange molecular dynamics to the UNRES force field: Tests with α and α + β proteins. *J. Chem. Theory Comput.* **5**(3): 627–640.

Czaplewski, C., Liwo, A., Makowski, M. et al. 2010. Coarse-grained models of proteins: Theory and applications. In Kolinski, A. *Multiscale Approaches to Protein Modeling.* Berlin: Springer.

Czaplewski, C., Ołdziej, S., Liwo, A. et al. 2004. Prediction of the structures of proteins with the UNRES force field, including dynamic formation and breaking of disulfide bonds. *Protein Eng. Des. Sel.* **17**: 29–36.

Danielsson, U.H., Lundgren, M. and Niemi, A.J. 2010. Gauge field theory of chirally folded homopolymers with applications to folded proteins. *Phys. Rev. E* **82**: 021910.

Dev, K.K., Nishimune, A., Henley, J.M. et al. 1999. The protein kinase Cα binding protein PICK1 interacts with short but not long form alternative splice variants of AMPA receptor subunits. *Neuropharmacology* **38**(5): 635–644.

Duggan, A., Garcia-Anoveros, J. and Corey, D.P. 2002. The PDZ domain protein PICK1 and the sodium channel BNaC1 interact and localize at mechanosensory terminals of dorsal root ganglion neurons and dendrites of central neurons. *J. Biol. Chem.* **277**(7): 5203–5208.

El Far, O., Airas, J., Wischmeyer, E. et al. 2000. Interaction of the C-terminal tail region of the metabotropic glutamate receptor 7 with the protein kinase C substrate PICK1. *Eur. J. Neurosci.* **12**(12): 4215–4221.

Ellis, R. 2013. Assembly chaperones: A perspective. *Philos. Trans. R. Soc., B* **368**: 20110398.

Enz, R., and Croci, C. 2003. Different binding motifs in metabotropic glutamate receptor type 7b for filamin a, protein phosphatase 1C, protein interacting with protein kinase C (PICK) 1 and syntenin allow the formation of multimeric protein complexes. *Biochem. J.* **372**(1): 183–189.

Excoffon, K.J., Hruska-Hageman, A., Klotz, M. et al. 2004. A role for the PDZ-binding domain of the coxsackie B virus and adenovirus receptor (CAR) in cell adhesion and growth. *J. Cell Sci.* **117**(19): 4401–4409.

Faddeev, L.D., and Takhtajan, L.A. 1987. *Hamiltonian Methods in the Theory of Solitons.* Berlin: Springer Verlag.

Gay, J.G., and Berne, B.J. 1981. Modification of the overlap potential to mimic a linear site-site potential. *J. Chem. Phys.* **74**: 3316–3319.

Gołaś, E., Maisuradze, G.G., Senet, P. et al. 2012. Simulation of the opening and closing of Hsp70 chaperones by coarse-grained molecular dynamics. *J. Chem. Theory Comput.* **8**(5): 1750–1764.

He, Y., Maciejczyk, M., Ołdziej, S. et al. 2013a. Mean-field interactions between nucleic-acid-base dipoles can drive the formation of the double helix. *Phys. Rev. Lett.* **110**: 098101-1–098105.

He, Y., Mozolewska, M.A., Krupa, P. et al. 2013b. Lessons from application of the UNRES force field to predictions of structures of CASP10 targets. *Proc. Natl. Acad. Sci., U S A* **110**: 14936–14941.

He, Y., Xiao, Y., Liwo, A. et al. 2009. Exploring the parameter space of the coarse-grained UNRES force field by random search: Selecting a transferable medium-resolution force field. *J. Comput. Chem.* **30**: 2127–2135.

Hirbec, H., Francis, J.C., Lauri, S.E. et al. 2003. Rapid and differential regulation of AMPA and kainate receptors at hippocampal mossy fibre synapses by PICK1 and GRIP. *Neuron* **37**(4): 625–638.

Hruska-Hageman, A.M., Wemmie, J.A., Price, M.P. et al. 2002. Interaction of the synaptic protein PICK1 (protein interacting with C kinase 1) with the non-voltage gated sodium channels BNC1 (brain Na+ channel 1) and ASIC (acid-sensing ion channel). *Biochem. J.* **361**(3): 443–450.

Hughesman, C.B., Turner, R.F.B. and Haynes, C. 2011a. Correcting for heat capacity and 5′-TA type terminal nearest neighbors improves prediction of DNA melting temperatures using nearest-neighbor thermodynamic models. *Biochemistry* **50**: 2642–2649.

Hughesman, C.B., Turner, R.F.B. and Haynes, C. 2011b. Role of the heat capacity change in understanding and modeling melting thermodynamics of complementary duplexes containing standard and nucleobase-modified LNA. *Biochemistry* **50**: 5354–5368.

Hung, A.Y., and Sheng, M. 2002. PDZ domains: Structural modules for protein complex assembly. *J. Biol. Chem.* **277**(8): 5699–5702.

Ishida, T., Ohnishi, K., Dio, M. et al. 1989. Proton nuclear magnetic resonance study on the aromatic amino acid-guanine nucleotide system: Effect of base methylation on the stacking interaction with tyrosine and phenylalanine. *Chem. Pharm. Bull.* **37**: 1–4.

Izvekov, S., and Voth, G.A. 2005a. A multiscale coarse-graining method for biomolecular systems. *J. Phys. Chem. B* **109**: 2469–2473.

Izvekov, S., and Voth, G.A. 2005b. Multiscale coarse graining of liquid-state systems. *J. Chem. Phys.* **123**: 134105.

Jaulin-Bastard, F., Saito, H., Le Bivic, A. et al. 2001. The ERBB2/HER2 receptor differentially interacts with ERBIN and PICK1 PSD-95/DLG/ZO-1 domain proteins. *J. Biol. Chem.* **276**(18): 15256–15263.

Kamiichi, K., Doi, M., Nabae, M. et al. 1987. Structural studies of the interaction between indole derivatives and biologically important aromatic compounds. Part 19. Effect of base methylation on the ring-stacking interaction between tryptophan and guanine derivatives: A nuclear magnetic resonance investigation. *J. Chem. Soc. Perkin Trans. 2*, **12**: 1739–1745.

Khalili, M., Liwo, A., Jagielska, A. et al. 2005b. Molecular dynamics with the united-residue model of polypeptide chains. II. Langevin and Berendsen-bath dynamics and tests on model α-helical systems. *J. Phys. Chem. B* **109**(28): 13798–13810.

Khalili, M., Liwo, A., Rakowski, F. et al. 2005a. Molecular dynamics with the united-residue model of polypeptide chains. I. Lagrange equations of motion and tests of numerical stability in the microcanonical model. *J. Phys. Chem. B* **109**(28): 13785–13797.

Kilmister, C.W. 1967. *Lagrangian Dynamics: An Introduction for Students.* New York, NY: Plenum Press.

Kityk, R., Kopp, J., Sinning, I. et al. 2013. Structure and dynamics of the ATP-bound open conformation of Hsp70 chaperones. *Mol. Cell* **48**: 863–874.

Kleinerman, D.S., Czaplewski, C., Liwo, A. et al. 2008. Implementations of Nosé–Hoover and Nosé–Poincaré thermostats in mesoscopic dynamic simulations with the united-residue model of a polypeptide chain. *J. Chem. Phys.* **128**(24): 245103.

Koliński, A., and Skolnick, J. 1994. Monte Carlo simulations of protein folding. I. Lattice model and interaction scheme. *Proteins Struct. Funct. Genet.* **18**: 338–352.

Koliński, A., and Skolnick, J. 2004. Reduced models of proteins and their applications. *Polymer* **45**: 511–524.

Kozłowska, U., Liwo, A. and Scheraga, H.A. 2007. Determination of virtual-bond-angle potentials of mean force for coarse-grained simulations of protein structure and folding from ab initio energy surfaces of terminally-blocked glycine, alanine, and proline. *J. Phys. Condens. Matter* **19**: 285203.

Kozłowska, U., Maisuradze, G.G., Liwo, A. et al. 2010. Determination of side-chain-rotamer and side-chain and backbone virtual-bond-stretching potentials of mean force from AM1 energy surfaces of terminally-blocked amino-acid residues, for coarse-grained simulations of protein structure and folding. 2. Results, comparison with statistical potentials, and implementation in the UNRES force field. *J. Comput. Chem.* **31**: 1154–1167.

Krokhotin, A., Liwo, A, Maisuradze, G.G. et al. 2014. Kinks, loops, and protein folding, with protein A as an example. *J. Chem. Phys.* 140: 025101.

Krokhotin, A., Niemi, A.J. and Peng, X. 2012. Towards quantitative classification of folded proteins in terms of elementary functions. *Phys. Rev. E* **85**: 031906.

Krupa, P., Sieradzan, A.K., Rackovsky, S. et al. 2013. Improvement of the treatment of loop structures in the UNRES force field by inclusion of coupling between backbone- and side-chain-local conformational states. *J. Chem. Theory Comput.* **9**: 4620–4632.

Kubo, R. 1962. Generalized cumulant expansion method. *J. Phys. Soc. Jpn.* **17**: 1100–1120.

Kumar, S., Bouzida, D., Swendsen, R.H. et al. 1992. The weighted histogram analysis method for free-energy calculations on biomolecules. I. The method. *J. Comput. Chem.* **13**(8): 1011–1021.

Lam, A.R., Rodriguez, J.J., Rojas, A. et al. 2013. Tracking the mechanism of fibril assembly by simulated two-dimensional ultraviolet spectroscopy. *J. Phys. Chem. A* **117**: 342–350.

Landau, K.D., and Lifshitz, E.M. 1960. *Electrodynamics of the Continuous Media.* New York, NY: Pergamon Press.

Laskowski, M., Jr., and Scheraga, H.A. 1954. Thermodynamic considerations of protein reactions. I. Modified reactivity of polar groups. *J. Am. Chem. Soc.* **76**: 6305–6319.

Lee, J., Liwo, A., Ripoll, D.R. et al. 2000. Hierarchical energy-based approach to protein-structure prediction: Blind-test evaluation with CASP3 targets. *Int. J. Quantum Chem.* **77**(1): 90–117.

Lee, J., Ripoll, D.R., Czaplewski, C. et al. 2001. Optimization of parameters in macromolecular potential energy functions by conformational space annealing. *J. Phys. Chem. B* **105**: 7291–7298.

Leonard, A.S., Yermolaieva, O., Hruska-Hageman, A. et al. 2003. cAMP-dependent protein kinase phosphorylation of the acid-sensing ion channel-1 regulates its binding to the protein interacting with C-kinase-1. *Proc. Natl. Acad. Sci. U S A* **100**(4): 2029–2034.

Lill, R. 2009. Function and biogenesis of iron–sulphur proteins. *Nature* **460**: 831–838.

Lin, S.H., Arai, A.C., Wang, Z. et al. 2001a. The carboxyl terminus of the prolactin-releasing peptide receptor interacts with PDZ domain proteins involved in A- amino-3-hydroxy-5-methylisoxazole-4-propionic acid receptor clustering. *Mol. Pharmacol.* **60**(5): 916–923.

Lin, W.J., Chang, Y.F., Wang, W.L. et al. 2001b. Mitogen-stimulated TIS21 protein interacts with a protein-kinase-Cα-binding protein rPICK1. *Biochem. J.* **354**(3): 635–643.

Liwo, A. 2013. Coarse graining: A tool for large-scale simulations or more? *Phys. Scr.* **87**: 058502.

Liwo, A., Arłukowicz, P., Czaplewski, C. et al. 2002. A method for optimizing potential-energy functions by a hierarchical design of the potential-energy landscape: Application to the UNRES force field. *Proc. Natl. Acad. Sci. U S A* **99**: 1937–1942.

Liwo, A., Czaplewski, C., Ołdziej, S. et al. 2008. Simulation of protein structure and dynamics with the coarse-grained UNRES force field. In Voth, G.A. *Coarse-Graining of Condensed Phase and Biomolecular Systems*. Farmington, CT: CRC Press, Taylor and Francis Group, pp. 107–122.

Liwo, A., Czaplewski, C., Pillardy, J. et al. 2001. Cumulant-based expressions for the multibody terms for the correlation between local and electrostatic interactions in the united-residue force field. *J. Chem. Phys.* **115**: 2323–2347.

Liwo, A., He, Y. and Scheraga, H.A. 2011. Coarse-grained force field: General folding theory. *Phys. Chem. Chem. Phys.* **13**(38): 16890–16901.

Liwo, A., Kaźmierkiewicz, R. and Czaplewski, C. 1998. United-residue force field for off-lattice protein-structure simulations; III. Origin of backbone hydrogen-bonding cooperativity in united-residue potentials. *J. Comput. Chem.* **19**: 259–276.

Liwo, A., Khalili, M., Czaplewski, C. et al. 2007. Modification and optimization of the united-residue (UNRES) potential energy function for canonical simulations. I. Temperature dependence of the effective energy function and tests of the optimization method with single training proteins. *J. Phys. Chem. B* **111**: 260–285.

Liwo, A., Khalili, M. and Scheraga, H.A. 2005. Ab initio simulations of protein-folding pathways by molecular dynamics with the united-residue model of polypeptide chains. *Proc. Natl. Acad. Sci. U S A* **102**: 2362–2367.

Liwo, A., Lee, J., Ripoll, D.R. et al. 1999. Protein structure prediction by global optimization of a potential energy function. *Proc. Natl. Acad. Sci. U.S.A.* **96**: 5482–5485.

Liwo, A., Ołdziej, S., Czaplewski, C. et al. 2004. Parameterization of backbone-electrostatic and multibody contributions to the UNRES force field for protein-structure prediction from ab initio energy surfaces of model systems. *J. Phys. Chem. B* **108**: 9421–9438.

Liwo, A., Ołdziej, S., Pincus, M.R. et al. 1997a. A united-residue force field for off-lattice protein-structure simulations. I. Functional forms and parameters of long-range side-chain interaction potentials from protein crystal data. *J. Comput. Chem.* **18**: 849–873.

Liwo, A., Pincus, M.R., Wawak, R.J. et al. 1993. Prediction of protein conformation on the basis of a search for compact structures; test on avian pancreatic polypeptide. *Protein Sci.* **2**: 1715–1731.

Liwo, A., Pincus, M.R., Wawak, R.J. et al. 1997b. A united-residue force field for off-lattice protein-structure simulations. II: Parameterization of local interactions and determination of the weights of energy terms by Z-score optimization. *J. Comput. Chem.* **18**: 874–887.

Maisuradze, G.G., Liwo, A. and Scheraga, H.A. 2009a. Principal component analysis for protein folding dynamics. *J. Mol. Biol.* **385**: 312–329.

Maisuradze, G.G., Liwo, A. and Scheraga, H.A. 2009b. How adequate are one- and two-dimensional free energy landscapes for protein folding dynamics? *Phys. Rev. Lett.* **102**: 238102(1–4).

Maisuradze, G.G., Liwo, A. and Scheraga, H.A. 2010a. Relation between free energy landscapes of proteins and dynamics. *J. Chem. Theory Comput.* **6**: 583–595.

Maisuradze, G.G., Liwo, A., Senet, P. et al. 2013. Local vs global motions in protein folding. *J. Chem. Theory Comput.* **9**: 2907–2921.

Maisuradze, G.G., Senet, P., Czaplewski, C. et al. 2010b. Investigation of protein folding by coarse-grained molecular dynamics with the UNRES force field. *J. Phys. Chem. A* **114**: 4471–4485.

Makowski, M., Liwo, A., Maksimiak, K. et al. 2007a. Simple physics-based analytical formulas for the potentials of mean force for the interaction of amino acid side chains in water. 2. Tests with simple spherical systems. *J. Phys. Chem. B* **111**: 2917–2924.

Makowski, M., Liwo, A. and Scheraga, H.A. 2007b. Simple physics-based analytical formulas for the potentials of mean force for the interaction of amino acid side chains in water. 1. Approximate expression for the free energy of hydrophobic association based on a gaussian-overlap model. *J. Phys. Chem. B* **111**: 2910–2916.

Makowski, M., Liwo, A. and Scheraga, H.A. 2011a. Simple physics-based analytical formulas for the potentials of mean force of the interaction of amino-acid side chains in water. VI. Oppositely charged side chains. *J. Phys. Chem. B* **115**: 6130–6137.

Makowski, M., Liwo, A., Sobolewski, E. et al. 2011b. Simple physics-based analytical formulas for the potentials of mean force of the interaction of amino-acid side chains in water. V. Like-charged side chains. *J. Phys. Chem. B* **115**: 6119–6129.

Makowski, M., Sobolewski, E., Czaplewski, C. et al. 2007c. Simple physics-based analytical formulas for the potentials of mean force for the interaction of amino acid side chains in water. 3. Calculation and parameterization of the potentials of mean force of pairs of identical hydrophobic side chains. *J. Phys. Chem. B* **111**: 2925–2931.

Makowski, M., Sobolewski, E., Czaplewski, C. et al. 2008. Simple physics-based analytical formulas for the potentials of mean force for the interaction of amino acid side chains in water. IV. Pairs of different hydrophobic side chains. *J. Phys. Chem. B* **112**: 11385–11395.

Meyer, G., Varoqueaux, F., Neeb, A. et al. 2004. The complexity of PDZ domain-mediated interactions at glutamatergic synapses: A case study on neuroligin. *Neuropharmacology* **47**(5): 724–733.

Molkenthin, N., Hu, S. and Niemi, A.J. 2011. Discrete nonlinear Schrödinger Equation and polygonal solitons with applications to collapsed proteins. *Phys. Rev. Lett.* **106**: 078102.

Momany, F.A., McGuire, R.F., Burgess, A.W. et al. 1975. Energy parameters in polypeptides. VII. Geometric parameters, partial atomic charges, nonbonded interactions, hydrogen bond interactions, and intrinsic torsional potentials for the naturally occurring amino acids. *J. Phys. Chem.* **79**: 2361–2381.

Monticelli, L., Kandasamy, S.K., Periole, X. et al. 2008. The MARTINI coarse-grained force field: Extension to proteins. *J. Chem. Theory. Comput.* **4**: 819–834.

Mozolewska, M., Krupa, P., Scheraga, H.A. et al. 2015. Molecular modeling of the binding modes of the iron-sulfur protein to the Jac1 co-chaperone from Saccharomyces cerevisiae by all-atom and coarse-grained approaches. *Proteins: Struct. Func. Bioinfo.* **83**: 1414–1426.

Murzin, A.G., Brenner, S.E., Hubbard, T. et al. 1995. SCOP: A structural classification of proteins database for the investigation of sequences and structures. *J. Mol. Biol.* **247**(4): 536–540.

Nanias, M., Czaplewski, C. and Scheraga, H.A. 2006. Replica exchange and multicanonical algorithms with the coarse-grained united-residue (UNRES) force field. *J. Chem. Theory Comput.* **2**(3): 513–528.

Némethy, G., and Scheraga, H.A. 1962. The structure of water and hydrophobic bonding in proteins. III. The thermodynamic properties of hydrophobic bonds in proteins. *J. Phys. Chem.* **66**, 1773–1789.

Némethy, G., and Scheraga, H.A. 1965. Theoretical determination of sterically allowed conformations of a polypeptide chain by a computer method. *Biopolymers* **3**: 155–184.

Némethy, G., Steinberg, I.Z. and Scheraga, H.A. 1963. The influence of water structure and of hydrophobic interactions on the strength of side-chain hydrogen bonds in proteins. *Biopolymers* **1**: 43–69.

Niemi, A.J. 2003. Phases of bosonic strings and two dimensional gauge theories. *Phys. Rev. D* **67**: 106004.

Ołdziej, S., Czaplewski, C., Liwo, A. et al. 2005. Physics-based protein-structure prediction using a hierarchical protocol based on the UNRES force field: Assessment in two blind tests. *Proc. Natl. Acad. Sci. U S A* **102**(21): 7547–7552.

Ołdziej, S., Kozłowska, U., Liwo, A. et al. 2003. Determination of the potentials of mean force for rotation about Cα...Cα virtual bonds in polypeptides from the ab initio energy surfaces of terminally-blocked glycine, alanine, and proline. *J. Phys. Chem. A* **107**: 8035–8046.

Ołdziej, S., Łagiewka, J., Liwo, A. et al. 2004. Optimization of the UNRES force field by hierarchical design of the potential-energy landscape. 3. Use of many proteins in optimization. *J. Phys. Chem. B* **108**: 16950–16959.

Orengo, C.A., Michie, A.D., Jones, S. et al. 1997. Analysis and assessment of ab initio three-dimensional prediction, secondary structure, and contacts prediction. *Structure* **5**: 1093.

Pauling, L., Corey, R.B. and Branson, H.R. 1951. The structure of proteins—2 hydrogen-bonded helical configurations of the polypeptide chain. *Proc. Natl. Acad. Sci. U S A* **43**: 205–211.

Pearlman, D.A., Case, D.A., Caldwell, J.W. et al. 1995. AMBER, a package of computer programs for applying molecular mechanics, normal mode analysis, molecular dynamics and free energy calculations to simulate the structural and energetic properties of molecules. *Comput. Phys. Commun.* **91**: 1–45.

Penzes, P., Johnson, R.C., Sattler, R. et al. 2001. The neuronal Rho-GEF Kalirin-7 interacts with PDZ domain-containing proteins and regulates dendritic morphogenesis. *Neuron* **29**(1): 229–242.

Perroy, J., El Far, O., Bertaso, F. et al. 2002. PICK1 is required for the control of synaptic transmission by the metabotropic glutamate receptor 7. *EMBO. J.* **21**(12): 2990–2999.

Pillardy, J., Czaplewski, C., Liwo, A. et al. 2001. Recent improvements in prediction of protein structure by global optimization of a potential energy function. *Proc. Natl. Acad. Sci. U S A* **98**(5): 2329–2333.

Reuben, J., and Polk, F.E. 1980. Nucleotide-amino acid interactions and their relation to the genetic code. *J. Mol. Evol.* **15**: 103–112.

Reymond, N., Garrido-Urbani, S., Borg, J.P. et al. 2005. PICK-1: A scaffold protein that interacts with nectins and JAMs at cell junctions. *FEBS Lett.* **579**(10), 2243–2249.

Rojas, A., Liwo, A., Browne, D. et al. 2010. Mechanism of fiber assembly; treatment of Aβ-peptide aggregation with a coarse-grained united-residue field force. *J. Mol. Biol.* **404**: 537–552.

Rojas, A.V., Liwo, A. and Scheraga, H.A. 2011. A study of the α-helical intermediate preceding the aggregation of the amino-terminal fragment of the β amyloid peptide (Aβ1-28). *J. Phys. Chem. B* **115**: 12978–12983.

Ruth, R.D. 1983. A canonical integration technique. *IEEE Trans. Nucl. Sci.* **30**(4): 2669–2671.

Sanz-Serna, J.M., and Calvo, M.P. 1994. *Numerical Hamiltonian Problems* (Vol. 7). London: Chapman & Hall.

Scheraga, H.A. 2004. The thrombin-fibrinogen interaction. *Biophys. Chem.* **112**: 117–130.

Shen, H., Czaplewski, C., Liwo, A. et al. 2008. Implementation of a serial replica exchange method in a physics-based united-residue (UNRES) force field. *J. Chem. Theory Comput.* **4**(8): 1386–1400.

Sheng, M., and Sala, C. 2001. PDZ domains and the organization of supramolecular complexes. *Annu. Rev. Neurosci.* **24**: 1–29.

Sieradzan, A.K., Hansmann, U.H.E., Scheraga, H.A. et al. 2012a. Extension of UNRES force field to treat polypeptide chains with D-amino-acid residues. *J. Chem. Theory Comput.* **8**: 4746–4757.

Sieradzan, A.K., Krupa, P., Scheraga, H.A. et al. 2015. Physics-based potentials for the coupling between backbone- and side-chain-local conformational states in the United Residue (UNRES) force field for protein simulations. *J. Chem. Theory Comput.*, **11**: 817–831

Sieradzan, A.K., Liwo, A. and Hansmann, U.H.E. 2012b. Folding and self-assembly of a small protein complex. *J. Chem. Theory Comput.* **8**: 3416–3422.

Sieradzan, A.K., Niadzvedtski, A., Scheraga, H.A. et al. 2014. Revised backbone-virtual-bond-angle potentials to treat the L- and D-amino acid residues in the coarse-grained united residue (UNRES) force field. *J. Chem. Theory Comput.* **10**: 2195–2203.

Sieradzan, A.K., Scheraga, H.A. and Liwo, A. 2012c. Determination of effective potentials for the stretching of Cα...Cα virtual bonds in polypeptide chains for coarse-grained simulations of proteins from ab initio energy surfaces of N-methylacetamide and N-acetylpyrrolidine. *J. Chem. Theory Comput.* **8**: 1334–1343.

Skolnick, J., Arakaki, A.K., Seung, Y.L. et al. 2009. The continuity of protein structure space is an intrinsic property of proteins. *Proc. Natl. Acad. Sci. U S A* **106**: 15690–15695.

Solis, A.D., and Rackovsky, S. 2000. Optimized representations and maximal information in proteins. *Proteins Struct. Funct. Bioinform.* **38**: 149–164.

Srivastava, P., and Zihai, L. 2003. Heat-shock proteins. *Curr. Protoc. Immunol.* **58-1T**: A.1T.1–A.1T.6.

Staudinger, J., Lu, J. and Olson, E.N. 1997. Specific interaction of the PDZ domain protein PICK1 with the COOH terminus of protein kinase Cα. *J. Biol. Chem.* **272**(51): 32019–32024.

Staudinger, J., Zhou, J., Burgess, R. et al. 1995. PICK1: A perinuclear binding protein and substrate for protein kinase C isolated by the yeast two-hybrid system. *J. Cell Biol.* **128**(3): 263–271.

Takei, K., Slepnev, V.I., Haucke, V. et al. 1999. Functional partnership between amphiphysin and dynamin in clathrin-mediated endocytosis. *Nat. Cell Biol.* **1**(1): 33–39.

Takeya, R., Takeshige, K. and Sumimoto, H. 2000. Interaction of the PDZ domain of human PICK1 with class I ADP-ribosylation factors. *Biochem. Biophys. Res. Commun.* **267**(1): 149–155.

Taylor, T.J., Bai, H., Tai, C.H. et al. 2013. Score functions to evaluate CASP10 predictions for free-modeling targets. *Prot. Struct. Funct. Bioinform.* **81**(Suppl. 11): 57–83.

Torres, G.E., Yao, W.D., Mohn, A.R. et al. 2001. Functional interaction between monoamine plasma membrane transporters and the synaptic PDZ domain-containing protein PICK1. *Neuron* **30**(1): 121–134.

Torres, R., Firestein, B.L., Dong, H. et al. 1998. PDZ proteins bind, cluster, and synaptically colocalize with Eph receptors and their ephrin ligands. *Neuron* **21**(6): 1453–1463.

Tozzini, V. 2005. Coarse-grained models for proteins. *Curr. Opin. Struct. Biol.* **15**: 144–150.

Tuckerman, M., Berne, B.J. and Martyna, G.J. 1992. Reversible multiple time scale molecular dynamics. *J. Chem. Phys.* **97**(3): 1990–2001.

Tycko, R. 2006. Molecular structure of amyloid fibrils: Insights from solid-state NMR. *Q. Rev. Biophys.* **39**: 1–55.

van Gunsteren, W.F., and Berendsen, H.J.C. 1987. *Groningen Molecular Simulation Package* (GROMOS). Groningen, the Netherlands: Biomos.

Wales, D.J. 2003. *Energy Landscapes*. Cambridge: Cambridge University Press.

Weiner, S.J., Kollman, P.A., Nguyen, D.T. et al. 1986. An all atom force field for simulations of proteins and nucleic acids. *J. Comput. Chem.* **7**(2): 230–252.

Williams, M.E., Wu, S.C., McKenna, W.L. et al. 2003. Surface expression of the netrin receptor UNC5H1 is regulated through a protein kinase C-interacting protein/protein kinase-dependent mechanism. *J. Neurosci.* **23**(36): 11279–11288.

Yin, Y., Maisuradze, G.G., Liwo, A. et al. 2012. Hidden folding pathways in free-energy landscapes uncovered by network analysis. *J. Chem. Theory Comput.* **8**: 1176–1189.

Yin, Y., Sieradzan, A.K., Liwo, A. et al. 2015. Physics-based potentials for coarse-grained modeling of protein–DNA interactions. *J. Chem. Theory Comput.*, **11**: 1792–1808.

Zeng, Y., Montrichok, A. and Zocchi, G. 2004. Bubble nucleation and cooperativity in DNA melting. *J. Mol. Biol.* **339**: 67–75.

CHAPTER 4

AWSEM-MD

From Neural Networks to Protein Structure Prediction and Functional Dynamics of Complex Biomolecular Assemblies

Garegin A. Papoian and Peter G. Wolynes

CONTENTS

4.1 Introduction 121
4.2 Protein Physics from the Energy Landscape Perspective 125
4.3 The Neural Network Origin of the AWSEM Hamiltonian 135
4.4 The Backbone Potential 145
4.5 Pairwise and FM-Based Contact Potentials 151
4.6 Water-Mediated Interactions 154
4.7 Hydrogen Bond Potentials 158
4.8 Sculpting Protein Folding Funnels as a Guide to Parameter
     Optimization 161
4.9 Applications 171
4.10 Future Outlook 180
     Acknowledgments 182
     References 182

4.1 INTRODUCTION

Proteins come in a myriad of shapes, exposing specific chemically hetero-geneous surfaces, and exhibiting contrasting mechanical properties rang-ing from the relative rigidity of globular forms and fibrous assemblies to high flexibility in disordered chains that nevertheless can specifically recognize appropriate targets. Proteins are the primary molecular engines that drive the mechanochemical processes needed to sustain contemporary life forms. Reliable, accurate, and predictive computational modeling of protein dynamics promises to revolutionize biology and medicine, speeding the rational development of therapeutic agents targeting many currently untreatable diseases. The modeling of protein dynamics has been historically developed along two divergent routes. In one route, atomistic protein force fields have been developed via a "bottom-up" approach, starting from quantum chemical calculations of small molecules, adding parametrized interactions to fit small molecule structural data, and, iteratively, adjusting the model to describe experimental protein dynamics in a few systems [1–4].

In the other route, coarse-grained force field models are mostly constructed "top down" and use a "Big Data" approach to exploit the "wisdom of the database" [5,6]. It is true that coarse-grained models also start from known quantum chemical constraints, like those employed by Pauling in his proposals of basic protein structures [7] but coarse-grained models also purposefully leave out some detailed physicochemical information (like exactly how many atoms are in an amino-acid residue!). Such models make up for their microscopic shortcomings by taking advantage of the large amounts of existing protein structural and sequence data both to determine the parameters in the model and to explore the form of the force field.

Fully atomistic simulations have come a long way and show great promise in tackling many important problems in protein biophysics. Doubtless, as computers continually improve, all biological molecular problems will become accessible to simulation by such fully atomistic approaches. It is good to recall, however, that the route to fully atomistic force fields which often seems straightforward sweeps many issues under the rug in their development. We do not yet know in what contexts these issues are important. These issues include the nonadditivity of the forces that arise from the internal structure of the atoms themselves (e.g., polarizability [8], hydrogen bond cooperativity [9]). Also it is clear that fully atomistic modeling eventually must confront the quantum nature of the local vibrational motions in protein, which has not yet been examined in all its important details [10–13].

In any event, the currently most pressing biological problems involve proteins and their assemblies containing thousands of amino-acid residues and functional motions occurring on timescales that start from the microsecond scale but continue on to milliseconds, minutes, and beyond. As a consequence, statistically meaningful, fully atomistic simulations of such large systems at biologically relevant timescales remain a computational challenge. Addressing the major problems of modern biology today using computational modeling requires coarse-grained protein force fields.

Another aspect of the philosophy behind coarse-grained modeling is, however, not completely understood by the larger community. Although it is true that one strong motivation for developing coarse-grained modeling is to leap-frog the slow development of computers, expedience is not the whole story. In addition, we should recognize that coarse-grained models try to explain protein dynamics at a level which is conceptually appropriate for the motions themselves. Large-scale protein motions such as folding and functional transitions, precisely because they are complex, are "emergent" phenomena [14]. As emergent phenomena, many (but not all!) details of the interatomic forces and motions will end up being averaged over during the motion: it is unnecessary to design ships using molecular dynamics; accurate continuum hydrodynamic modeling suffices. Likewise, understanding folding and function and then communicating the insight gleaned from computer simulations are both easier when you start already at an appropriately coarser resolution.

The emergent nature of the physics owing to the large system size is not the only reason for the power of coarse-grained models. Those aspects of protein dynamics that are most relevant to biology presumably are those that have been sculpted by evolution and are adaptive for the organism. Evolution changes proteins residue-by-residue, not atom-by-atom. Biologically, evolution thus has only limited need for the most precise descriptions. Of course, since evolution happens in the physical world that is based on atoms, natural selection does "know" about the exact consequences of interresidue forces in great and subtle detail. These "details" can involve very small free energy differences. Such small free energy differences often come from a fine balance between opposing interactions between atoms in the protein and those in the solvent water. Trying to infer these balances directly from the deeper, all-atom model is nontrivial to do [2]. It is fortunately also unnecessary in most cases.

By approximately averaging over unnecessary atomic detail then, once a sufficiently accurate coarse-grained model has been constructed, the coarse-grained model can give a much more compact and understandable description of the mechanism of biomolecular dynamics by employing energy landscape theory [15] along with its tools such as frustration analysis [16–19]. The more efficient and comprehensible description available from a coarse-grained model then can act as an intellectual springboard for developing new experiments.

We see then that using coarse-grained force field models should not be viewed simply as an expedient compromise forced on us by limited computational resources but brings out the most biologically important features of protein motion.

While it is clear that many details of the fully atomistic models can be averaged over, it is not clear exactly what the irrelevant degrees of freedom are, nor what is the best way to describe the remaining degrees of freedom nor how to simulate them efficiently. These ambiguities have led to a diverse set of coarse-grained models. Comparing the usefulness of different coarse-grained models is difficult also because investigators employ a wide range of computational resources to simulate the models. While this wide range of computational capabilities allows the study of a correspondingly wide range of problems, making one on one comparisons of algorithms or models is tricky.

In this chapter, we review one particular class of useful coarse-grained protein models that has been developed over the last 25 years in our laboratories. The contemporary form of the resulting protein energy function is called AWSEM (the associative memory, water mediated, structure and energy model). The AWSEM force field is the heir to earlier force fields that started as simple potentials, that were motivated by the idea of employing the theory of neural networks to extract information from structural databases. Originally, these models focused on recognizing a native protein fold among several hundred decoy folds [20,21]. The next phase of this research relied on powerful optimization ideas from energy landscape theory that connected the functional form and parameterization of coarse-grained Hamiltonians to new understanding from the physics of protein folding [22–24]. Development in the subsequent 15 years continued using energy landscape ideas, where the main focus became discovering the microscopic physics of interactions within and between proteins in their aqueous environment [25–27]. Over time, the force field has transitioned from being dominated by specific bioinformatic terms (structural memories in the neural network model) to one relying more on many broadly transferable physical terms, such as cooperative hydrogen bonding in alpha helices and beta sheets. A key idea that emerged during this development was the importance of water-mediated interactions where landscape theory plus the database showed us that water's interaction with protein goes far beyond the hydrophobic effect [25]. Both the bioinformatic, neural-network-inspired aspects and the more purely physical aspects of the force field contribute to the practical successes achieved using the AWSEM potential.

This chapter is organized as follows. Because AWSEM is based on leveraging our modern understanding of physics of protein folding using energy landscape ideas, in the next section we briefly review protein folding theory, clarifying how various fundamental ideas from the theories of spin glasses and polymer physics underlie the folding process. After this, we discuss a series of coarse-grained protein force fields that have been used for predicting protein structure from amino-acid sequences either with or without knowledge about homologous proteins.

In later sections, we elaborate on the details of the AWSEM potential and also describe the parameter optimization strategy that follows from energy landscape theory. We then discuss a number of applications of AWSEM to predicting the folding of globular proteins, and also to predicting mechanisms of protein association and formation of protein-DNA complexes, such as nucleosomes. The successful engineering of protein structure prediction codes may seem to be a separate problem from understanding protein dynamics. Nevertheless, developing a coarse-grained force field that reliably folds many unrelated proteins has provided a very stringent test of our understanding of the underlying folding physics and has been a strong motivation for decoding folding mechanisms as studied in the laboratory. Since this challenge has now largely been met, we believe the resulting coarse-grained force field can be applied with some confidence to many problems that go beyond protein folding and structure prediction to studying biomolecular function and evolution. We thus conclude this chapter by providing our outlook on the anticipated development of AWSEM in the near term.

4.2 PROTEIN PHYSICS FROM THE ENERGY LANDSCAPE PERSPECTIVE

Protein folding is a wonderfully deep intellectual problem at the intersection of structural biology, genetics, evolution, polymer physics, and statistical mechanics. Starting from these diverse areas, researchers have queried the protein folding phenomenon from numerous perspectives and points of view. This intellectual diversity endows the "folding problem" with almost limitless richness, and, hence, makes the quest for understanding rather open-ended. At its most general, the folding problem can be seen as the question of how biological systems achieve functional structure at the molecular level. With that said, many physicists had historically first become interested in the folding problem because it was posed as a seeming paradox of search, namely, a brute force exploration of the conformational space to find the uniquely folded structure should take astronomically long timescales: How is this random search avoided [28]? Less philosophically, physicists and physical chemists were also challenged with the practical structure prediction problem: How does one develop a general algorithm to predict structures of globular proteins based only on the one-dimensional protein sequence as an input? Anfinsen's refolding experiment suggests this should be possible, since by folding *in vivo*, proteins decode their sequence themselves spontaneously [29]. The availability of such an algorithm opens up many prospects of bringing structural physicochemical understanding to vast expanses of biology since the number of known genetic sequences vastly outnumbers the number of known protein structures. While structural determination techniques continue to improve both in resolution and efficiency, they remain relatively slow compared to techniques for acquiring sequence information by itself.

The first of the protein folding problems, the philosophical search paradox, has been resolved within the energy landscape theory of protein folding, as elaborated in recent reviews [30] and discussed below. While the paradox has by now been dissipated, many interesting additional questions have emerged from the energy landscape viewpoint, making the study of protein folding still a very exciting area of research in molecular biophysics. The second, practical structure prediction problem is not as fully resolved, in the sense that no algorithm has been yet developed that both quickly and reliably predicts the structure of any protein from sequence alone to the level available from crystallographic resolution when that technique can be employed. Nevertheless, enormous strides have been made in the last two decades, where a significant fraction of proteins can be structurally well predicted but still at somewhat lower resolution than crystallography, with the coverage of successes quickly increasing with time with respect to classes of protein folds and sequences. Purely computational biology is therefore becoming a useful tool in elucidating biological function. We highlight below some of the recent successes of the AWSEM force field [31] in this regard. Interestingly, AWSEM has already achieved high speed, reliability, and precision for the related binding structure prediction problem, namely, finding how two proteins associate into a larger complex when only their monomeric structures are known [32]. In any event, access to even moderate resolution information can be very enlightening and is often sufficient to address many biological questions even while quantitative predictions like those needed for drug design remain a challenge even with high-resolution structures.

For the remainder of this section, we take turns in discussing the protein folding problem from the points of view of pure thermodynamics, polymer physics, and the dynamics of spin glasses, subsequently synthesizing these notions into a coherent theoretical framework within the energy landscape theory.

Numerous experiments and computer simulations along with theoretical analyses have revealed that many small proteins show a simple two-state thermodynamic behavior, suddenly making a transition from a disordered to an ordered phase as the thermodynamic control variables are smoothly changed, for example, by lowering the solution temperature or by removing a denaturant, such as urea [33]. Phenomenologically, these studies thus point to a weak first-order thermodynamic phase transition in a finite system, showing a high degree of cooperativity because even small proteins are still pretty big molecules. Furthermore, it turns out that at least two types of disordered phases need to be considered: (1) a random coil (RC) phase, where the protein chain is swollen and highly fluctuating much like a polymeric gas, and (2) a collapsed molten globule (MG) phase, but where the protein has not yet adopted its unique native (N) structure, still structurally fluctuating like a polymeric liquid [34–36]. Hence, we should consider the following phase equilibria:

$$\text{RC} \rightleftharpoons \text{MG} \rightleftharpoons N$$

Theory predicts and experiment largely confirms that the first of these transitions for homopolymers is second order, where large self-similar fluctuations are expected near the transition critical point [34–36]. Interestingly, many natural proteins operate under physiological conditions where the two transition points appear to be proximal. Hence, folding occurs near a tricritical point. The corresponding implications of this confluence of phase coexistence behavior are further elaborated below.

Here, we first concentrate on the transition between the molten globule phase, which is entropically rich but high in energy, to an entropy-poor, but low-energy native state. Since the folding order parameter, such as the fraction of native contacts, changes discontinuously between the MG and N phases, this transition will generally be first order [37]. As is customary in the protein folding field, we use the term "energy" colloquially, to mean the solvent-averaged free energy of any given protein chain conformation.

Calorimetric experiments of course also include energy changes of the solvent. The basic thermodynamics of the globule-to-native transition can be empirically understood by applying the paradigm of Landau's mean-field theory [37], where, first, an order parameter is introduced to describe the process for folding [36] (the fraction of native contacts, Q, serves well in that role for folding). Then, near the folding midpoint, a two-well free energy profile is expected, with a barrier separating them, with the two minima becoming more asymmetrically disposed as the temperature is lowered below the folding temperature [36–38].

When macroscopic phase changes are considered from a kinetic viewpoint, first-order phase transitions occur either via nucleation processes or through spinodal decomposition, which dominates when the disordered phase is highly unstable [39,40]. Presumably, the thermodynamic drive to folding need only be moderate under biological conditions, so the nucleation mechanism would be expected to dominate. Detailed theoretical analyses and simulations indeed demonstrate the analogy of folding to the nucleation process [35,41–43], however, several important differences from the macroscopic situation unique to proteins need to be considered. First, proteins are relatively small systems not very much larger than the range of the interresidue forces, hence, potentially the critical nucleus size could equal or even exceed the size of the whole protein [41]. In that case, it may be better to either drop the classical nucleation picture altogether, and instead rely on a spatially-averaged global mean-field theory, or, on the contrary, adopt a spatially resolved field-theoretical treatment that interpolates between the two pictures [44]. Second, as opposed to the crystallization of simple pure liquids, protein chains are highly heterogeneous systems. This heterogeneity, coupled with the polymeric nature of the proteins, greatly complicates the simple homogeneous nucleation picture. Heterogeneous interactions between different parts of the protein would roughen the critical nucleus, and can generate multiple coexisting nuclei or lead to many other complexities [41].

Notice that the classical thermodynamic analysis does not really point to any paradoxes yet. Indeed, the molecular search problem for an ice nucleus to form in supercooled water is hard but obviously not insuperable. Indeed, from the current, accumulated wealth of knowledge about protein folding kinetics in the laboratory, the above discussed mean-field picture of two-state equilibrium seems to be largely self-consistent and indeed very useful by itself. However, the price one pays for using only the formidable power of thermodynamic phenomenology is a nearly complete opaqueness about folding mechanisms at the microscopic level. Thermodynamics alone does not tell us how exactly any given protein chain folds. Similarly, the fact that proteins are polymers, made from 20-letter alphabet amino acids, with special sequences evolved over billions of years, is still largely ignored by the simplest nucleation paradigm. Was evolution even needed for folding to be possible? We start to consider these topics next.

The fact that proteins are polymers [45], long connected chains, determines many of their most remarkable properties. Proteins, are, however, not like most artificially prepared polymers, where the monomeric building blocks are either all-identical (homopolymers) or are arranged in some predictable patterns of two or three types of monomers (di- and triblock copolymers, respectively) [45]. Instead proteins are made from a large range of different monomeric types [46]. These monomer units, amino acids are, however, nearly identical at the backbone level, with the exception of proline [46]. The side-chain groups on the 20 different amino acids also do not vary a great deal in steric size but they do have quite different chemistry, some charged, some polar, others apolar [36,46]. Not only are many monomeric types used to make a protein, but the monomers are not even arranged in simple patterns. Highly repetitive sequences, like those seen in artificial polymers, often do not seem to fold at all, but are considered to yield "intrinsically disordered" proteins [47–49]. In contrast, a typical foldable protein sequence appears nearly random when considered as a one-dimensional character string [50]. The apparent randomness is taken, in fact, as the signature of a coding sequence for a structured protein. That such "random" sequences can fold into unique three-dimensional shapes is a rather remarkable achievement that energy landscape theory suggests required evolution to come about.

Unlike a random walk, the protein backbone cannot adopt any trajectory in space, but is highly restricted by several factors. First, as with other real polymers, the protein chain cannot intersect itself because of excluded volume steric interactions. Furthermore, the backbone stereochemistry greatly restricts the phi and psi dihedral angles of each residue, reducing correspondingly the available conformational space [33]. As with other sp^3 hybridized single carbon–carbon bonds, three discrete rotameric states emerge at each dihedral angle. Not all these rotamers can be freely realized because they are constrained additionally by local steric interactions. Interestingly, even if only two rotameric states were allowed per residue, which is a likely underestimate, the overall number of available backbone conformations would be 2^{100} (approximately 10^{30}) for a hundred residue protein. Directly sampling all these conformations to search for the native state, if that were to be the search strategy, is not conceivable at any timescale of interest to biological organisms. This is the famous Levinthal paradox [51]. Here, it should be noted that successful coarse-grained protein force fields should account for the restricted stereochemistry of real protein chains, otherwise, too flexible backbone models would lead to an enormous explosion of the number of unphysical conformations. On the contrary, too rigid backbones would disallow many native conformations and would also create serious kinetic traps. The backbone model of AWSEM [31], which is somewhat complex, assumes an ideal peptide geometry (i.e., it does not allow bending motions, e.g., which are in fact quite stiff compared to dihedral angle charges) and imposes potential terms that guide the phi and psi dihedral angles to small regions in the Ramachandran plot [33,46], as observed for real proteins.

What are the consequences of the interactions between side chains? We separately consider homopolymeric and heteropolymeric effects. Studying the homopolymer might initially seem completely out of place, since proteins contain many types of amino acids. Nevertheless, when a broad ensemble of structures is considered, the structurally averaged physical properties of the protein chain are well described for single sequences by averaging over sequences of similar overall composition (they are self-averaging, to use the jargon of glass physics [52]). Consider, for example, the role of the overall hydrophobicity of the protein. If a sequence is more hydrophobic on average, many of its residues will have a tendency to avoid solvent and thus to stick to each other. This tendency leads to collapse from a random coil phase to a molten globule phase [53]. Similarly, a chain having many positively charged residues would resist collapse because of the desolvation penalty of the charges, and furthermore, the chain would swell because of the like-charge repulsion. Hence, by merely looking at the composition of any given sequence, reasonable conclusions can usually be drawn about its behavior [25], especially in the unfolded state: for example, whether a protein is "intrinsically disordered" can be predicted with some reliability [54]. The average properties of the chain are largely sufficient for understanding whether the unfolded state prefers to collapse into a liquid-like molten globule or to stay as a more entropically rich gas-like random coil.

The most interesting physical effects associated with proteins being heteropolymers are due to "frustration" [52,55–57]. Since some side chains will attract each other in the solvent, while other pairs of side chains would rather stay apart, a random sequence, once it is collapsed to a random compact con-formation [58], would yield structures not only having many favorable inter-actions, but also having many unfavorable interactions because the chain must remain covalently connected. Differently arranged compact chain con-formations will have distinct energies. A number of these collapsed con-formations might end up by chance being extremely low in energy, hence, thermodynamically dominating at low enough temperatures due to their corresponding Boltzmann weights. The emerging set of nearly degenerate candidate ground states will be structurally very dissimilar to each other for a truly random sequence. Once a chain has found such a low-energy min-imum, there is no driving force for it to explore configurational space any further. Because of this lack of guidance the Levinthal paradox reemerges: the chain does not have enough time on biologically relevant timescales to find all the degenerate minima because it will take a long time to escape from any given one. Trapping instead of search itself becomes the difficulty for a random chain to find its best structure, the deepest minimum.

The argument just outlined, due to Bryngelson and Wolynes [55], breathed new life into the Levinthal argument, showing its validity at least for long random sequences. The refined conundrum of trap escape has, however, been resolved by bringing ideas of statistical Hamiltonians from spin-glass theory to explain quantitatively exactly how hard the trapping search problem in protein folding physics really is [55,56,59]. Energy land-scape theory shows that the new Levinthal paradox can most elegantly be resolved only if protein sequences have evolved to create energetic corre-lations among chain conformations at a global scale [34]. Conformations that are similar to the very low-energy native conformation will also be low in energy in some (potentially nonlinear) proportionality to the structural overlap between these two states [60]. The correlations lead to a guided or funneled search, not a random process of aimless repeated trap escape. For any folded structure, one can vary the overlaid sequence in such a way that unfavorable, frustrated interactions, which would ordinarily on a sta-tistical basis be ubiquitous for the random sequences, can be nearly elim-inated. Such a design which is even possible by a random evolutionary search through sequence space would lead to a landscape globally funneled to the native basin. The resulting "principle of minimum frustration" [55] resolves the original Levinthal paradox by saying evolved proteins are spe-cial. In fact, statistical landscape theory allows this degree of specialness to be quantified.

This more careful analysis of the folding dynamics via the energy landscape theory raises a new problem: even in a globally funneled landscape, local traps might still be occasionally very deep, leading to kinetic trapping while the ensemble of conformations "slides down" the funnel. One way to quantify the difficulty of the trapping issue is to stratify the funnel, namely, to consider isosurfaces of the structural order parameter, such as the fraction of native contacts, Q [60–61] (see Figure 4.1). Then, a diffusive dynamics along the longitudinal direction toward the folded state will be slowed transiently by the trapping, while the huge number of the remaining transverse degrees of freedom would be sampled at a nearly constant value of Q [62–63] (see Figure 4.2). Hence, any particular constant Q hypersurface may be thought of itself as a dynamically useful statistical ensemble [34,60], where, in particular, the statistics of traps would define a low enough temperature where all large-scale motions nearly freeze. By repeating this calculation at different strata of the funnel, one obtains the following interesting quantities: $E(Q)$—the average (nonthermal) energy of the isosurface Q, $dE(Q)$—the standard deviation of energy minima at a given stratum, and $S(Q)$—the total number of conformations available to the stratum at Q [34,60,64]. Taking these data as an input, the stratified random energy model (REM) [65] allows one to calculate the expected glass transition temperature at any hyper surface corresponding to Q, and also other related thermodynamics quantities, such as average thermal energy and free energy [66–68].

The resulting free energy profile, $F(Q)$, allows one not only to predict the equilibrium populations of various conformations at various degrees of folding but also to understand the directionality of flow through the configuration space [34]. Furthermore, simulations or experiments can be used to estimate $D(Q)$, the position-dependent diffusion constant that measures how rapidly structures reconfigure [62]. This rate depends on the proximity to the glass transition temperature, T_G. A Kramers theory based on the free energy profile $F(Q)$ and the diffusion rate $D(Q)$ yields detailed kinetic predictions about folding timescales [62]. If the folded protein is thermodynamically stable at a temperature T_F substantially greater than T_G, trapping in the misfolded state is not a serious issue. This analysis shows that the minimal frustration principle can be expressed in quantitative terms as $T_F > T_G$. While in this part of our exposition, we have focused on the transition from the liquid-like molten globule to the folded state, many proteins are found to be near the collapse transition under physiological conditions. Both the "homopolymeric" average composition, indicating the sequence's overall hydrophobicity, as well as heteropolymeric

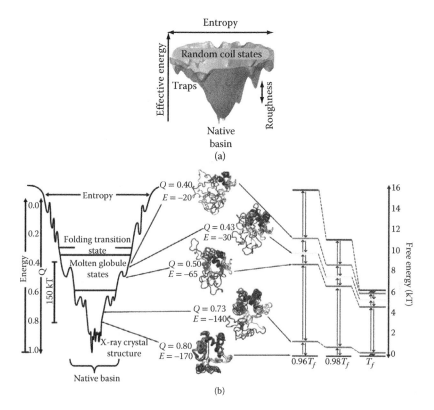

FIGURE 4.1   The native state of a globular protein should be not only thermodynamically dominant under physiological conditions, but also kinetically accessible. Generic heteropolymers consisting of a sequence of 20 different amino acids are predicted to have rough energy landscapes, where folding experiments would be initial-condition dependent, leading to a broad array of very different final conformations. Protein sequences have evolved to avoid this problem by organizing their conformational landscape into a funnel with small or moderate ruggedness, which allows reliable folding to the native basin on biologically relevant timescales. The upper portion of the funnel contains very large number of random coil conformations that grow exponentially with chain length. For some proteins, the mid portion might include compact molten globule states, where the tertiary contacts are still strongly fluctuating. The native state usually has a sharply defined topology, however, when viewed at high structural resolution, it can be further partitioned into either distinct allosteric states or other types of generic substates. (a) Conceptual aspects of the folding funnel are depicted. (b) Computer simulations show more quantitative insights into the organization of the energy landscape of cytochrome c, a protein frequently probed in folding studies. (Panel (b): from JA Hegler et al., *HFSP J.*, 2, 307–313, 2008.)

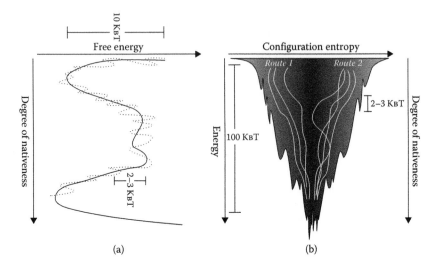

FIGURE 4.2 The folding funnel can be projected into a one-dimensional reaction coordinate, Q. This can be followed by solving the Kramers problem of a fictitious particle diffusing on the obtained $F(Q)$ profile, obtaining first passage time distributions and other kinetic observables of interest. (From JN Onuchic and PG Wolynes, *Curr Opin Struct Biol.*, 14, 70–75, 2004.)

interactions, enter into determining the exact collapse thermodynamics and kinetic course [34,35,58]. Because of this, complete theoretical models of folding kinetics should include all three phases, RC, MG, and N. Analysis can become rather involved because of the complexity of the many phase changes occurring simultaneously.

We conclude this section by summarizing how various findings from protein folding physics have helped the engineering of high-quality protein force fields. In fact, energy landscape ideas have been used to tune both coarse-grained and atomistic models, but here we will only elaborate on the former where the theory has been more explicitly employed. As discussed above, one important choice in coarse-grained models that will determine search kinetics concerns the representation of the protein chains backbone. Using the most simplistic representation may first seem desirable for reasons of computational efficiency, since fewer dynamical equations need to be integrated. Coarse graining, however, can lead to an unrealistic expansion of the chain conformation space or, on the contrary, a severe overrestriction on the possible structures; either of these issues could lead to either too big a search or an overly slow trapping dominated search. In AWSEM [31], even though three beads are used per residue, the assumption of an ideal peptide geometry and many backbone interaction constraints in fact leads to backbones rather similar to those in all-atom representation: for the backbone, AWSEM is essentially atomistically accurate, while having many fewer atoms to follow dynamically in the computer. In terms of treating the side chains, many choices exist for selecting different degrees of fine- or coarse-grained representation for the various chemical groups. For the side chains, AWSEM is quite coarse-grained: each amino acid is represented by only a single bead at the C_β position. The price that is paid for this simplicity is that a relatively complicated set of potentials between the C_βs emerges. In order to gain back the information loss, a number of additional terms in AWSEM, elaborated below, are introduced, that also take into account backbone hydrogen bonding, which may be cooperative, and also water-mediated interactions among the solvent-exposed residues. The latter interactions, defining a new *hydrophilic* code of amino acids [25], play a particularly important role in larger proteins containing internal interfaces or many charged groups and also play a special role in protein binding.

With all these potential forms come a large number of parameters that must be determined in an objective way. AWSEM parameters have been largely optimized self-consistently using the principles of the energy landscape theory of protein folding outlined above. The key is that the principle of minimum frustration can be recast mathematically as the maximization of the T_F/T_G ratio. By carrying out this optimization, the glass transition temperature is made as low as possible in units of the folding temperature so that simulated annealing can avoid traps even at temperatures low enough for the native fold to be stable. With all these ingredients, developed over 25 years, AWSEM has achieved remarkable successes both in folding and binding predictions, as documented in later sections. However, next, we present an overview of the origins of the AWSEM force field, providing a historic context.

4.3 THE NEURAL NETWORK ORIGIN OF THE AWSEM HAMILTONIAN

The task of finding a three-dimensional structure which corresponds to a given sequence may be fruitfully viewed in the context of information theory. In particular, it has been recognized that the number of unique protein folds is not very large, perhaps on the order of a few thousand [69]. Hence, if almost all natural folds were to be discovered, then low-resolution structure prediction could be based on matching any given input sequence to one of the known structural folds (potentially followed by further structural refinement, if a higher resolution prediction is desired). Such a mapping, from sequence to a known structure, is somewhat analogous to memory retrieval or recognition, where the memories indicate (not necessarily perfect) associations between sequences and structures, learned and stored in some mathematical way [70]. Biological organisms use neuronal networks for analogous tasks, and a variety of mathematical models of neural networks have been invented to mimic biological information processing [70,71]. Ideas for this development have already been carried over to other areas of inquiry and application. Especially in the last decade, computational neural networks have transitioned from being a largely research curiosity to mainstream products that we use every day on the Internet, ranging from character and speech recognition to real-time translations between languages and image recognition search engines [72].

In this spirit, in the context of matching protein sequence to structure, Friedrichs and Wolynes in 1989 proposed using neural networks to store a number of sequence-structure memories and then retrieve a particular structure if one of these sequences, either intact or partially degraded, because of mutations or insertions and deletions, were given as an input [20]. At that point, a choice of a neural network model needed to be made.

It turns out that one important class of neural networks, belonging to the so-called class of *recurrent artificial neural networks* [70], may be mathematically modeled as involving the dynamical evolution of a specially constructed quenched spin-glass Hamiltonian, invented by John Hopfield in 1982 [73]. The Hopfield network model consists of N microscopic spin variables that have two states, for example, $s = \pm 1$. These spins interact via pairwise coupling constants, J_{ij}, that are symmetric. The time evolution of each spin is asynchronous, attempted at some mean rate of W, with the following transition rules. If the local field, $\sum_j J_{ij} s_j$, is greater than some threshold U_i (e.g., all thresholds could be set to zero), then the state of the s_i spin is switched to $+1$. If the same sum is less than the threshold, then the s_i spin is switched to -1. To complete the model, an *ansatz* for the coupling constants, J_{ij} is needed. This is where memories come in. If J_{ij}s were to be completely randomly distributed, then the local minima found through time evolution would correspond to random spin configurations. Instead, the following simple prescription for the coupling elements, $J_{ij} = \sum_\mu s_i^\mu s_j^\mu$, may be used to store memories, where $\{s_i^\mu\}$ indicate the spin configurations of the memory state μ. If we assume we would like to store n memories, μ would run from 1 to M.

It can be shown that the Hopfield construction produces an energy landscape with basins of attractions that correspond to the input stored memories [71,73]. In particular, if we were to start dynamics with an initial spin configuration different from but sufficiently close to one of the memories, μ', the temporal evolution will generally lead to recall of the complete memory, where the ultimate spin configuration is $\{s_i^{\mu'}\}$, perhaps with only a few spins being incorrectly flipped (see Figure 4.3). The recall property of the network is a result of the "orthogonality" of the distinct input memories. Indeed, let's assume we prepare the system in the $\{s_i^{\mu'}\}$ spin configuration, and evaluate the following sum that determines whether the spin s_i should be flipped:

$$\sum_j J_{ij} s_j^{\mu'} = \sum_\mu s_i^\mu \left(\sum_j s_i^\mu s_j^{\mu'} \right). \tag{4.1}$$

The term in the brackets, $\sum_j s_i^\mu s_j^{\mu'}$, will have an expectation value of zero for uncorrelated distinct memories μ and μ', deviating from zero only because of noisy interference between the memories. On the other hand, when the index of the outer sum becomes equal to μ', the inner summation evaluates in N, producing an overall result of $s_i N$, i.e., $-N$ or N, depending on the sign of s_i. Hence, unless the interference noise is large, which may happen

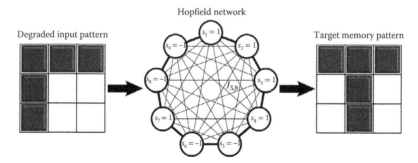

FIGURE 4.3   Some of the salient features of the Hopfield network operation are schematically shown. If only one target memory (right panel) is stored in the Hopfield network (middle panel), the coupling constants are given by $J_{ij} = s_i s_j$, where $\{s_i\}$ spins characterize the particular memory. With these coupling constants, the Hopfield network can recognize numerous degraded input configurations (e.g., as shown in the left panel) as related to the stored memory by following asynchronous temporal evolution of network spins, that, in the case of a single stored memory, are nearly guaranteed to converge to the target in a finite number of steps. However, as additional memories are stored in the network, in particular above the network capacity, the recall capability degrades with the dynamics starting to lead to spurious minima.

if too many memories are stored compared to network's storage *capacity*, the memory, $\{s_i^{\mu'}\}$, should be stable under the rules of evolution. Many interesting properties of the Hopfield network, including an intuitive estimate of network's storage capacity and occurrence of recurrent dynamics when J_{ij}s are nonsymmetric (i.e., corresponding to breaking of detailed balance), were already discussed in Hopfield's classic paper [73]. In the last decade, there has been a great resurgence in the application of neural networks based on Hopfield's model, where the spins evolve stochastically (thermally) rather than following the deterministic cellular automate type rules outlined above, that lead to steepest-descent-like dynamics. The resulting neural networks are called Boltzmann machines, and one particular type, restricted Boltzmann machines (RBMs), has already been widely used for learning from complex data, including the recognition of handwriting and speech [72,74].

Coming back to the problem of protein structure prediction, one quickly realizes one cannot directly apply Hopfield's Hamiltonian to the folding problem. In the Hopfield memory model, the initial state and the memories are the same type of objects (i.e., both are sets of spin configurations), while in structure prediction, the input is the sequence of amino acids in the protein while the output is something very different, the three-dimensional structure of the protein. Hence, the association is heterologous.

For pedagogical purposes, let's approach constructing an algorithm for association along two different avenues, followed by a subsequent synthesis. First, to achieve conceptual simplification, suppose we were to require only the recognition that two sequences are homologous. For example, sequence encoded by two letters, hydrophobic ($q = 1$) and hydrophilic ($q = -1$), $\{q_i^{\text{input}}\}$, we wish a network to find given another closely homologous memory, $\{q_i^{\mu}\}$, under conditions that the number of memories is below the storage capacity and $\{q_i^{\text{input}}\}$ is sufficiently similar to $\{q_i^{\mu}\}$. For this problem, we could introduce an explicit Hamiltonian of the Hopfield-network type [20]

$$\mathcal{H}^{\text{seq}} = -\sum_{i \neq j} J_{ij} q_i q_j \tag{4.2}$$

where $J_{ij} = \sum_{\mu} q_i^{\mu} q_j^{\mu}$ is summed over memory sequences as described above. Hence, the Hamiltonian may be explicitly rewritten as follows:

$$\mathcal{H}^{\text{seq}} = -\sum_{\mu} \sum_{i \neq j} q_i^{\mu} q_j^{\mu} q_i q_j \tag{4.3}$$

Notice this is an unwanted symmetry when all spins of the sought memory $\{q_i^{\mu}\}$ are flipped. To break this symmetry, the following modification may be made [20]:

$$\mathcal{H}^{\text{seq}} = -\sum_{\mu}\sum_{i \neq j}\left(q_i^{\mu}q_j^{\mu}q_iq_j + q_i^{\mu}q_i + q_j^{\mu}q_j\right) \qquad (4.4)$$

This Hamiltonian may be used to discover homologues of a given sequence in the memory database [12,20–22]. One could generalize this approach in several ways. First, instead of relying only on a single attribute of each amino acid, $\{q_i\}$, we could rely on multiple attributes, and label them, as $\{q_{m,i}\}$, where the mth subscript indicates the mth attribute (e.g., electrostatic charge or solvent exposure). Then, the generalized Hamiltonian would become

$$\mathcal{H}^{\text{seq}} = -\sum_{\mu}\sum_{m,n}\sum_{i \neq j}q_{m,i}^{\mu}q_{n,j}^{\mu}q_{m,i}q_{n,j} \qquad (4.5)$$

Hence, one could study, for example, the solvent-exposure patterns of alphahelical segments versus beta sheets using such a Hamiltonian or investigate to what degree various attributes are correlated [21]. Notice that from the structural perspective, an attribute such as solvent exposure has an emergent character, originating from many-body structural interactions and packing. Therefore, the Hamiltonian (Equation 4.5) may already be used to find interesting interrelations among structures and sequences in a protein database. In the original papers, the amino-acid attributes, $\{q_{m,i}\}$, were called "charges" [12,20–22], motivated by the idea of conserved quantities arising from various symmetries in theoretical physics.

Another route of applying neural networks to proteins concerns three-dimensional structure recognition. Namely, given several different polymeric conformations (and completely ignoring the sequence information), we would like an input polymer conformation to be recognized if it happens to be structurally close enough to one of the memories. Here, the first problem is to choose microscopic structural variables that can play the role of spins in the Hopfield network. The Cartesian coordinates themselves are not suitable, because we need to be able to recognize structures even if they are translated in space or rotated in orientation. To address this invariant recognition problem, a set of the scalar interbead distances, for example, between C_{α} atoms, r_{ij}, may be used. Let's introduce the following function, $\theta(r_{ij} - r_{ij}^{\mu})$, which is 1 if $\text{abs}(r_{ij} - r_{ij}^{\mu}) < \sigma$ and zero, otherwise. Here, σ indicates distance tolerance, which could be, for example, on the order of 1 Å.

Then, the following structural Hamiltonian would allow recognition of a set of conformational memories:

$$\mathcal{H}^{\text{str}} = -\sum_{\mu}\sum_{i\neq j}\Theta(r_{ij} - r_{ij}^{\mu}) \tag{4.6}$$

Finally, we can combine the sequence and structural recognition by requiring that a given protein sequence in a specific conformation be low energy only when both sequence and structure are well matched to one of the database memories. This logical AND statement is accomplished by simple multiplication of the corresponding terms in the two Hamiltonians [20]:

$$\mathcal{H}^{\text{str-seq}} = -\sum_{\mu}\sum_{m,n}\sum_{i\neq j}\gamma_{ij}^{\mu}(q_i, q_j, q_i^{\mu}, q_j^{\mu})\Theta(r_{ij} - r_{ij}^{\mu}) \tag{4.7}$$

where we have introduced a residue attribute correlation function,

$$\gamma_{ij}^{\mu}(q_i, q_j, q_i^{\mu}, q_j^{\mu}) \equiv q_i^{\mu}q_j^{\mu}q_iq_j + q_i^{\mu}q_i + q_j^{\mu}q_j \tag{4.8}$$

which indicates whether the target and memory residue pairs, i and j, are similar or not. If the attributes, $\{q_i\}$, may only adopt values ± 1, then, γ_{ij}^{μ} is $+3$ if there is an exact match between the target and memory pair attributes, otherwise it is -1. Figure 4.4a shows an example of structure prediction using this Hamiltonian.

When simulating the folding protein chain, the residues cannot move independently, hence, it is also necessary to introduce interactions that constrain the connectivity of neighboring residues along the polymer, either by using harmonic distance constraints or by employing a significantly more elaborate backbone Hamiltonian like that discussed in the next section. Hence, the associative memory Hamiltonian (AMH) family of potentials minimally contains the following terms [21]:

$$\mathcal{H}^{\text{AMH}} = \mathcal{H}^{\text{bb}} - \sum_{\mu}\sum_{m,n}\sum_{i\neq j}\gamma_{ij}^{\mu}(q_i, q_j, q_i^{\mu}, q_j^{\mu})\Theta(r_{ij} - r_{ij}^{\mu}) \tag{4.9}$$

where $\mathcal{H}^{\text{bb}}$ indicates polymer backbone preferences. Early applications of the Hamiltonian (Equation 4.9) revealed that stochastic dynamics would allow a storage capacity on the order of $0.5N$, where N is the target protein chain length [75]. In particular, using input sequences which were obtained by a modest degradation of one of the memory protein sequences via mutations, molecular dynamics, or Monte Carlo search would still lead to the recall of near-native structure of the input protein sequence.

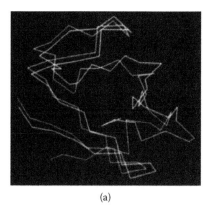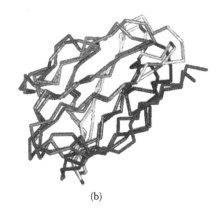

(a)   (b)

FIGURE 4.4   (a) A structure prediction for *Desulfovibrio vulgaris* rubredoxin is shown from the work [20] that introduced associative memory Hamiltonians for reconstructing protein tertiary structures from distantly homologous sequences given as input. Only one out of 80 sequences in the memory database was homologous to this protein, at approximately 50% sequence similarity, showing this Hamiltonian's (i.e., Equations 4.7 and 4.8) capacity to recognize related sequences and associate them with specific protein structures. ((a) reproduced from MS Friedrichs and PG Wolynes, *Science*, 246, 371–373, 1989.) (b) A structure prediction for the variable fragment from a Bence-Jones immunoglobulin is shown, using variational parameter optimization [22], discussed below in this section. ((b) from RA Goldstein et al., *Proc Natl Acad Sci USA*, 89, 4918–4922, 1992.)

A natural generalization of the AMH approach was to enlarge the types of residue attributes, for example, by using previously computed mutation propensity among various groups of amino acids [21]:

$$\gamma_{ij}^{\mu}(A_i, A_j, A_i^{\mu}, A_j^{\mu}) \equiv \xi(A_i, A_i^{\mu})\xi(A_j, A_j^{\mu}) - \bar{\gamma} \tag{4.10}$$

where $\xi(A_i, A_i^{\mu})$ is the probability that an amino acid of type A_i would mutate in a certain evolutionary timeframe to an amino acid of type A_i^{μ} of the memory protein and $\bar{\gamma}$ is the same quantity averaged over unrelated proteins. Clustering of observed interconversions among various amino acids suggests the partitioning of amino acids into five classes. A five-letter amino-acid code, with γ_{ij}^{μ}s given by Equation 4.10, increases the storage capacity by a factor of 5.6 [21]. Additional enhancement of storage was obtained by allowing limited insertions and deletions between the target and memory sequences, where, interestingly, the Hamiltonian dynamically aligns the target sequences to various memories during the folding run, as the trajectory evolves. Filtering was also found to be helpful, where more sophisticated structural criteria were used to ascertain the similarity between the target and memory pairs, i and j. For example, if the ith residue of the memory protein was located in an alpha helix and the jth residue was located in a beta sheet, the same was required from the target i, j pair for the γ_{ij}^{μ} to be favorable. Mathematically, filtering was achieved by multiplying γ_{ij}^{μ} terms described above by $\Delta(\Gamma_i, \Gamma_i^{\mu})\Delta(\Gamma_j, \Gamma_j^{\mu})$, which is 1 if the structural properties Γ_i and Γ_j of the target match the analogous properties of the memory, μ, and otherwise 0. With these and other enhancements, the maximum capacity was increased on the order of a hundred-fold, from approximately 30 memories for a 50-residue target to over 3500, nearly equaling the number of known protein folds.

In terms of protein structure prediction algorithms, neural network modeling with a Hamiltonian (Equation 4.9) relies on the ability of the Hamiltonian to recognize the target sequence among a finite number of database memories, even if the sequence match is imperfect. With the γ_{ij}^{μ}s given by Equation 4.8 or even Equation 4.10, this approach has a limited potential for learning and generalization, emphasizing mainly the ability to recall based on finding a partial sequence match. Nevertheless, when this is feasible, highly accurate structural prediction can be made. Practically speaking, however, providing coverage of all available protein folds and major families of sequence variations went beyond the capacity of AMHs of the form given above. Hence, to achieve broad transferability, the best insights from the AMH approach needed to be blended with alternative approaches to modeling proteins.

The ability to learn and generalize is predicated on finding common symmetries and patterns among the database proteins. For example, hydrophobic residues tend to be found in protein cores, or alpha-helical chain segments often have specific sequence patterns. Some of these biological symmetries, for example, the mutation propensities preserving structure, were included in the earlier papers employing the AMH. The original Hopfield ansatz for the coupling constants, $\gamma_{ij}^{\mu}(q_i, q_j, q_i^{\mu}, q_j^{\mu}) = q_i^{\mu} q_j^{\mu} q_i q_j$, is most suited for efficient recall of memories; however, because it is prescriptive, it does not leave a lot of room for learning general patterns in the database. To achieve the latter goal, the $\gamma_{ij}^{\mu}(q_i, q_j, q_i^{\mu}, q_j^{\mu})$s may be treated as variational parameters [22,76], which can then be optimized so that some set of target proteins would fold efficiently. Hence, one would be seeking to obtain a unique set of $\gamma(q_i, q_j, q_i^{\mu}, q_j^{\mu})$s, which are the same for all pairs of i and j, i.e., where the dependence on the residue pair indices and memories is implicit, appearing only via the arguments determining γ's value, $\{q_i, q_j, q_i^{\mu}, q_j^{\mu}\}$. With this approach [22,76], we may expect that the learning algorithm, by virtue of achieving efficient folding of a number of target proteins (using a large, representative protein memories database) would also discover, for example, physically reasonable patterns of amino-acid substitutions, for example, of a charged residue by a polar one on a protein's surface or the propensity of hydrophobic amino acids to associate versus the frailty of hydrophilic contacts.

Learning of physically intuitive interactions by optimizing $\gamma(q_i, q_j, q_i^\mu, q_j^\mu)$s nudges AMH toward another style of coarse-grained modeling of proteins, namely methods relying on statistical potentials derived from analyzing frequencies of various interactions and structural features found in a database of a large number of proteins. This procedure resembles inverse statistical mechanics using potentials of mean force (PMF) in coarse-grained modeling of nonbiological liquids and polymers [6]. However, even with variationally optimized $\gamma(q_i, q_j, q_i^\mu, q_j^\mu)$s, where the latter start to become physically meaningful, the memory proteins still explicitly enter the Hamiltonian, revealing the neural network origin of the framework. For a given target protein, let's assume we rely only on a single memory protein, whose sequence, however, does not completely coincide with the target sequence (i.e., these are two homologous proteins). In that case, only a partially funneled energy landscape might result, with rugged features, where the target structure identical to the memory conformation is low energy, but other competing conformations are also rather stable, interfering with folding. However, as additional memories are added to the database, other parts of the target sequence would become matched, deepening the funnel for the native basin and eliminating or at least smoothing out unrelated deep traps. Similarly, simultaneous optimization of the T_f/T_g ratios [22] for multiple targets is expected to further eliminate spurious patterns, while reinforcing the coherent generalizable interactions. This dual strategy of including many memory proteins (but not exceeding the overall capacity) and many targets eventually leads to the development of a fully transferrable potential, which could fold not only proteins from the training set but also unrelated test proteins [77].

To further nudge the force field in the PMF direction, in the hope of maximally extending the transferability, one could add a number of additional physical terms to the AM potential, including residue pair-dependent hydrogen bonding, a contact potential, and water-mediated interactions, among others. This was the historical route of the development of AMH [23,24,26,27,77], which finally resulted in the AWSEM Hamiltonian [31], with the following form:

$$\mathcal{H}^{\text{AWSEM}} = \mathcal{H}^{\text{bb}} + \mathcal{H}^{\text{AM}} + \mathcal{H}^{\text{PMF}} \tag{4.11}$$

In the next section, we elaborate on the $\mathcal{H}^{\text{bb}}$ and $\mathcal{H}^{\text{PMF}}$ terms. The $\mathcal{H}^{\text{AM}}$ term has undergone an interesting transformation over the last 20 years. First, instead of global matching or alignment of target proteins to memory proteins, much shorter fragments are now used in AWSEM, on the order of 10 residues. The intuition behind this choice originates from addressing the great difficulty of modeling many-body local interactions (e.g., local packing or polarizability) via largely pairwise PMF terms. Furthermore, since it is usually easy to find 10-residue fragments in the database of memory proteins (where the latter may be potentially completely unrelated to the target protein) with nearly complete matching of target and memory sequences, all $\gamma(q_i, q_j, q_i^\mu, q_j^\mu)$s are currently set to 1. Hence, the fragment memory (FM) term in AWSEM reverts to a simple Hopfield-like form (in structure space—compare with Equation 4.6)

$$\mathcal{H}^{\text{FM}} = -\sum_\mu \sum_{i \neq j} \theta(r_{ij} - r_{ij}^\mu) \tag{4.12}$$

introducing matched memory structures as mini-funnels on the energy landscape, which weakly trap the fragment when the molecular dynamics trajectory happens to cruise near a particular FM's basin of attraction. A key idea here is the harmony between the local and tertiary structures [28]: the local chain conformation induced due to a particular memory will be greatly stabilized if simultaneously favorable interactions occur between the fragment and its surrounding chain. Although currently [31] sequence information is only used to select FMs, it is possible that in the future development of the $\mathcal{H}^{\text{FM}}$ term, for example, when longer fragments are used or explicitly includes the structural context of the FM (such as its surface accessibility) as well, sequence correlations could again be explicitly taken into account by optimizing the corresponding $\gamma(q_i, q_j, q_i^\mu, q_j^\mu)$s.

Finally, we would like to raise the following question: Are neural network approaches to studying protein folding merely empirical, or does this perspective bring fundamental insights that otherwise might be overlooked? This question reveals some conflict between the reductionist view that folding is driven by highly specific, microscopic physical–chemical interactions versus the holistic view that would consider folding as a generic information-theoretic problem of translating a one-dimensional sequence into a three-dimensional structure that was solved by evolution. From the latter perspective, folding is a robust computation. However, it is fundamentally dissimilar from many other prominent examples of biological computation, such as symbolic translation of DNA genetic sequence to messenger RNA and then to protein sequence [75]. The latter processes can be viewed as sequential digital operations that could be easily programmed into von Neumann machines. They involve many far from equilibrium steps that are essentially irreversible. Kinetic proofreading arising from the breakdown of detailed balance allows each of these steps to be nearly error free. On the other hand, translating a protein's sequence into a unique three-dimensional structure is an extremely nonlinear operation, that is achieved via numerous parallel pathways, exhibiting many degeneracies of both various sequences that fold to nearly the same structure and also residual structural substates that are present after folding of a specific sequence [75]. Local errors are not a problem: folding is robust. Hence, from the information-theoretic viewpoint, folding indeed is more akin to analog, parallel, highly nonlinear, fault-tolerant computations performed by neural networks as opposed to traditional von Neumann digital computer programs. The energy landscape theory of protein folding, the funnel picture, provides the physical foundation for this emergent information-theoretic perspective. The neural network paradigm of protein structure prediction and folding, where memories are correlated in complex ways over sequences and conformations, brings fundamental new insights from an angle that might not be expected by the majority of physical scientists. Although this connection between the folding problem and neural networks has been explored, it is likely many interesting discoveries using this perspective still lie ahead.

4.4 THE BACKBONE POTENTIAL

As mentioned above, in the earliest works using the AMH, the backbone was represented by C_α beads connected via harmonic springs [21,75]. However, this potential allows for too much chain flexibility, raising the entropy of the unfolded state. This excess entropy, in turn, lowers the folding temperature more than the glass transition temperature so trapping becomes a problem. At another extreme, every backbone atom could be explicitly represented, to fully take into account all the stereochemical constraints of the peptide backbone, which hold to greater accuracy for allowable configurations.

After the original Friedrichs and Wolynes papers [20,75], a not elaborate but still efficient backbone model was introduced [21], having three beads per residue, located at the C_α, C_β, and O atomic positions. Furthermore, by leveraging the broad stiffness spectrum of a protein chain, where bond lengths and angle fluctuate much less compared to very dynamic dihedral angles, an ideal peptide geometrical model had been profitably assumed, allowing on-the-fly calculation of the positions of the remaining backbone atoms [21],

$$\begin{aligned}
\mathbf{r}_{N_i} &= 0.48318\mathbf{r}_{C_{\alpha_{i-1}}} + 0.70328\mathbf{r}_{C_{\alpha_i}} - 0.18643\mathbf{r}_{O_{i-1}} \\
\mathbf{r}_{C'_i} &= 0.44365\mathbf{r}_{C_{\alpha_i}} + 0.23520\mathbf{r}_{C_{\alpha_{i+1}}} + 0.32115\mathbf{r}_{O_i} \\
\mathbf{r}_{H_i} &= 0.84100\mathbf{r}_{C_{\alpha_{i-1}}} + 0.89296\mathbf{r}_{C_{\alpha_i}} - 0.73389\mathbf{r}_{O_{i-1}}
\end{aligned} \quad (4.13)$$

Here, i indexes the residue position along the chain. H_i refers to the amide hydrogen. As elaborated below, the amide nitrogen and hydrogen positions are used to construct directional hydrogen bond potentials.

We would like to make a clear distinction between the all-atom representation of the backbone in the energy functions in the AWSEM potential versus only three explicit beads that move during molecular dynamics. Hence, interactions mediated by virtual beads are transmitted as corresponding forces to the real beads using Equation 4.13 and the chain rule.

The backbone potential consists of a number of terms, developed over many years, both systematically and by trial and error, that endow the polymer chain with protein-like stereochemistry [31]:

$$V_{\text{backbone}} = V_{\text{con}} + V_{\text{chain}} + V_\chi + V_{\text{rama}} + V_{\text{excl}} \quad (4.14)$$

Next, we briefly discuss each of these terms. For additional information, interested readers are advised to consult the AWSEM publication [31], in particular its Supplemental Information section, as well as some of the earlier papers where these potentials were introduced or extensively discussed [12, 21–23,25–27,76,77].

To maintain an ideal peptide geometry, one could constrain various bonds and angles of the protein backbone. However, it is more efficient to use a carefully chosen set of harmonic distance constraints to achieve the same effect. This, in particular, allows one to avoid the computational cost from using three-body bending potentials and the computational burden of trigonometric functions. Two sets of harmonic distance constraints, shown

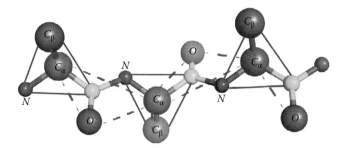

FIGURE 4.5 The three-beads-per-residue backbone model of the AWSEM and related earlier potentials is shown. C_α, C_β, and O are explicitly specified and serve as dynamical variables, while the remaining backbone atoms, N (blue), C' (green), and H (amide proton, not shown) are computed on the fly assuming an ideal peptide geometry. Magenta dashed lines indicate the connectivity potential, Equation 4.15, while brown solid lines indicate the chain potential, Equation 4.16.

with dashed and solid lines in Figure 4.5, have been historically called connectivity and chain potentials:

$$V_{\mathrm{con}} = \lambda_{\mathrm{con}} \sum_{i=1}^{N} \left[\left(\mathbf{r}_{C\alpha_i O_i} - \mathbf{r}^0_{C\alpha_i O_i} \right)^2 + \left(\mathbf{r}_{C\alpha_i C_{\beta_i}} - \mathbf{r}^0_{C\alpha_i C_{\beta_i}} \right)^2 \right]$$
$$+ \lambda_{\mathrm{con}} \sum_{i=1}^{N-1} \left[\left(\mathbf{r}_{C\alpha_i C\alpha_{i+1}} - \mathbf{r}^0_{C\alpha_i C\alpha_{i+1}} \right)^2 + \left(\mathbf{r}_{O_i C\alpha_{i+1}} - \mathbf{r}^0_{O_i C\alpha_{i+1}} \right)^2 \right],$$

$$(4.15)$$

and

$$V_{\mathrm{chain}} = \lambda_{\mathrm{chain}} \left[\sum_{i=2}^{N} \left(\mathbf{r}_{N_i C_{\beta_i}} - \mathbf{r}^0_{N_i C_{\beta_i}} \right)^2 + \sum_{i=1}^{N-1} \left(\mathbf{r}_{C'_i C_{\beta_i}} - \mathbf{r}^0_{C'_i C_{\beta_i}} \right)^2 \right.$$
$$\left. + \sum_{i=2}^{N-1} \left(\mathbf{r}_{N_i C'_i} - \mathbf{r}^0_{N_i C'_i} \right)^2 \right]$$

$$(4.16)$$

As can be seen from Figure 4.5, the V_{con} potentials link neighboring amino-acid residues, while the V_{chain} term is responsible for maintaining ideal bond angles around the $C\alpha$ atoms of each residue. The equilibrium distances entering Equations 4.15 and 4.16 are provided in Ref [31], where one can also find the other constants appearing in the subsequent discussion of many other AWSEM terms below. The harmonic spring constants, λ_{chain} and λ_{con}, are 5 and 10 kcal/Å^2, respectively, allowing thermal fluctuations of the corresponding distances only on the order of 0.3 Å. This can be seen by comparing $\lambda(\Delta r)^2$ to $k_B T$.

Early AMH simulations, with only harmonic bonds connecting C_αs, often exhibited frequent errors with inverted chirality at different chain scales. To address this problem, a chirality potential was introduced for all amino acids except glycine, such that the natural left-handedness of amino acids is maintained during molecular dynamics. Let's first define χ,

$$\chi_i = \left(\mathbf{r}_{C'_i C_{\alpha_i}} \times \mathbf{r}_{C_{\alpha_i} N_i} \right) \cdot \mathbf{r}_{C_{\alpha_i} C_{\beta_i}} \tag{4.17}$$

which is the scalar product of a vector normal to the $C' - C_\alpha - N$ plane and the $C_\alpha - C_\beta$ vector. Then, the following potential prevents chirality inversion of the C_α atom:

$$V_\chi = \lambda_\chi \sum_{i=2}^{N-1} (\chi_i - \chi_0)^2. \tag{4.18}$$

Prior studies have shown that preserving chirality at the residue level prevents the chain from falling into unphysical traps with defective chiralities along the chain, especially at smaller and moderate spatial scales. However, sometimes, low-energy chain conformations are still produced that are near mirror images of the native topology, indicating that the local amino-acid chirality does not completely determine the global chirality of the whole chain.

The two most salient dynamic degrees of freedom of a protein chain are the ϕ and ψ dihedral angles of each residue, determined by the following sets of four atoms, correspondingly, $C'_{i-1} - N_i - C_{\alpha i} - C'_i$ and $N_i - C_{\alpha i} - C'_i - N_{i+1}$. Structural analyses of experimentally determined protein conformations show that the ϕ, ψ angles are strongly clustered around three or four centers in the so-called Ramachandran plot [33,46]. The major restriction on the allowed values for ϕ, ψ angles originates from steric constraints. Biasing the protein chain conformations toward the allowed Ramachandran regions eliminates a large number of unphysical protein chain conformations, hence, reducing the unfolded state entropy:

$$V_{\mathrm{rama}} = -\lambda_{\mathrm{rama}} \sum_{i=2}^{N-1} \sum_{j} W_j e^{-\sigma_j (\omega_{\phi_j} (\cos(\phi_i - \phi_{0j}) - 1)^2 + \omega_{\psi_j} (\cos(\psi_i - \psi_{0j}) - 1)^2)} \tag{4.19}$$

Here, the $\{\phi_0, \psi_0\}$ tuples point to locations on Ramachandran plot where attractive wells are introduced to bias the corresponding dihedral angles. The resulting default potential is shown in Figure 4.6a. The Ramachandran potential (Equation 4.19) is not residue specific (except for proline, Figure 4.6b, which has a unique backbone chemistry). However, sequence-specific biasing turns can be added to V_{rama} if a protein's secondary structure

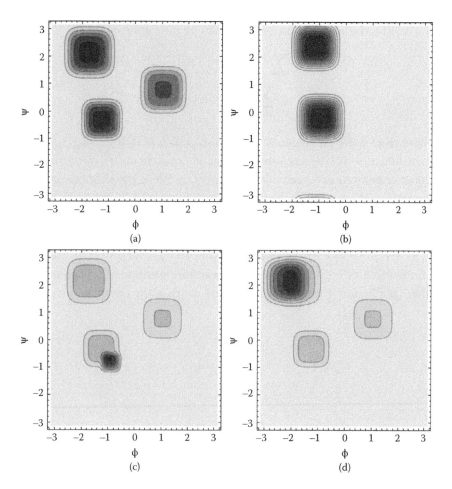

FIGURE 4.6 Various forms of the Ramachandran biases are shown. (a) Default; (b) proline specialization; (c) alpha-helical bias; and (d) beta-sheet bias. (Reprinted with permission from A Davtyan et al., *J Phys Chem B*, 116, 8494–8503. Copyright 2012 American Chemical Society.)

is predicted before the simulation using purely bioinformatic approaches. State-of-the-art predictors of secondary structure typically achieve 60%–80% success rate. If one is interested in studying proteins that are only alpha helical or contain only beta sheets, V_{rama} parameters can be correspondingly fine-tuned (see Figure 4.6c and d).

The final backbone term takes into account excluded volume interactions. Without this term, the chain would tend to overcollapse, and also unphysical entanglements of the chain may be produced. In addition, again, the entropy would be too high without excluded volume constraints. Hence, in AWSEM, the steric repulsions are given by the following form:

$$V_{\text{excl}} = \lambda_{\text{excl}} \sum_{ij} \Theta(r_{\text{ex}}^C - r_{C_i,C_j})(r_{C_i,C_j} - r_{\text{ex}}^C)^2$$
$$+ \lambda_{\text{excl}} \sum_{ij} \Theta(r_{\text{ex}}^O - r_{O_i,O_j})(r_{O_i,O_j} - r_{\text{ex}}^O)^2, \qquad (4.20)$$

where $\Theta(x)$ is the Heaviside step function, which is zero when $x < 0$ and one when $x > 0$. The excluded volume parameters are $r_{\text{ex}}^C = 3.5$ Å (for sequence separations greater than 4) and $r_{\text{ex}}^O = 3.5$ Å. For shorter sequence separations, r_{ex}^O remains the same, while r_{ex}^C becomes 4.5 Å. In Equation 4.20, the subscripts in r_{C_i,C_j} refer to both alpha and beta carbons. The shape of this potential is a parabolic ramp, where the potential is zero when, for example, $r_{O_i,O_j} > r_{\text{ex}}^O$, while it increases parabolically as r_{O_i,O_j} goes to zero, reaching a finite value of $\lambda_{\text{excl}}(r_{\text{ex}}^O)^2$ when two beads are positioned on top of each other (see Figure 4.7a). At this complete overlap, the energetic penalty is

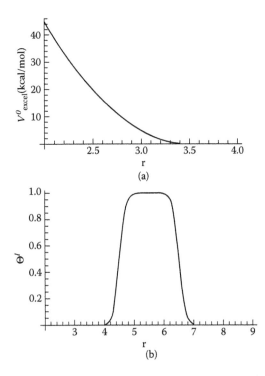

FIGURE 4.7 (a) Excluded volume interaction potential between two oxygen atoms is shown (see Equation 4.20). (b) The function introduced in Equation 4.23, Θ_{ij}^I, is plotted. (Reprinted with permission from A Davtyan et al., *J Phys Chem B*, 116, 8494–8503. Copyright 2012 American Chemical Society.)

prohibitive, approximately 50 kcal/mol. The parameter λ_{excl} is chosen so that the thermal energy is not enough for two beads to approach each other more closely than approximately 3.0 Å (when $r_{ex} = 3.5$ Å).

4.5 PAIRWISE AND FM-BASED CONTACT POTENTIALS

We would like to remind the reader of the basic potential for AMH, developed in early 1990s, given by Equation 4.9, which is repeated here for convenience (replacing H by V),

$$V^{AMH} = V^{bb} - \sum_{\mu} \sum_{m,n} \sum_{i \neq j} \gamma_{ij}^{\mu}(q_i, q_j, q_i^{\mu}, q_j^{\mu}) \theta(r_{ij} - r_{ij}^{\mu}) \qquad (4.21)$$

where the backbone potential was already elaborated in the previous section. The distance comparison function, $\theta(r_{ij} - r_{ij}^{\mu})$, treats short- and long-range interactions, both in sequence and spatially, on equal footing. This mean-field treatment can be improved by creating separate sets of $\gamma_{ij}^{\mu}(q_i, q_j, q_i^{\mu}, q_j^{\mu})$s for residues in certain ranges of sequence separation and also by their spatial proximity [77,78]. Hence, interactions were subsequently grouped into short-, medium-, and long-range classes based on interresidue sequence separation (i.e., $|j - i| < 4$, $4 < |j - i| < 12$, and $|j - i| > 12$, respectively, where i and j are residue indices). Similarly, three spatial distance classes were introduced, called first, second, and third wells, which were delineated in the following way: the first well corresponds to contacts where 4.5 Å $< r_{ij} < 6.5$ Å, the second well to 6.5 Å $< r_{ij} < 9.5$ Å, and the third well to 9.5 Å $< r_{ij} < 15$ Å. Nine sets of γ_{ij}s were optimized for the resulting 3×3 combinations of sequence separation and spatial distance classifications.

Another concurrent idea was to treat the *long range in sequence* contacts via PMF-type interactions as opposed to using sets of protein memories [76]. Especially in the context of *de novo* structure prediction, where even distinct homologs are not expected to be present in the memories' database, target sequence to memory sequence correlations can only be sequence local in scope. From the physical viewpoint, contacts that are far in sequence presumably occur because of global cooperativity of all interactions and also the direct attraction between any particular pair of amino acids, where these indirect and direct interactions are coupled in very complicated ways. However, it may not be necessary to unentangle completely the corresponding many-body effects. Instead, we may simply postulate direct pairwise interactions among amino acids, and then optimize the corresponding parameters by requiring that the target proteins be maximally funneled, i.e., again by maximizing T_f/T_g. In AWSEM [31], which is the most recent reincarnation of the AM family of Hamiltonians, direct contact interactions described in this paragraph have the following form:

$$V_{\text{direct}} = -\lambda_{\text{direct}} \sum_{j-i>9}^{N} \gamma_{ij}(q_i, q_j) \Theta_{ij}^I \qquad (4.22)$$

where

$$\Theta_{ij}^I = \frac{1}{4} \left(1 + \tanh \left[\eta \left(r_{ij} - r_{\min}^I\right)\right]\right) \left(1 + \tanh \left[\eta \left(r_{\max}^I - r_{ij}\right)\right]\right) \qquad (4.23)$$

The functional form of Θ_{ij}^I, based on hyperbolic tanh functions, approximates multiplication of two Heaviside step functions, creating a continuously defined square well, between $r_{\min} = 4.5$ Å and $r_{\max} = 6.5$ Å, which goes to zero with an interfacial width of $1/\eta$, on the order of 0.5 Å (see Figure 4.7b). The distances, r, are computed between C_β atoms, except for glycine, where C_α is used instead. The depth of each well is $-\lambda_{\text{direct}}\gamma_{ij}(q_i, q_j)$, where λ_{direct} scales the overall potential. Note that Equation 4.22 is memoryless, i.e., has a PMF form. When the corresponding 20×20 matrix of $\gamma_{ij}(q_i, q_j)$s were optimized, having 210 unique elements because pairwise interactions are symmetric, some well-known features of basic physical chemistry of amino acids naturally emerge, such as an effective attraction between hydrophobic residues [26]. In the prior works, it was found that the related 20×20 interaction matrix derived by Miyazawa and Jernigan [79], using the quasichemical approximation, can be diagonalized, and that it then shows only a single nontrivial eigenvector (hence, it is highly reducible). In contrast, the direct interaction matrix, derived by the energy landscape optimization strategy, contains on the order of 10 nontrivial eigenvectors, corresponding to finer details of physically meaningful interactions, such as hydrophobicity, side-chain size, charge, polarizability, and modulating the pairwise interactions when one or both of the side chains are branched [26]. In AWSEM, contact interactions are utilized for residues that are at least nine residues apart in sequence, as shown by the summation indices in Equation 4.22.

Residue pairs belonging to short- and medium-range classes, hence, residing on the same local protein segment, were treated with the correspondingly optimized γ_{ij}s in the 1990s and early 2000s. In the recent AWSEM code, the shorter sequence separation interactions are replaced by a simpler

FM potential, given by Equation 4.12, which is written out here in a more explicit form:

$$V_{FM} = -\lambda_{FM} \sum_{\mu} \omega_{\mu} \sum_{ij} \gamma_{ij} \exp \left[-\frac{(r_{ij} - r_{ij}^{\mu})^2}{2\sigma_{ij}^2} \right] \qquad (4.24)$$

In the above equation, the i, j summation is not over residues but is more fine-grained, over beads (i.e., over three atoms per residue). The tolerance factor for the distance similarity is sequence separation dependent, given by $\sigma_{ij} = |I - J|^{0.15}$. λ_{FM} allows adjustment of the overall strength of the FM term. Although, currently, all residue pair-dependent interactions strengths are identical, $\gamma_{FM}(q_i, q_j) = 1$, and memories also enter with equal weights, $\omega_{\mu} = 1$, these parameters could be optimized in the future based on the ideas from neural network learning and energy landscape theory of protein folding, discussed throughout this chapter.

We finish this section by comparing how sequence-structure memory potentials were used before the introduction of FMs and afterward. In the earlier versions of AMH, from 1989 to the mid-2000s, the target protein sequence was globally aligned, with the allowance of insertions and deletions, to the sequences of memory proteins in the databank. The alignments could be performed not only in a usual way, by directly matching sequences, but also by aligning target protein's sequence to memory protein's structure using some coarse-grained energy function that is minimized at the best alignment [78]. This approach would be expected to work well if some of the memory proteins are homologous to the target protein, at least distantly (focusing, in particular, in the "twilight" zone, from 25% to 40% homology). However, for completely *de novo* structure prediction, there could be scenarios when no memories can be found that are globally related to the target sequence in any meaningful way. In that case, relying too strongly on the neural network approach may seem inappropriate, since we do not have any global memories to recall or potentially even generalize from. However, at shorter fragment levels, many correlated sequences can still be found in the database, even in the absence of global sequence correlations. Hence, the potential can "learn" from those local sequence-structure motifs. The caveat is that the FMs most likely exist in a potentially unrelated structural context, surrounded by residues unlike the chemical details of target fragment's surroundings. Nevertheless, there is hope, borne out by subsequent extensive simulations and analyses, that the local fragment's conformations are indicative of the fragment's propensities to form various local structures, and, if enough variety of memories are present (but not exceeding the network's capacity), the consistency between the target's global and local structures leads to picking of the right memory during a molecular dynamics run. Hence, for a given target protein, different local structural motifs along the chain can be picked out from unrelated memory proteins; they are self-consistently molded together to form a coherent tertiary structure. Interestingly, the FM term in AWSEM is simply a more restricted version of the Associative Memory term introduced in 1989, paralleling in some way the introduction of structural restrictions on Hopfield-type neural networks as modern deep-learning algorithms have been gradually developed.

Overall, as documented in the later sections, the last decade of development has lead to an AWSEM potential that is both widely transferrable and reasonably accurate, both when globally homologous proteins are used to generate FMs and also in the complete absence of any known, even distant homology.

4.6 WATER-MEDIATED INTERACTIONS

While in atomistic simulations of proteins, water molecules were explicitly included starting from 1980s and 1990s, they needed to be integrated out in coarse-grained protein simulations, replaced by some effective implicit-solvent interactions. The main contribution of water was considered to be mediated by the hydrophobic effect, a line of thought started by Kauzmann in the 1950s [53]. Hence, "simple" contact potentials were developed, of the Miyazawa–Jernigan type [79], where the interaction energies among hydrophobic residues were stabilizing and among hydrophilic residues were neutral or destabilizing. These interaction patterns indeed lead to a partial microphase separation, sending a higher fraction of hydrophobic residues to the globule's core, with the surface being highly enriched with hydrophilic residues. One might colloquially call these types of contact potentials the hydrophobic code (hydrophobic/polar [HP]) for protein folding.

While extending the energy landscape theory of protein folding to binding problems, Papoian and Wolynes found that a significant fraction of protein dimeric interfaces, on the order of a third of the total, were energetically "antifunneled" when using such conventional Miyazawa–Jernigan type contact potentials [68]. A closer inspection indicated that these dimers contained largely hydrophilic interfaces, with an abundance of crystallographic water molecules mediating interprotein interactions [68]. Motivated by this observation, the intermediate range in distance contacts (from 6.5 to 9.5 Å between Cβs) was partitioned into water-mediated and protein-mediated interactions [25]. The corresponding two 20 × 20 interaction matrices (see Figure 4.8 that was reproduced from a recent review [80]) were learned either by the T_b/T_g optimization using a database of several hundred protein home- and heterodimers, where T_b is formally the binding temperature, or by optimizing T_f/T_g over a set of protein monomer folding [25,26]. The interaction potentials obtained from requiring either efficient binding or folding turned out to be nearly identical, indicating that knowledge-based learning may be leading to the discovery of fundamental physicochemical interactions. For example, water-mediated

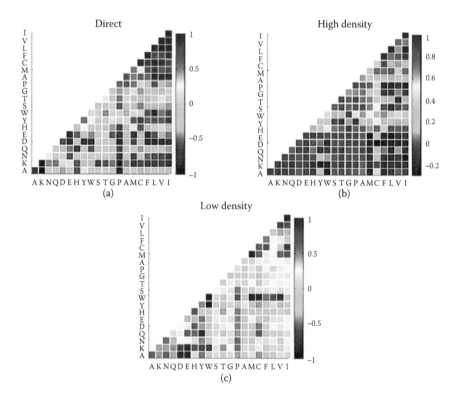

FIGURE 4.8   20 × 20 parameter tables, derived in [26], are graphically shown for direct, protein-mediated, and water-mediated interactions. The interactions are color-coded in such a way that red interactions are attractive and blue interactions are less favorable. When two residue Cβs are found at distances between 4.5 and 6.5 Å, the corresponding *direct* interactions are based on the parameters from (a). At the next distance range, from 6.5 to 9.5 Å between Cβs, amino acids interact primarily either via protein-mediated (b) or water-mediated (c) parameters, depending on their respective instantaneous coordination numbers. The latter indicate whether any particular residue is located near a protein's surface, exposed to solvent, or buried deeply inside. (From NP Schafer et al., *Isr J Chem.*, 54, 1311–1337, 2014. With permission.)

interactions were found to be favorable between two arginines, which was rather unexpected because of naive anticipation of electrostatic repulsion between same charged residues [25]. However, later bioinformatic and molecular dynamics studies have confirmed this finding, showing that the hydrophobic parts of arginines may indeed prefer to associate, while the repulsion of charged groups is diminished by bridging water molecules and mobile counterions [81,82].

A context-dependent potential, which changes its value as a pair of residues transitions from a protein's core to its water-exposed surface, is straightforward to implement in energy-based dynamic methods, such as Monte Carlo sampling, via conditional statements in the algorithm. However, in molecular dynamics simulations, a continuous and differentiable energy function is needed which would "automatically" switch the interaction potential from protein-mediated to water-mediated or vice versa, as the residue pair's local environment fluctuates. To accomplish this, we introduced a function [26], σ_{ij}^{wat}, which basically computes the instantaneous coordination number of each residue, and is around 1 if both residues are water exposed and around 0 otherwise (see Figure 4.9).

$$\sigma_{ij}^{\text{wat}} = \frac{1}{4}\left(1 - \tanh\left[\eta_\sigma\left(\rho_i - \rho_0\right)\right]\right)\left(1 - \tanh\left[\eta_\sigma\left(\rho_j - \rho_0\right)\right]\right)$$
$$\sigma_{ij}^{\text{prot}} = 1 - \sigma_{ij}^{\text{wat}} \tag{4.25}$$

Here, ρ type variables indicate the coordination number of the residue in the given conformation, calculated from $\rho_i = \sum_{j=1}^{N} \Theta_{ij}^I$, hence counting the number of contacts in the first well. With this notation, the second-well potential in AWSEM is written as

$$V_{\text{water}} = -\lambda_{\text{water}} \sum_{j-i>9}^{N} \Theta_{ij}^{II}\left(\sigma_{ij}^{\text{wat}}\gamma_{ij}^{\text{wat}}(a_i, a_j) + \sigma_{ij}^{\text{prot}}\gamma_{ij}^{\text{prot}}(a_i, a_j)\right) \tag{4.26}$$

where, analogous to Θ_{ij}^I, Θ_{ij}^{II} creates a square well bracketed between $r_{\min}^{II} = 6.5$ Å and $r_{\max}^{II} = 9.5$ Å. Note that because of the form of the potential, Equation 4.26, a pair of residues in the right distance range interact either via a protein-mediated, $\gamma_{ij}^{\text{prot}}$, or water-mediated, γ_{ij}^{wat}, interaction depending on whether they are located in the protein's interior or on its water-exposed surface.

Water-mediated interactions represent a new, hydrophilic code for protein binding and folding [26]. Indeed, in folding studies, prediction of highly

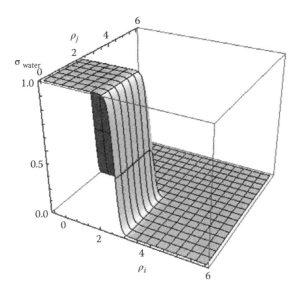

FIGURE 4.9 The switching function from Equation 4.25, σ_{ij}^{wat}, is shown, as a function of coordination numbers of two interaction residues, ρ_i and ρ_j. (Reprinted with permission from A Davtyan et al., *J Phys Chem B*, 116, 8494–8503. Copyright 2012 American Chemical Society.)

charged proteins greatly improved because of better treatment of polar interactions [26]. However, the most dramatic impact of water-mediated interactions has been in binding predictions [32]. Indeed, many protein complex interfaces are organized with a central hydrophobic region, imparting affinity but with low specificity, and with an outer ring of polar residues coupled by water-mediated interactions, which may not contribute much affinity but leads to high specificity in the recognition of the correct binding partner [83]. AWSEM's successes in binding predictions are briefly reviewed in Section 4.9 of this chapter.

Yet another way to look at residue's hydrophobicity is to consider its tendency to prefer the protein's core or to be water-exposed or perhaps to reside in an intermediate layer of the globule [25]. Direct contacts already encode these preferences; however, an explicit burial term would allow further fine-tuning. Hence, piggy-backing on the machinery of water-mediated interactions, we introduced a many-body burial term [25,26],

$$V_{\text{burial}} = -\frac{1}{2}\lambda_{\text{burial}} \sum_{i=1}^{N} \sum_{\mu=1}^{3} \gamma_{\text{burial}}(a_i, \rho_\mu)$$
$$\times \left(\tanh\left[\eta\left(\rho_i - \rho_{\min}^\mu\right)\right] + \tanh\left[\eta\left(\rho_{\max}^\mu - \rho_i\right)\right] \right) \quad (4.27)$$

where each residue, again denoted by a_i, has a specific preference, $\gamma_{\text{burial}}(a_i, \rho_\mu)$, for low-, medium-, and high-density wells, with ρ_i indicating its instantaneous coordination number, $\rho_i = \sum_{j=1}^{N} \Theta_{ij}^I$.

4.7 HYDROGEN BOND POTENTIALS

Even at the level of a single hydrogen bond, quantum chemical calculations show that while electrostatic interactions are dominant, orbital overlap also plays a role, explaining the sharper angular dependence of hydrogen bond energy compared to what would be expected from interactions involving point charges [84]. In the context of a protein immersed in water, one has to further consider cooperative effects and complicated desolvation processes. Hence, how to reasonably model hydrogen bonds represents a challenge even in detailed atomistic simulations. Because of these reasons, in particular, in many coarse-grained protein force fields, explicit hydrogen bonds are not included at all. In contrast, AWSEM represents the other extreme, where hydrogen bonds are represented in a rather elaborate way, to account for the underlying physicochemical complexity and richness of these interactions [31].

First, it is useful to differentiate hydrogen bonds in alpha helices versus beta sheets [24,85–87]. In addition, it has been recognized that in coarse-grained simulations of protein folding or binding, short range and highly directional hydrogen bonds may lead to premature trapping and subsequently slow exploration dynamics [24]. Hence, for hydrogen bonds comprising beta sheets, it is useful to introduce a second, coarser length scale, such that chain segments would have a tendency to approximately align in parallel or antiparallel fashion, without completely locking in place too early. These alignment inducing interactions, dubbed the "liquid-crystal" potential, help the beta sheets to efficiently anneal during the early stages of protein folding. This potential has the following form [24,31,86]:

$$V_{P-AP} = -\gamma_{\text{APH}} \sum_{i=1}^{N-13} \sum_{j=i+13}^{\min(i+16,N)} \nu_{i,j}\nu_{i+4,j-4}$$

$$- \gamma_{\text{AP}} \sum_{i=1}^{N-17} \sum_{j=i+17}^{N} \nu_{i,j}\nu_{i+4,j-4} - \gamma_P \sum_{i=1}^{N-13} \sum_{j=i+9}^{N-4} \nu_{i,j}\nu_{i+4,j+4}$$

(4.28)

where

$$\nu_{i,j} = \frac{1}{2}\left(1 + \tanh\left[\eta\left(r_0 - r_{C\alpha_i,C\alpha_j}\right)\right]\right)$$

(4.29)

is near 1 when the C_α atoms of two residues are found less than the cut-off distance, $r_0 = 8.0$ Å and near 0, otherwise. The first term in Equation 4.28, weighted by γ_{APH}, aligns continuously connected chain segments into beta-hairpins and the second and third terms induce antiparallel and parallel alignments of two chain segments, respectively. The quadratic terms in Equation 4.28, such as $\nu_{i,j}\nu_{i+4,j+4}$, give rise to cooperativity, where formation of a bond between residues $i + 4, j + 4$ is enhanced by the already existing bond between the i, j residues.

In addition to low-resolution "liquid-crystal" interactions approximately aligning chain segments, AWSEM also includes structurally high-resolution hydrogen bonds, based on the following elementary interaction form [24,31,85–87]:

$$\theta_{i,j} = \exp\left[-\frac{\left(r_{ij}^{ON} - \langle r^{ON} \rangle\right)^2}{2\sigma_{NO}^2} - \frac{\left(r_{ij}^{OH} - \langle r^{OH} \rangle\right)^2}{2\sigma_{HO}^2}\right] \tag{4.30}$$

where r^{ON} indicates the instantaneous distance between the carbonyl oxygen of residue i to the nitrogen of residue j and, r^{OH}, analogously corresponds to the distance between the carbonyl oxygen of residue i to the amide hydrogen of residue j. Distances in angular brackets indicate the corresponding equilibrium values, obtained from statistical analysis of protein structures. Similar to other contact-based terms in AWSEM, σ_{NO} and σ_{HO} allow for a distance variation, on the order of 0.7 Å, such that structural distortions within that range would still count toward contributing to the hydrogen bond potential. Because both ON and OH distances are taken into account, $\theta_{i,j}$ contains angular dependence.

To avoid the possibility of spurious hydrogen bond formation, which could be detrimental both thermodynamically and kinetically, we would like to "turn on" hydrogen bonds only if both chain segments are sufficiently extended, as observed previously from structural surveys of known proteins. Hence, the following term (used below) helps to identify and filter out the chain segments that are too coiled [31]:

$$\nu_i^\mu = \frac{1}{2}\left(1 + \tanh\left[\eta^\mu\left(r_{i-2,i+2}^{C\alpha} - r_c^{HB}\right)\right]\right) \tag{4.31}$$

Here, given the residue i that is considered for hydrogen bonding to residue j, the $C\alpha - C\alpha$ distance between the $i - 2, i + 2$ residues are measured, and if that distance is less than a cutoff distance, r_c^{HB}, equal to 12 Å, then ν_i^μ becomes nearly zero. A similar computation is carried out for the partner residue, j. This ensures that both residues i and j are centered in five-residue chain segments that are sufficiently extended.

With these formulas established, the finer-grained hydrogen bonding in AWSEM can be written as consisting of three terms:

$$V_\beta^{ij} = V_{\beta,1}^{ij} + V_{\beta,2}^{ij} + V_{\beta,3}^{ij} \tag{4.32}$$

where the first term is given by the following noncooperative interaction potential:

$$V_{\beta,1}^{ij} = -\Lambda_1(|j - i|)\theta_{i,j}\nu_i^I\nu_j^{II} \tag{4.33}$$

In Equation 4.33, $\Lambda_1(|j - i|)$ provides different weights to hydrogen bonds between the i and j residues depending on the sequence separation. Three different sequence separation categories have been defined, from 4 to 18, from 18 to 45, and more than 45 residues apart. ν_i^I and ν_j^{II} ensure that both interacting strands are in extended enough conformations (see Equation 4.31).

The other two terms of the hydrogen bonding potential engender cooperativity, enhancing the likelihood of hydrogen bond formation if a neighboring bond is already formed. For the antiparallel beta strands, the following potential captures simultaneous emergence of two proximate hydrogen bonds:

$$V_{\beta,2}^{ij} = -\Lambda_2(a_i, a_j, |j - i|)\theta_{i,j}\theta_{j,i}\nu_i^I\nu_j^{II} \tag{4.34}$$

where

$$\begin{aligned}
\Lambda_2(a_i, a_j, |j - i|) = {} & \lambda_2(|j - i|) - 0.5\alpha_1(|j - i|)\ln P_{HB}(a_i, a_j) \\
& - 0.25\alpha_2(|j - i|)[\ln P_{NHB}(a_{i+1}, a_{j-1}) \\
& + \ln P_{NHB}(a_{i-1}, a_{j+1})] \\
& - \alpha_3(|j - i|)[\ln P_{anti}(a_i) + \ln P_{anti}(a_j)]
\end{aligned} \tag{4.35}$$

As elaborated by Equation 4.35, this interaction is modulated by a number of bioinformatically derived parameters, including $P_{HB}(a_i, a_j)$ and $P_{NHB}(a_i, a_j)$, which indicate the propensity for residues of type a_i and a_j to be hydrogen bonded or not. Similarly, $P_{anti}(a_i)$ indicates the likelihood for a residue of type a_i to form an antiparallel-hydrogen bond.

Finally, the third term gives rise to cooperativity of hydrogen bonds in beta strands that are aligned in parallel:

$$V_{\beta,3}^{ij} = -\Lambda_3(a_i, a_j, |j - i|)\theta_{i,j}\theta_{j,i+2}\nu_i^I\nu_j^{II} \tag{4.36}$$

where

$$\Lambda_3(a_i, a_j, |j - i|) = \lambda_3(|j - i|) - \alpha_4(|j - i|) \ln P_{\text{parHB}}(a_{i+1}, a_j)$$

$$- \alpha_5(|j - i|) \ln P_{\text{par}}(a_{i+1}) + \alpha_4(|j - i|) \ln P_{\text{par}}(a_j)$$

(4.37)

The set of coefficients, $\{\lambda_i\}$ and $\{\alpha_i\}$, have been obtained by maximizing T_f / T_g, as discussed earlier (and further elaborated in the next section). Their specific values, along with all other AWSEM parameters, are listed in the Supplemental Information of reference [31].

In addition to hydrogen bonding in beta sheets, AWSEM contains a separate hydrogen bonding potential for alpha helices, with the following form [31,85]:

$$V_{\text{helical}} = -\lambda_{\text{helical}} \sum_{i=1}^{N-4} (f(a_i) + f(a_{i+4}))(\gamma_{\text{prot}} \sigma_{i,i+4}^{\text{prot}} + \gamma_{\text{wat}} \sigma_{i,i+4}^{\text{wat}})$$

$$\times \exp\left[-\frac{\left(r_{i,i+4}^{\text{ON}} - \langle r^{\text{ON}} \rangle\right)^2}{2\sigma_{\text{ON}}^2} - \frac{\left(r_{i,i+4}^{\text{OH}} - \langle r^{\text{OH}} \rangle\right)^2}{2\sigma_{\text{OH}}^2}\right]$$

(4.38)

The strength of a hydrogen bond interaction between residues i, j is modulated by $f(a_i) + f(a_{i+4})$, based on the propensities of residues of type a_i and a_{i+4} to be found in alpha helices. The corresponding propensities, $f(a_i)$, have been obtained from the statistical analysis of the protein data bank. The term in the second parenthesis of Equation 4.38 allows tuning of the strength of the alpha-helical hydrogen bonds depending on whether they are (instantaneously) located on protein's surface or the core. σ^{prot} and σ^{wat} have been reused from Equation 4.25. The intuition behind this distinction goes back to Pauling, who recognized that hydrogen bonds are more important in the protein's core than on its surface because of the desolvation of carbonyl oxygens and amide hydrogens in the former environment [7].

4.8 SCULPTING PROTEIN FOLDING FUNNELS AS A GUIDE TO PARAMETER OPTIMIZATION

Developing a coarse-grained force field is comprised of two important steps. First, the mathematical form of a set of potential energy functions needs to be proposed. This step is based mainly on physical intuition. In the previous several sections, we have motivated and introduced the functional forms of the components of the AWSEM force field that is targeted for applications in protein structure prediction and for studying protein functional dynamics. The next crucial step is parameter optimization. There are a multitude of approaches from inverse statistical mechanics, often related to each other, that may be used to determine the best set of parameters for the given form of potential so as to satisfy some chosen optimization criteria. For example, for relatively simpler molecules, renormalization group calculations, force matching, or reproducing structural correlations could serve as the avenues for determining the Hamiltonian parameters [88–90]. For proteins, represented by a 20-letter amino-acid alphabet, hundreds of parameters need to be optimized even when only an interaction of pairwise form is used; the number of parameters quickly may grow to thousands or even tens of thousands, if three- or higher body interactions are also considered. Not surprisingly, brute force trial and error cannot handle this combinatorial explosion of the number of parameters in the context of a complex biomolecular Hamiltonian. The energy landscape sculpting approach, described next, represents a powerful approach to solving the parameter optimization problem, resulting in highly predictive coarse-grained protein force fields. Energy landscape sculpting is a kind of machine learning.

To motivate the main ideas, consider some typical globular protein that is experimentally known to fold to a unique structure, notwithstanding the conformational substates of the native basin. Then, if we were to apply a force field of the AWSEM form to model this protein, but used completely random parameters in that force field, a very rugged conformational landscape would certainly result. The landscape would have many deep traps, structurally unrelated to each other and likely unrelated to the actual native state. Hence, a natural question is: how difficult is the problem of finding a set of parameters that would reliably fold this protein? For one protein, this is not very difficult; for example, within the structure-based protein coarse-graining methods, native contacts are identified as low in energy. Ensuring this native contact dominance for a single structure is basically sufficient to create a smooth and deep folding funnel [91]. Such a "bespoke" potential would be almost completely nontransferrable, especially with respect to proteins that are neither structurally nor sequence-wise related to this particular training protein.

Hence, instead of focusing on a single protein, we must choose several dozen or hundreds of diverse training proteins, and try to derive a set of parameters that creates deep, smooth folding funnels for all of these proteins simultaneously. This task is now a significantly more difficult problem, requiring first of all that the potential functions be reasonably complete. If the potential is of a sufficiently complete flexible form, which is presumably the case with the AWSEM potential, the next step is the formulation of an easy-to-solve optimization problem that would lead, with one set of parameters, to simultaneous sculpting of folding funnels for the large number of proteins in the training set. In the absence of serious overlearning, analogous to the acquiring of a "superstition," the same set of parameters would be expected to perform well when the force field is subsequently applied to an unrelated collection of test proteins, hence, effectively resolving the challenging parameter transferability problem.

So it is necessary to uncover a quantitative criterion, which is easily computed, that when optimized, would produce well-funneled energy landscapes for each of the training proteins, that is landscapes with funnels that are nearly devoid of energetically low-lying conformational traps. It is natural to rely on the characteristic temperatures underlying protein folding to construct such a criterion. In the first part of this chapter, we discussed three important characteristic temperatures for a heteropolymeric protein chain, T_f (folding), T_g (glass transition), and T_c (collapse transition). For efficient folding, the collapse transition should be near the folding temperature [35,58]. Therefore, the folding criterion depends largely on only the two remaining independent temperatures (although we still have to carefully keep track of the collapse transition, as elaborated below). For any particular protein, the most efficient folding landscape is created when glass transition temperature for the landscape is low, while its folding temperature is high. This enables the protein chain to glide toward its native basin by simulating at T_f without being significantly slowed down by energetically deeply lying random trap configurations, which start to dominate at T_g. Using this notion, maximizing the ratio of T_f over T_g has been used to find the optimal AWSEM force field parameters that sculpt the smoothest folding funnels for the training set of proteins [22,77]. Interestingly, it turns out that yet another but equivalent approach, one that relies on maximizing the free energy difference, ΔF, between the folded and unfolded ensembles just above the glass transition temperature, also leads to the same equations for funnel-sculpting as is derived by maximizing T_f/T_g [92]. This agreement between a spin-glass model of protein folding and the cumulant expansion analysis of computer simulations of heteropolymeric chain collapse, freezing, and folding highlights the internal thermodynamic consistency of energy landscape theory.

Another general important concern about force field parameterization deals with the question of accuracy. Namely, how must the accuracy of the force field scale with the protein chain size in order to fold not only small but also large proteins? If strong scaling of accuracy with size were needed, it might be nearly impossible to find Hamiltonians that are accurate enough to fold large proteins. This would be the case if one were to try to find the lowest energy state of a completely random sequence. Luckily, unlike random heteropolymers, natural proteins have funneled landscapes. The force field accuracy needed for evolved proteins with funneled energy landscape turns out to be only weakly dependent on protein's length [93]. Hence, we might reasonably expect that protein coarse-grained force fields that are complete enough, and are optimized using a limited amount of structural data, would still lead to robust folding funnels for many training and test proteins of different sizes.

The following optimization algorithm (with some variations) has been traditionally used to obtain the parameters for AWSEM and the preceding force fields [22,23,26,77,80,86]. First, an initial guess is made for the parameters, based, for example, on the likelihoods of finding specific types of interactions in the structural database of proteins. After this selection, this initial Hamiltonian is then used to generate an ensemble of structures for each training protein, which either represent denatured states or which are constrained to belong to the native basin. Then, one calculates both the average energy and the energy variance for the simulated denatured and native ensembles. Here, if we can write the Hamiltonian in the following form:

$$V = \sum_i \gamma_i \phi_i, \qquad (4.39)$$

where γ_i represents the interaction strength for the functional form given by ϕ_i, these two statistical quantities can be written as an explicit function of the γ_i. The functional form of components ϕ_i, for example, could correspond to one of the wells of the direct interaction potentials, Equation 4.22. Alternatively, ϕ_i could represent a grouping of potentials, for example, all interactions among residues being close to each other in sequence. Such a grouping would allow an optimization of the relative weights of Hamiltonian's major classes of interactions, for example, balancing FM terms with long-range interactions.

To construct the denatured and folded ensembles, a biasing potential may be added, $V_{loc}(Q)$, typically of a harmonic form, $k(Q-Q_0)^2$, to generate native ($Q_0 \approx 1.0$) and denatured ($Q_0 \approx 0.25$) conformational ensembles. From these sets of structures, an auxiliary vector, $\vec{A}$, is constructed:

$$A_i = \langle \phi_i \rangle_d - \langle \phi_i \rangle_n \qquad (4.40)$$

and a matrix, **B**, is computed:

$$B_{i,j} = \langle \phi_i \phi_j \rangle_d - \langle \phi_i \rangle_d \langle \phi_j \rangle_d \tag{4.41}$$

In these expressions, the angular brackets with n and d subscripts indicate averaging over the simulated native and denatured ensembles, respectively. The energy gap between the native and denatures states for any vector of coefficients, $\{\gamma_i\}$, can then be written as a dot product,

$$\delta E = \vec{\gamma} \cdot \vec{A} \tag{4.42}$$

while the energy variance of the denatured state becomes a matrix quadratic form:

$$\Delta E^2 = \vec{\gamma} \cdot \mathbf{B} \cdot \vec{\gamma}. \tag{4.43}$$

The folding and glass transition temperatures are computed as functions of the stability gap, δE, and the energetic ruggedness of the denatured ensemble, ΔE, and are explicitly expressed as functions of the γ_is. Below we briefly highlight one particular route to obtaining the T_f/Tg ratio, referring the interested reader to recent comprehensive reviews and original papers for the further details and alternative derivations [22,23,26,34,77,80,80,86].

We can estimate the folding and glass transition temperatures by relying on a phenomenological model of protein folding using ideas from spin-glass theory. The statistical properties of the fully denatured ensemble energy landscape, if it has little native structure in it, are approximated by those of a random energy model (REM) [65], where the energies of individual microstates are assumed to be uncorrelated. Even when correlations are taken into account, the REM results are only slightly modified. In the REM, the probability distribution of the energies of unfolded states, which is the sum of many terms, will be nearly Gaussian:

$$P(E)dE = c_0 e^{-\frac{(E-\bar{E})^2}{2\Delta E^2}} dE \tag{4.44}$$

where c_0 is the normalization constant, $\bar{E}$ represents the average energy of the denatured ensemble, and the energetic variance, ΔE^2, was introduced above.

The total number of chain configurations in the unfolded state is denoted as, Ω_0, with the corresponding maximal configurational entropy being

$$S_0 = k_B \log \Omega_0 \tag{4.45}$$

These variables allow one to define the corresponding density of states via

$$\Omega(E) = \Omega_0 P(E) \tag{4.46}$$

with the corresponding microcanonical entropy being given by

$$S(E) = S_0 - \frac{(E - \bar{E})^2}{2\Delta E^2} \tag{4.47}$$

Using the thermodynamic identity, $\frac{\partial S(E)}{\partial E} = \frac{1}{T}$, the thermal energy is obtained as

$$E(T) = \bar{E} - \frac{\Delta E^2}{k_B T} \tag{4.48}$$

while the canonical entropy becomes

$$S(T) = S_0 - \frac{\Delta E^2}{2k_B T} \tag{4.49}$$

The last equation allows us to estimate the glass transition temperature, since at this temperature configurational entropy vanishes, $S(T_g) = 0$. Solving this equation yields

$$T_g = \sqrt{\frac{\Delta E^2}{2k_B S_0}} \tag{4.50}$$

Equation 4.50 sensibly predicts that either increasing the overall *a priori* number of protein chain conformations or decreasing the energetic ruggedness should lower the glass transition temperature.

The free energy of the denatured ensemble, F_d, then becomes

$$F_d = E_d(T) - TS_d(T) = \bar{E}_d - TS_0 - \frac{\Delta E^2}{2k_B T} \tag{4.51}$$

In the simplest approximation, the free energy of the native ensemble is given by

$$F_n = \bar{E}_n \tag{4.52}$$

Hence, at the folding temperature, corresponding to the midpoint of the folding transition, $F_d = F_n$,

$$\bar{E}_n = \bar{E}_d - T_f S_0 - \frac{\Delta E^2}{2k_B T_f} \tag{4.53}$$

which rearranges to the following equation:

$$\delta E - -T_f S_0 - \frac{\Delta E^2}{2k_B T_f} = 0 \tag{4.54}$$

that should be solved for T_f:

$$T_f = \frac{\delta E + \sqrt{\delta E^2 - 2S_0 \Delta E^2 / k_B}}{2S_0} \tag{4.55}$$

For proteins that robustly fold to their native conformations, the stability gap, δE, is expected to completely dominate the second term in Equation 4.55; hence, greatly simplifying it in this limit, one finds

$$T_f = \frac{\delta E}{S_0} \tag{4.56}$$

Collecting the results for the folding temperature and the glass transition temperature, we find $\frac{T_f}{T_g}$ is maximized when we maximize the following ratio:

$$\frac{T_f}{T_g} = \frac{\delta E}{S_0} \sqrt{\frac{2k_B S_0}{\Delta E^2}} = \sqrt{\frac{2k_B}{S_0}} \frac{\delta E}{\Delta E} \tag{4.57}$$

Since S_0 is nearly independent of the parameters, γ, maximization of $\frac{T_f}{T_g}$ is achieved when the stability gap is maximized in units of the standard deviation of energy in the denatured state. The maximization of $\frac{\delta E}{\Delta E}$ can next be formulated in terms of the auxiliary linear algebra objects $\vec{A}$ and $\mathbf{B}$ defined above. Each of these involves obtaining the averages, $\langle \phi_i \rangle_d - \langle \phi_i \rangle_n$ and $\langle \phi_i \phi_j \rangle_d - \langle \phi_i \rangle_d \langle \phi_j \rangle_d$, respectively, using the corresponding simulations of training proteins' native and denatured ensembles, but employing the original trial parameters, $\vec{\gamma}_0$:

$$R \equiv \frac{\delta E}{\Delta E} = \frac{\vec{\gamma} \cdot \vec{A}}{\sqrt{\vec{\gamma} \cdot \mathbf{B} \cdot \vec{\gamma}}} \tag{4.58}$$

Notice that if all $\{\gamma_i\}$ are simultaneously scaled by a constant, α, the variational function, R, stays invariant. By taking advantage of this invariance, the conditions for variationally maximizing R, obtained by solving the following set of equations, $\{\partial R / \partial \gamma_k = 0\}$, turns out to be equivalent to another constrained optimization equation:

$$\frac{\partial}{\partial \gamma_k} \left(\vec{\gamma} \cdot \vec{A} - \mu \sqrt{\vec{\gamma} \cdot \mathbf{B} \cdot \vec{\gamma}} \right) = 0 \tag{4.59}$$

where μ plays the role of an undetermined Lagrange multiplier. As seen in Equation 4.59, we seek to maximize the energy gap between the folded and unfolded states while holding the energy variance in the denatured ensemble fixed as the individual elements of $\vec{\gamma}$ are varied. After the partial derivatives in Equation 4.59 are found (which are elaborated in the Appendix of reference [80], for example), the following linear equations are obtained:

$$\mu' \sum_j B_{j,k}\gamma_j = A_k \tag{4.60}$$

where μ has been recast as another undetermined Lagrange multiplier, μ'. In the matrix notation, Equation 4.60 can be written as

$$\mu' \mathbf{B} \cdot \vec{\gamma} = \vec{A} \tag{4.61}$$

which is solved by

$$\vec{\gamma} = \frac{1}{\mu'}\mathbf{B}^{-1} \cdot \vec{A} \tag{4.62}$$

Maximizing T_f/T_g without employing any additional constraints, as shown above, may have some unwanted consequences, since the dynamics of the denatured ensemble depends on both local and global degrees of order in it. First, it is desirable to tune the average energy of the denatured ensemble, $\bar{E}_d$, in such a way so to keep the collapse temperature close to the folding temperature. Using a simple mean-field theory [77], the collapse temperature can be estimated to occur at $T_c \approx \delta E_{\text{collapse}}/Nk_B \approx \sum_i \gamma_i \langle \phi_i \rangle_d/Nk_B \equiv \vec{A}' \cdot \gamma$, where the elements of $\vec{A}'$ are computed by averaging over the denatured ensemble conformations, $A'_i = \langle \phi_i \rangle_d/N$. Then, a new variational functional, which constrains the degree of collapse of the denatured ensemble, becomes

$$R' = \vec{\gamma} \cdot \vec{A} + \mu_2 \vec{\gamma} \cdot \vec{A}' - \mu_1 \sqrt{\vec{\gamma} \cdot \mathbf{B} \cdot \vec{\gamma}} \tag{4.63}$$

We see the Lagrange multiplier, μ_2, can be used to tune the collapse temperature T_c (e.g., to nudge T_c close to T_f), while μ_1 still sets the scale of energetic ruggedness in the denatured ensemble. Next, we address the question of how to determine μ_2, since it is needed to solve Equation 4.63.

The formal solution of the set of equations, $\{\partial R'/\partial \gamma_k = 0\}$, results in

$$\vec{\gamma} = \frac{1}{\mu'_1}\mathbf{B}^{-1} \cdot \left(\vec{A} + \mu'_2 \vec{A}'\right) \tag{4.64}$$

Premultiplying both sides of Equation 4.64 by $\vec{A}'$, we obtain

$$T_c = \vec{A}' \cdot \vec{\gamma} = \frac{1}{\mu_1'} \vec{A}' \cdot \mathbf{B}^{-1} \cdot \left(\vec{A} + \mu_2' \vec{A}' \right) \qquad (4.65)$$

which may then be solved for μ_2':

$$\mu_2' = \frac{\mu_1' T_c - \vec{A}' \cdot \mathbf{B}^{-1} \cdot \vec{A}}{\vec{A}' \cdot \vec{A}'} \qquad (4.66)$$

Therefore, we first choose μ_1' to set the energy scale and then compute μ_2' from Equation 4.66, substituting it back into Equation 4.64 to solve for the optimized parameters, $\vec{\gamma}$.

Additional constraints may be reasonably imposed on the optimization of parameters, again to take into account some of the physics of folding beyond the uncorrelated REM. For example, if the short range in sequence interactions happens to be too strong, local segments of the polymer chain can become trapped in low-energy configurations, thus rigidifying various chain segments, which, in turn, would greatly slow down the global chain rearrangement dynamics. To avoid this locally glassy dynamics, we separately compute the contributions of the short range in sequence forces to the energetic ruggedness, $\mathbf{B}'$, where its elements are defined as $B_{i,j}' = \langle \phi_i' \phi_j' \rangle_d - \langle \phi_i' \rangle_d \langle \phi_j' \rangle_d$, with the prime notation indicating the relevant Hamiltonian terms (i.e., short range in sequence interactions), and ensure that it does not grow too large. Introducing a new Lagrange multiplier, the optimization functional then becomes

$$R'' = \vec{\gamma} \cdot \vec{A} + \mu_2 \vec{\gamma} \cdot \vec{A}' - \mu_1 \sqrt{\vec{\gamma} \cdot \mathbf{B} \cdot \vec{\gamma}} - \mu_3 \sqrt{\vec{\gamma} \cdot \mathbf{B}' \cdot \vec{\gamma}} \qquad (4.67)$$

By tuning μ_3, we can control the desired ratio between the energetic variance of short-range interactions to the overall energetic variance, $\vec{\gamma} \cdot \mathbf{B}' \cdot \vec{\gamma} / \vec{\gamma} \cdot \mathbf{B} \cdot \vec{\gamma}$. In practice, we follow steps analogous to the ones shown in Equations 4.64 through 4.66, first formally solving for $\vec{\gamma}$,

$$\vec{\gamma} = \frac{1}{\mu_1'} \left(\mathbf{B} + \mu_3' \mathbf{B}' \right)^{-1} \cdot \left(\vec{A} + \mu_2' \vec{A}' \right) \qquad (4.68)$$

then obtaining a system of two quadratic equations to be solved for μ_2' and μ_3' (with μ_1' setting the overall energy scale), and then numerically obtaining $\vec{\gamma}$ from Equation 4.68.

Several numerical and statistical issues need to be addressed to achieve robust optimization that is transferrable. First, the finite amount of sampling in the input ensembles introduces noise into the elements of $\vec{A}$ and **B**. The latter is especially problematic, because upon inversion the smallest eigenvalues of **B**, which are the most sensitive to noise, make the largest contribution to $\mathbf{B}^{-1}$. If these small eigenvalues are too noisy, the optimization scheme would become unstable. To rectify this problem, a singular value decomposition of the matrix **B** is carried out, by omitting the smallest eigenvalues that lack sufficient statistics from the subsequent inversion [23]. To determine which specific eigenvalues need to be omitted, they are statistically analyzed either by splitting the simulation trajectory into several groups or by adding a small amount of test noise to the elements of the **B** matrix [26,86].

The optimization scheme elaborated above has been couched in the context of learning the potentials for a single training protein. When multiple training proteins are employed, as is needed for transferability, a question then arises as to how to do the averaging over these proteins. One approach is simply to average $\vec{A}$ and **B** with uniform weights over the training proteins. Alternatively, a weighted average of these quantities over training examples can be used, with $\omega_m = (T_g/T_f)_m$ being the weight of protein numbered m among overall M training proteins [77]. Then, Equation 4.68 becomes

$$\vec{\gamma} = \hat{\mathbf{B}}^{-1}\vec{\hat{A}} \qquad (4.69)$$

where

$$\hat{\mathbf{B}} = \frac{1}{M}\sum_m \omega_m \left(\mathbf{B} + \mu_3'\mathbf{B}'\right)_m \qquad (4.70)$$

$$\hat{\mathbf{A}} = \frac{1}{M}\sum_m \omega_m \left(\vec{A} + \mu_2'\vec{A}'\right)_m \qquad (4.71)$$

With the above-mentioned choice of ω_m, the more "problematic" proteins, those having higher glass transition temperatures compared with their folding temperatures, are given more weight [77]. Yet another alternative is to write out a unique set of equations (Equation 4.69) separately for each training protein and then solve all these systems simultaneously (as a single overdetermined system) using a least-square approach [26].

Finally, once a new set of force field parameters is obtained, the resulting Hamiltonian must be used for further simulations of the denatured and native ensembles. The resulting denatured state conformations and their energies as well as the native energy are changed from what they were with the initial parameters. An iterative procedure must be then used to obtain a new parameter set, $\vec{\gamma}_{n+1}$, using the Hamiltonian parameterized with $\vec{\gamma}_n$, repeating the procedure until a self-consistent solution is found. In practice, to improve numerical stability, the parameter sets of consecutive rounds, $\vec{\gamma}_n$ and γ_{n+1}, are mixed:

$$\vec{\gamma}_{n+1} = (1 - \epsilon)\vec{\gamma}'_{n+1} + \epsilon\vec{\gamma}_n \tag{4.72}$$

where $\vec{\gamma}'_{n+1}$ is obtained from Equation 4.69 using a Hamiltonian from the previous round of optimization, parameterized by $\vec{\gamma}_n$. The mixing constant, ϵ, is typically chosen between 0.05 and 0.25.

Overall, the self-consistent parameter optimization approach based on energy landscape theory finally produces the most challenging low-energy structural decoys: those that would be generated by the Hamiltonian itself. These are the traps that must be discriminated against to favor the native structures. The optimization scheme is quite sophisticated. It enables careful statistical analysis to prevent overlearning. It also employs extra physical constraints on the landscape that prevent unwanted freezing of various collective degrees of freedom and also control the coil-to-globule collapse temperature and other important physical characteristics important in protein folding and heteropolymer dynamics. When this optimization strategy is followed and the set of training proteins is chosen to be large and diverse, the resulting optimized potential is rendered transferrable to many other proteins that are not found among or even resemble the training proteins, neither structurally nor even sequence-wise.

4.9 APPLICATIONS

As elaborated above, the AMH family of protein structure prediction force fields was initially designed to recognize homologous structures among database memories, given a protein sequence as an input. The early work led to establishing the "capacity" limits of recalling memories from databases. These studies demonstrated that synonymous substitutions of amino acids allowed for significant generalization and learning. In particular, by suitably grouping of amino acids, a structure better than any of those of the database memories could be predicted. Subsequent evolution of the AMH force fields focused on adding many physically motivated interaction potentials, including hydrogen bonds for alpha helices and beta sheets, stereochemical fine-tuning of the backbone dihedral angle potentials, and, importantly, several classes of water-mediated potentials that complemented the traditional hydrophobic code for protein folding with a largely independent hydrophilic one. Several historical overviews of the performance of these potentials with respect to the quality of structure prediction for mainly single-domain proteins have been recently elaborated [30,80]. While we refer the interested readers to these reviews for more details, here we conclude this chapter by briefly overviewing the performance of the latest instantiation of the AMH family of potentials, the AWSEM force field. We do this not only in assessing its use for conventional protein structure prediction, but also much beyond that, showing how the technology can be used in predicting structures of protein complexes, both protein homo- and heterodimers, and protein–DNA complexes, and how it can be used to study protein oligomerization and aggregation, both when it is abnormal and pathogenic and functional.

AWSEM coherently combined several developments of the AMH/AMW force fields occurring in the last 10 years, including water-mediated interactions [26], hydrogen bonding specific to alpha helices [85], and local conformational biasing by relying on memory fragments [94]. To systematically evaluate the performance of various structure prediction algorithms, a panel of 10–15 alpha-helical proteins were used as test targets, ranging in size from approximately 60 to 200 residues [31] (see Figure 4.10). Importantly,

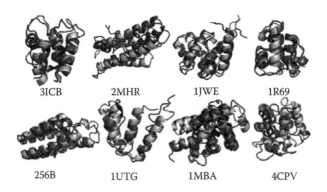

FIGURE 4.10 Protein structure prediction results using AWSEM-MD for a set of test proteins are compared with the corresponding x-ray structures. The following PDB codes are used: 3ICB: vitamin D-dependent calcium-binding protein; 2MHR: myohemerythrin; 1JWE: N-terminal domain of E. coli DNAB helicase; 1R69: aminoterminal domain of phage 434 repressor; 256bB: cytochrome B562; 1utg: uteroglobin; 1MBA: aplysia limacina myoglobin; and 4CPV: carp parvalbuminCARP. (The results are taken from A Davtyan et al., *J Phys Chem B*, 116, 8494–8503, 2012, while the images are from PG Wolynes, *Biochimie*, 119, 218–230, 2015. With permission.)

none of these diverse set of proteins were used in training the force field, hence, they serve as a rigorous performance benchmark. For structure prediction studies, a common simulation protocol uses temperature annealing, where an unfolded chain is initially simulated at very high temperatures, where vigorous chain rearrangement is allowed, followed by continuously lowering the temperature all the way to way below the folding temperatures. It has been estimated that the chance of finding a chain conformation within around 6 Å root-mean-square deviation (RMSD) of the native structure, for a moderate size protein, is extremely small [95] (from 10^{-5} for an 80-residue chain to 10^{-12} for a 200-residue chain) unless the force field captures correctly many salient aspects of folding physics. Hence, in a relatively short annealing simulation, which takes only at most several hours on a single modern CPU, finding a trajectory snapshot that is native-like is very significant, indicating a successful potential. On a side note, especially in earlier steps of protein folding, when the chain does not resemble much the native conformation, RMSD is not an informative measure of the progress of the folding reaction. Instead, it is better to use either the fraction of native contacts or, more ambitiously, the fraction of native distances, Q, as folding order parameters [96]. For 13 alpha-helical proteins, the best Q obtained from 20 independent temperature annealing runs for each target protein are shown in Figure 4.11. For those readers who are more familiar with the RMSD measure, a Q value of 0.7 for a small protein corresponds to approximately 2 Å RMSD. As a rule, Q values above 0.4 indicate a recognizably correct native fold, while Q values above approximately 0.6 indicate almost atomistic resolution predictions.

We further considered three separate real-life scenarios for using AWSEM [31]. There is still a need, occasionally, to make a prediction for a sequence for which no discernible homology can be found. In that case, which would be by far the most challenging task for any prediction algorithm, one is forced to prepare a database of memories for the FM portion of AWSEM, where there are no homologues, even distant ones. We call such a memories database setup "homologues excluded" (HE). At the other extreme, a sequence might already be known to have many known structural homologues, perhaps distant ones, and, in this case, one would be advised to include FMs only from those homologous protein structures ("homologues only" [HO]). In the third, intermediate scenario, where homologues exist but they have not been recognized by pure sequence analysis, both homologous and nonhomologous proteins would be allowed in the fragments

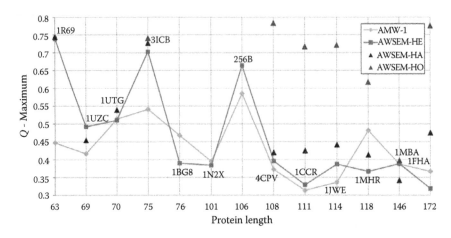

FIGURE 4.11 Maximum Q-scores obtained from a series of temperature annealing simulations for each protein, sorted by chain length, for various fragment memory choices: "homologs excluded" (AWSEM-HE, light squares), "homologs allowed" (AWSEM-HA, black triangles), and "homologs only" (AWSEM-HO, gray triangles). (AMW-1 indicates results from GA Papoian et al., *Proc Natl Acad Sci USA*, 101, 3352–3357, 2004; figure from A Davtyan et al., *J Phys Chem B*, 116, 8494–8503. Copyright 2012 American Chemical Society.)

database (HA). In the most challenging case of de novo structure prediction (HE in Figure 4.11), AWSEM accurately predicts structures for most proteins having less than 100 residues, while it captures only some aspects of native features for larger proteins [31]. This latter test is a bit unfair for a physics-based approach because most of the larger proteins in the panel actually contain a heme cofactor, which is not taken into account in the current AWSEM. The lack of heme likely explains a large part of the discrepancies for larger proteins. Interestingly, when apomyoglobin is used as a target instead of the heme-containing myoglobin, the best predicted structure becomes significantly more native-like (Q 0.44, RMSD 3.9 Åversus Q 0.37 for myoglobin) [97]. For the better calculations, Chen et al. examined yet another FM strategy, based on running explicit-solvent atomistic simulations of short (approximately 20 residue) peptides [97]. This latter approach, at significantly added computational expense, allows one to completely forgo bioinformatic component of the AWSEM potential.

After the introduction of water-mediated interaction potentials in 2004 [26], (initially named AMW), the subsequent AWSEM predictions for protein binding have been shown to be remarkably accurate in a structural sense [32] (see Figure 4.12). The accuracy likely arises because the monomer structure is being treated as a homologue and the available conformational

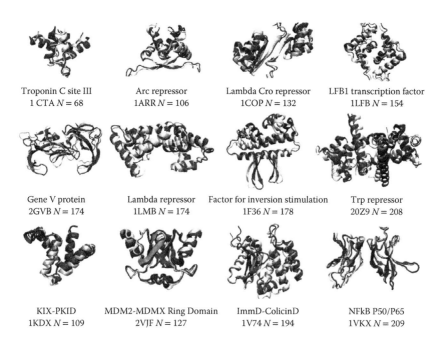

FIGURE 4.12 Binding structure predictions start with an initial condition of two unfolded monomers, without any native contact bias between the monomers, that are allowed to interact via the AWSEM potential as the simulation temperature is gradually lowered. The best predicted structures and the corresponding x-ray structures are shown in light gray and dark gray, respectively. (From W Zheng et al., *Proc Natl Acad Sci USA*, 109, 19244–19249, 2012. With permission.)

space for states that are merely misbound is small compared to more extensive space for misfolded states. Nevertheless, this shows that AWSEM indeed captures the right physics by having an important contribution of polar contacts at many binding interfaces. In using AWSEM to predict binding, it is being employed as a highly flexible docking algorithm, in that one starts with unfolded chains. To systematically evaluate the binding capabilities of the AWSEM force field, eight homodimers and four heterodimers were simulated using temperature annealing, where the initial chain conformations of monomers were completely unfolded [32]. As can be seen in Figure 4.12, the predicted structures for all 12 dimeric complexes agree with the crystallographic structures of the complexes to within atomistic resolution, which might be somewhat unexpected because AWSEM is a coarse-grained force field. Furthermore, two-dimensional free energy profiles, computed using the structural similarity to individual monomers, Q_A and Q_B for two monomers A and B, predict properly whether there is two-state or three-state binding for the various dimers in the set, in agreement with the related experimental observations [32].

These latter results, which shed light on various possible binding mechanisms, exemplify a major advantage of using AWSEM, beyond its original roots in structure prediction. Mechanistic issues can be addressed by constructing reasonably accurate free energy profiles and examining the specific structures in various regions of order parameters. Many exciting problems in molecular biology today lie at the interface of purely structural issues and dynamic biophysical mechanisms. AWSEM and its predecessors (AMH and AMW) have already been extensively used to study many mechanistic issues.

AWSEM having been instantiated in the LAMMPS open source code [98] can take advantage of other coarse-grained potentials used to model other biomolecules. For instance, AWSEM's coarse-grained protein force field has recently been combined in a simple way with the three-site DNA model that was invented by de Pablo and coworkers [99,100]. In the hybrid model [101–104], the protein and DNA beads interact in a sequence-independent way using screened electrostatic interactions of the Debye–Hückel type. In an interesting application of this combined protein–DNA potentials, Potoyan et al. investigated the NF-κB/IκB/DNA ternary system [102,104]. NF-κB is one of the most important human proteins, playing the role of a master regulator of hundreds of other genes, modulating immune and anticancer responses. It was previously thought that *in vivo* NF-κB's binding to DNA would simply follow a mass-action law, where the nonequilibrium control of the genetic circuits, in which it's involved, arises only from protein synthesis and degradation processes. However, using the AWSEM/DNA hybrid, Potoyan et al. showed that NF-κB's natural regulator, IκB, can actively dislodge an NF-κB molecule bound to DNA in a process called molecular stripping. To strip NF-κB from the DNA, IκB acts as a very large allosteric effector for NF-κB binding to the DNA. IκB changes the conformation of the NF-κB upon binding to the NF-κB/DNA complex so that the NF-κB–DNA interactions become disrupted, leading to subsequent unbinding. This disruption is mainly of electrostatic origin, where effectively the negatively charged PEST sequence from the IκB competes with negative charges on DNA. The biophysical mechanism of molecular stripping seen in this system hints at yet another level of kinetic control over cell-biological circuits beyond synthesis and degradation.

The AWSEM models have been employed to study nucleosomes, which are made up of histones that package the nuclear DNA of higher organisms [101]. DNA folding inside nuclei is mediated by many proteins, but histones play the primary role. Eight histone proteins form a hockey-puck-like octamer, which is highly positively charged, around which the negatively charged 147 DNA base pairs wrap, forming a nucleosomal particle. Some of the genomic DNA, continuing as remaining unwrapped linker DNA, follows to form another neighboring nucleosome and so on, giving rise to a polynucleosomal chain called chromatin. The fundamental biology and physics of chromatin folding and regulation by posttranscriptional modifications are at the forefront of modern science. Many fundamental aspects of histone and nucleosomal dynamics have been studied using molecular dynamics simulations, both atomistic and coarse-grained. The AWSEM force field has been used recently to make predictive calculations of histone dimers, their interactions with chaperones, as well as nucleosomal stability and disassembly, as elaborated next [105].

Although all histones share the same stereotypical "histone fold," some histones, such as H3, have biologically important variants, such as the centromere-specific H3 variant, CENP-A, which is essential for mitosis. Interestingly, the only histone which is evolutionarily invariant is H4. Zhao et al. applied a combination of explicit-solvent atomistic simulations, at the microsecond timescale, and, independently, carried out separate coarse-grained AWSEM simulations of the H3/H4 dimer and CENP-A/H4 dimer and as well as the ternary complex with a chaperone, HJURP [105]. This allowed then to investigate conformational dynamics underlying these important nucleosomal building blocks. They discovered that the H4 histone is indeed the most rigid histone, serving as a reinforcing structural framework for assembling higher order super-structures [105]. In contrast, they found that the CENP-A/H4 dimer, which is needed to assemble the kinetochore during cell division, is significantly more dynamic than its canonical counterpart, the H3/H4 dimer (see Figure 4.13). Furthermore, the molecular dynamics simulations indicated that the centromere-specific chaperone HJURP stabilizes the CENP-A/H4 dimer by forming a specific electrostatic interaction network with the latter. These results from both atomistic and coarse-grained simulations led to the hypothesis that during the CENP-A/H4 deposition process, the HJURP chaperone protects various substructures of the dimer, so as to serve both as a folding and binding chaperone [105]. This study also showed broad consistency between atomistic and AWSEM simulations, when both approaches are applied to the same molecular systems. As expected, AWSEM calculations have the advantage of allowing vastly more sampling, while atomistic simulations

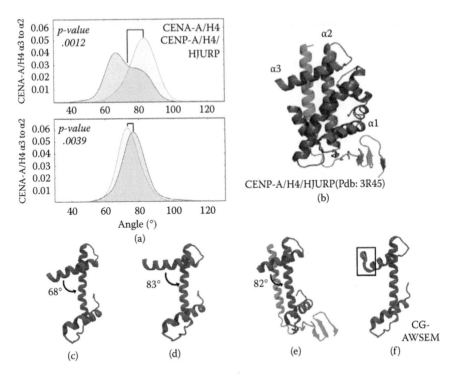

FIGURE 4.13 (a)–(f) AWSEM simulations of the native-state dynamics of histone proteins and their chaperones have been shown to be broadly consistent with the corresponding atomistic simulations in explicit solvent [105]. Without the chaperone HJURP, AWSEM simulations show that α_3 helix of the centromeric protein CENP-A, in a dimeric form, is highly dynamic, stochastically switching between two basins around 68° and 82°. Their chaperone, HJURP, when bound to the CENP-A dimer, structurally stabilizes the α_3 helix, forming an intricate network of electrostatic interactions in the ternary complex. These calculations and the associated *in vivo* experiments show that HJURP serves both as a folding and a binding chaperone for CENP-A, by protecting various substructures of the CENP-A dimer in distinct biological and structural contexts. (Reprinted with permission from H Zhao et al., *J Am Chem Soc.*, 138, 13207–13218. Copyright 2016 American Chemical Society.)

provide finer structural details. Although energy landscape theory of protein dynamics, upon which AWSEM is built, does not necessarily imply that a coarse-grained force field such as AWSEM would accurately reproduce dynamics of the native basin of proteins at relatively high resolution, this, nevertheless, turns out to be the case for histone binary and ternary complexes.

The AWSEM/DNA hybrid described above, with Debye–Hückel electrostatic interactions between protein and DNA beads, has also been applied to investigate the native dynamics and disassembly pathways of the canonical nucleosome. Zhang et al. [101] have showed that, in agreement with experiments, unpeeling of nucleosomal DNA is coupled to dislodging first the H2A/H2B histones (see Figure 4.14). The latter do not leave immediately, but instead interact in a nonnative manner with the remaining (H3–H4)$_2$ tetramer. Furthermore, if the nucleosome is intact, containing flexible tails, DNA unwinding is asymmetric, in agreement with recent single molecule pulling experiments. The two DNA ends peel off in a more symmetric way for the tailless nucleosomes. Subsequently, free energy calculations were carried out to probe this difference, indicating that asymmetric unwinding arises directly from the electrostatic interactions between the DNA end segments and the H3 tails [101].

Finally, the AWSEM-MD can also be used to study protein aggregation, both functional and pathological. AWSEM-MD calculations have buttressed a proposal for the molecular mechanism responsible for formation of long-term memory (LTM) that involves aggregation [106]. It has long been recognized that LTMs are formed by modifying the number, strength, and structure of specific localized synaptic connections. Since most proteins in cells have relatively short half-lives, this raises the question of how LTMs persist over long timescales. Kandel and coworkers [107] and Heinrich and Lindquist [108] suggested that cytoplasmic polyadenylation element binding (CPEB) may form oligomerized prions that could maintain memories for long periods of time. Recently, AWSEM was used to develop a molecular model of CPEB oligomerization, finding that several metastable

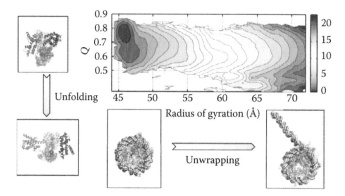

FIGURE 4.14 The AWSEM protein coarse-grained force field has been combined with a three-bead per base model of de Pablo and coworkers to study a number of protein–DNA processes, including the free energy landscape of a nucleosomal particle. (Reprinted with permission from B Zhang et al., *J Am Chem Soc.*, 138, 8126–8133. Copyright 2016 American Chemical Society.)

oligomers are energetically viable [106]. In the absence of external tension, oligomerization based on coiled-coil structures is thermodynamically favorable. But, under tension, which may be exerted by the cytoskeletal network, the oligomers transition to beta-sheet structures. This sets up a positive feedback between cytoskeletal filaments and CPEB oligomers, where CPEB beta-sheet formation is activated by the cytoskeletal dynamics, which, in turn, promotes actin mRNA activity, to stabilize LTM.

4.10 FUTURE OUTLOOK

In this chapter, we have guided the reader through an almost three-decade journey that lead to the development of current AWSEM and its extensions to protein–DNA interactions. In terms of the future of the force field, there is still a lot of room for improvement. Some obvious improvements are incremental and narrowly technical. On the other hand, serendipitous discoveries may also lie ahead, just as the water-mediated interactions were discerned by trying to determine whether protein dimers' binding landscapes are funneled or not. We next briefly outline some directions where we see the force field developing in the near future and then follow by pointing out some areas where the applications of AWSEM could make an impact.

As evidenced by the somewhat elaborate nature of this chapter, AWSEM has already evolved into quite a complicated force field, containing numerous terms. For those readers that prefer simplicity (we count ourselves in that group too) we would have been delighted to develop a simple protein modeling force field that is also accurate and predictive. However, the major information loss in removing atomistic degrees of freedom seems to necessitate derivation of a complex Hamiltonian if the ambition is to arrive at a broadly transferrable, predictive, and accurate model. Nevertheless, while it is possible that the AWSEM Hamiltonian could be simplified in the future, perhaps by omitting some of the terms, the converse is even more likely: additional protein physics need to be discovered and incorporated into AWSEM as additional potential terms. For example, many-body polarization interactions could play an important role. Similarly, although electrostatic interactions have recently been incorporated via the screened Debye–Hückel potential, we are planning to also develop explicit ion-based interactions to treat more realistically the complex many-body physics near highly charged surfaces, such as DNA and nucleosomes, as well as the effects of key divalent ions like Mg^{2+} and Ca^{2+}.

FMs is another area where innovative work could improve the performance of AWSEM. The readers who carefully went through the neural network origins of AWSEM in the beginning part of this chapter might have noticed the relatively simple functional form of the currently used FM potential, which assigns equal weights to all memories. Hence, some of the recent breakthroughs in neural network research and deep-learning algorithms could be used as inspiration to take the FM interactions to another level. In this regard, recent work on using conformational snapshots from explicit-solvent atomistic simulations of fragments provides another important avenue for new development. In addition, the problem that current AWSEM does not include cofactors, one of the major reasons why folding structure predictions are not fully successful for larger proteins, could also be addressed by including coarse-grained cofactor fragments parameterized from atomistic simulations. With all these new developments, as new potentials are introduced and new milestones are reached, it will be necessary to carry out "grand optimizations," as outlined in the earlier sections. Since the last such optimization on a large scale was performed by Papoian et al. in 2004, it is important to chart a path for periodic reoptimizations of AWSEM parameters every several years.

Arguably, the most powerful biophysical applications of AWSEM have so far have arisen from studying macromolecular complex formation processes, either among several protein monomers, or between proteins and DNA. In these simulations, the monomers are internally stabilized by including a single FM taken from the crystal structure of the monomer. As discussed above, alternatively, FMs could be sampled from snapshots of an atomistic simulation. With this strength of AWSEM in mind with regard to complicated binding problems, we envision almost a limitless number of interesting biophysical systems where AWSEM could be profitably applied either in its classical prediction mode, to predict the structure of macromolecular assemblies, or to study kinetic functional mechanisms, where interactions of one protein unit or domain with another influences how and whether the latter binds to another protein or DNA segment. In our own groups, we are currently applying AWSEM to study folding of membrane proteins, protein misfolding, and aggregation, including functional aggregation, interactions of histone proteins with each other, their chaperones, and DNA, and the role of posttranslational modifications, such as acetylation, in modulating these interactions. We are also extending AWSEM for simulating more accurately intrinsically disordered proteins. Hence, in summary, we feel justified in predicting that AWSEM will provide important contributions to understanding many new exciting bimolecular systems and processes.

ACKNOWLEDGMENTS

GAP would like to thank many members of the PGW's research group for delightful conversations about AMH, AMW, and AWSEM during his postdoctoral work at the University of California, San Diego, in particular Michael Eastwood and Johan Ulander, among others. He also would like to express his sincere gratitude to Aram Davtyan for his courage and dedication over many years in rederiving the highly complex and numerous force terms of the AWSEM potential and implementing AWSEM from scratch in C++, while he was a graduate student in GAP's group. He is grateful to the NSF (Grants: CAREER CHE-0846701, CHE-1363081 and POLS-1206060), Beckman Young Investigator program, and Camille Dreyfus Teacher-Scholar program for supporting the work on AWSEM by his research group initially at the University of North Carolina Chapel Hill and subsequently at the University of Maryland, College Park.

PGW happily acknowledges his many coworkers in protein structure prediction over the years, many of whose ideas and hard work are being put to practical use. AWSEM importantly benefited from the efforts of Weihua Zheng and Nick Schafer. PGW's work has been supported over the years both by NSF and NIH. Currently, he is supported by the Center for Theoretical Biological Physics (NSF Grants PHY-1308264 and PHY-1427654), the National Institute of General Medical Sciences PPG Grant P01 GM071862, and the D. R. Bullard-Welch Chair (Grant C-0016) at Rice University.

REFERENCES

1. Y Duan, C Wu, S Chowdhury, MC Lee, G Xiong, W Zhang, R Yang, P Cieplak, R Luo, T Lee et al. A point-charge force field for molecular mechanics simulations of proteins based on condensed-phase quantum mechanical calculations. *J Comput Chem*, 24(16):1999–2012, 2003.
2. RB Best, X Zhu, J Shim, PEM Lopes, J Mittal, M Feig, and AD MacKerell Jr. Optimization of the additive CHARMM all-atom protein force field targeting improved sampling of the back-bone φ, ψ and side-chain $\chi 1$ and $\chi 2$ dihedral angles. *J Chem Theory Comput*, 8(9):3257–3273, 2012.
3. D Frenkel and B Smit. *Understanding Molecular Simulation*. Orlando, FL: Academic Press, 2nd edition, 2001.

4. B Leimkuhler and C Matthews. *Molecular Dynamics: With Deterministic and Stochastic Numerical Methods*. London: Springer, 2015.

5. V Mayer-Schönberger and K Cukier. *Big Data: A Revolution That Will Transform How We Live, Work, and Think*. Boston, MA: Houghton Mifflin Harcourt, 2013.

6. GA Voth (ed.). *Coarse-Graining of Condensed Phase and Biomolecular Systems*. Boca Raton, FL: CRC Press, 2009.

7. L Pauling, RB Corey, and HR Branson. The structure of proteins: Two hydrogen-bonded helical configurations of the polypeptide chain. *Proc Natl Acad Sci*, 37(4):205–211, 1951.

8. TP Straatsma and JA McCammon. Molecular dynamics simulations with interaction potentials including polarization development of a noniterative method and application to water. *Mol Simul*, 5(3–4):181–192, 1990.

9. CJ Tainter, PA Pieniazek, Y-S Lin, and JL Skinner. Robust three-body water simulation model. *J Chem Phys*, 134(18):184501, 2011.

10. C Zheng, CF Wong, J McCammon, and PG Wolynes. Quantum simulation of ferrocytochrome-c. *Nature*, 334(6184):726–728, 1988.

11. CF Wong, C Zheng, J Shen, J McCammon, and PG Wolynes. Cytochrome c: A molecular proving ground for computer simulations. *J Phys Chem*, 97(13):3100–3110, 1993.

12. R Goldstein and PG Wolynes. Protein tertiary structure prediction using associative memory Hamiltonians—A progress report. In *Advances in Biomolecular Simulations*, vol. 239, pp. 200–209. Melville, NY: AIP Publishing, 1991.

13. DM Leitner and JE Straub. *Proteins: Energy, Heat and Signal Flow (Computation in Chemistry)*. Boca Raton, FL: CRC Press, 2009.

14. RB Laughlin, D Pines, J Schmalian, BP Stojković, and P Wolynes. The middle way. *Proc Natl Acad Sci USA*, 97(1):32–37, 2000.

15. M Oliveberg and PG Wolynes. The experimental survey of protein-folding energy landscapes. *Q Rev Biophys*, 38(03):245–288, 2005.

16. DU Ferreiro, JA Hegler, EA Komives, and PG Wolynes. Localizing frustration in native proteins and protein assemblies. *Proc Natl Acad Sci USA*, 104(50):19819–19824, 2007.

17. DU Ferreiro, JA Hegler, EA Komives, and PG Wolynes. On the role of frustration in the energy landscapes of allosteric proteins. *Proc Natl Acad Sci USA*, 108(9):3499–3503, 2011.

18. DU Ferreiro, EA Komives, and PG Wolynes. Frustration in biomolecules. *Q Rev Biophys*, 47(4):285–363, 2014.

19. M Jenik, RG Parra, LG Radusky, A Turjanski, PG Wolynes, and DU Ferreiro. Protein frustratometer: A tool to localize energetic frustration in protein molecules. *Nucleic Acids Res*, 40:W348–W351, 2012.

20. MS Friedrichs and PG Wolynes. Toward protein tertiary structure recognition by means of associative memory Hamiltonians. *Science*, 246(4928):371–373, 1989.

21. MS Friedrichs, RA Goldstein, and PG Wolynes. Generalized protein tertiary structure recognition using associative memory Hamiltonians. *J Mol Biol*, 222(4):1013–1034, 1991.

22. RA Goldstein, ZA Luthey-Schulten, and PG Wolynes. Optimal protein-folding codes from spin-glass theory. *Proc Natl Acad Sci USA*, 89(11):4918–4922, 1992.

23. C Hardin, MP Eastwood, Z Luthey-Schulten, and PG Wolynes. Associative memory Hamiltonians for structure prediction without homology: Alpha-helical proteins. *Proc Natl Acad Sci USA*, 97(26):14235–14240, 2000.

24. C Hardin, MP Eastwood, MC Prentiss, Z Luthey-Schulten, and PG Wolynes. Associative memory Hamiltonians for structure prediction without homology: α/β proteins. *Proc Natl Acad Sci USA*, 100(4):1679–1684, 2003.

25. GA Papoian, J Ulander, and PG Wolynes. Role of water mediated interactions in protein-protein recognition landscapes. *J Am Chem Soc*, 125(30):9170–9178, 2003.

26. GA Papoian, J Ulander, MP Eastwood, Z Luthey-Schulten, and PG Wolynes. Water in protein structure prediction. *Proc Natl Acad Sci USA*, 101(10):3352–3357, 2004.

27. C Zong, GA Papoian, J Ulander, and PG Wolynes. Role of topology, non-additivity, and water-mediated interactions in predicting the structures of α/β proteins. *J Am Chem Soc*, 128(15):5168–5176, 2006.

28. N Go. Theoretical studies of protein folding. *Ann Rev Biophys Bioeng*, 12(1):183–210, 1983.

29. CB Anfinsen. Studies on the principles that govern the folding of protein chains. *Science*, 181:223–30, 1972.

30. PG Wolynes. Evolution, energy landscapes and the paradoxes of protein folding. *Biochimie*, 119:218–230, 2015.

31. A Davtyan, NP Schafer, W Zheng, C Clementi, PG Wolynes, and GA Papoian. AWSEM-MD: Protein structure prediction using coarse-grained physical potentials and bioinformatically based local structure biasing. *J Phys Chem B*, 116(29):8494–8503, 2012.

32. W Zheng, NP Schafer, A Davtyan, GA Papoian, and PG Wolynes. Predictive energy landscapes for protein–protein association. *Proc Natl Acad Sci USA*, 109(47):19244–19249, 2012.

33. AV Finkelstein and O Ptitsyn. *Protein Physics: A Course of Lectures*. London: Academic Press, 2002.

34. JD Bryngelson, JN Onuchic, ND Socci, and PG Wolynes. Funnels, pathways, and the energy landscape of protein folding: A synthesis. *Proteins*, 21(3):167–195, 1995.

35. DK Klimov and D Thirumalai. Criterion that determines the foldability of proteins. *Phys Rev Lett*, 76(21):4070, 1996.

36. AV Finkelstein and AY Badretdinov. Rate of protein folding near the point of thermodynamic equilibrium between the coil and the most stable chain fold. *Fold Des*, 2(2):115–121, 1997.

37. LD Landau and EM Lifshitz. *Statistical Physics*, Part 1, vol. 5, Oxford, UK: Butterworth-Heinemann 1980.

38. JN Onuchic, Z Luthey-Schulten, and PG Wolynes. Theory of protein folding: The energy landscape perspective. *Ann Rev Phys Chem*, 48(1):545–600, 1997.

39. RAL Jones. *Soft Condensed Matter*, (Oxford Master Series in Condensed Matter Physics, Vol. 6). Oxford, UK: Oxford University Press, 2002.

40. J-L Barrat and J-P Hansen. *Basic Concepts for Simple and Complex Liquids*. Cambridge, UK: Cambridge University Press, 2003.

41. PG Wolynes. Folding funnels and energy landscapes of larger proteins within the capillarity approximation. *Proc Natl Acad Sci USA*, 94(12):6170–6175, 1997.

42. VI Abkevich, AM Gutin, and EI Shakhnovich. Specific nucleus as the transition state for protein folding: Evidence from the lattice model. *Biochemistry*, 33(33):10026–10036, 1994.

43. AR Fersht. Nucleation mechanisms in protein folding. *Curr Opin Struct Biol*, 7(1):3–9, 1997.

44. BA Shoemaker, J Wang, and PG Wolynes. Structural correlations in protein folding funnels. *Proc Natl Acad Sci USA*, 94(3):777–782, 1997.

45. M Rubinstein and RH Colby. *Polymer Physics*. Oxford, UK: Oxford University Press, 2003.

46. A Fersht. *Structure and Mechanism in Protein Science: A Guide to Enzyme Catalysis and Protein Folding*. New York: Macmillan, 1999.

47. AK Dunker, CJ Oldfield, J Meng, P Romero, JY Yang, JW Chen, V Vacic, Z Obradovic, and VN Uversky. The unfoldomics decade: An update on intrinsically disordered proteins. *BMC Genomics*, 9(Suppl 2):S1, 2008.

48. GA Papoian. Proteins with weakly funneled energy landscapes challenge the classical structure–function paradigm. *Proc Natl Acad Sci USA*, 105(38):14237–14238, 2008.

49. HJ Dyson. Making sense of intrinsically disordered proteins. *Biophys J*, 110(5):1013–1016, 2016.

50. O Weiss, MA Jimenez-Montano, and H Herzel. Information content of protein sequences. *J Theor Biol*, 206(3):379–386, 2000.

51. C Levinthal. How to fold graciously. *Mossbauer Spectrosc Biol Syst*, 67:22–24, 1969.

52. V Dotsenko. *Introduction to the Replica Theory of Disordered Statistical Systems*. Cambridge, UK: Cambridge University Press, 2005.

53. W Kauzmann. Some factors in the interpretation of protein denaturation. *Adv Protein Chem*, 14:1–63, 1959.

54. B Xue, RL Dunbrack, RW Williams, AK Dunker, and VN Uversky. PONDR-FIT: A meta-predictor of intrinsically disordered amino acids. *Biochim Biophys Acta*, 1804(4):996–1010, 2010.

55. JD Bryngelson and PG Wolynes. Spin glasses and the statistical mechanics of protein folding. *Proc Natl Acad Sci USA*, 84(21):7524–7528, 1987.

56. E Shakhnovich and AM Gutin. Formation of unique structure in polypeptide chains: Theoretical investigation with the aid of a replica approach. *Biophys Chem*, 34(3):187–199, 1989.

57. VS Pande, AY Grosberg, and T Tanaka. Statistical mechanics of simple models of protein folding and design. *Biophys J*, 73(6):3192, 1997.

58. JD Bryngelson and PG Wolynes. A simple statistical field theory of heteropolymer collapse with application to protein folding. *Biopolymers*, 30(1–2):177–188, 1990.

59. JD Bryngelson and PG Wolynes. Intermediates and barrier crossing in a random energy model (with applications to protein folding). *J Chem Phys*, 93(19):6902–6915, 1989.

60. SS Plotkin, J Wang, and PG Wolynes. Statistical mechanics of a correlated energy landscape model for protein folding funnels. *J Chem Phys*, 106(7):2932–2948, 1997.

61. JA Hegler, P Weinkam, and PG Wolynes. The spectrum of biomolecular states and motions. *HFSP J*, 2(6):307–313, 2008.

62. ND Socci, JN Onuchic, and PG Wolynes. Diffusive dynamics of the reaction coordinate for protein folding funnels. *J Chem Phys*, 104(15):5860–5868, 1996.

63. JN Onuchic and PG Wolynes. Theory of protein folding. *Curr Opin Struct Biol*, 14(1):70–75, 2004.

64. JN Onuchic, PG Wolynes, Z Luthey-Schulten, and ND Socci. Toward an outline of the topography of a realistic protein-folding funnel. *Proc Natl Acad Sci USA*, 92(8):3626–3630, 1995.

65. B Derrida. Random-energy model: An exactly solvable model of disordered systems. *Phys Rev B*, 24(5):2613, 1981.

66. SS Plotkin and JN Onuchic. Understanding protein folding with energy landscape theory. Part I: Basic concepts. *Q Rev Biophys*, 35(2):111–167, 2002.

67. SS Plotkin and JN Onuchic. Understanding protein folding with energy landscape theory. Part II: Quantitative aspects. *Q Rev Biophys*, 35(3):205–286, 2002.

68. GA Papoian and PG Wolynes. The physics and bioinformatics of binding and folding—An energy landscape perspective. *Biopolymers*, 68(3):333–349, 2003.

69. AL Cuff, I Sillitoe, T Lewis, AB Clegg, R Rentzsch, N Furnham, M Pellegrini-Calace, D Jones, J Thornton, and CA Orengo. Extending CATH: Increasing coverage of the protein structure universe and linking structure with function. *Nucleic Acids Res*, 39(Suppl 1):D420–D426, 2011.

70. B Müller, J Reinhardt, and MT Strickland. *Neural Networks: An Introduction.* Berlin, Germany: Springer Science, 2012.

71. H Sompolinsky. Statistical mechanics of neural networks. *Phys Today*, 41(21):70–80, 1988.

72. Y LeCun, Y Bengio, and G Hinton. Deep learning. *Nature*, 521(7553):436–444, 2015.

73. JJ Hopfield. Neural networks and physical systems with emergent collective computational abilities. *Proc Natl Acad Sci USA*, 79(8):2554–2558, 1982.

74. GE Hinton, S Osindero, and Y-W Teh. A fast learning algorithm for deep belief nets. *Neural Comput*, 18(7):1527–1554, 2006.

75. MS Friedrichs and PG Wolynes. Molecular dynamics of associative memory Hamiltonians for protein tertiary structure recognition. *Tetrahedron Comput Methodol*, 3(3):175–190, 1990.

76. RA Goldstein, ZA Luthey-Schulten, and PG Wolynes. Protein tertiary structure recognition using optimized Hamiltonians with local interactions. *Proc Natl Acad Sci USA*, 89(19):9029–9033, 1992.

77. KK Koretke, Z Luthey-Schulten, and PG Wolynes. Self-consistently optimized energy functions for protein structure prediction by molecular dynamics. *Proc Natl Acad Sci USA*, 95(6):2932–2937, 1998.

78. KK Koretke, Z Luthey-Schulten, and PG Wolynes. Self-consistently optimized statistical mechanical energy functions for sequence structure alignment. *Protein Sci*, 5(6):1043–1059, 1996.

79. S Miyazawa and RL Jernigan. Residue–residue potentials with a favorable contact pair term and an unfavorable high packing density term, for simulation and threading. *J Mol Biol*, 256(3):623–644, 1996.

80. NP Schafer, BL Kim, W Zheng, and PG Wolynes. Learning to fold proteins using energy landscape theory. *Isr J Chem*, 54(8–9):1311–1337, 2014.

81. D Pednekar, A Tendulkar, and S Durani. Electrostatics-defying interaction between arginine termini as a thermodynamic driving force in protein-protein interaction. *Proteins*, 74(1):155–163, 2009.

82. J Vondrášek, PE Mason, J Heyda, KD Collins, and P Jung-wirth. The molecular origin of like-charge arginine–arginine pairing in water. *J Phys Chem B*, 113(27):9041–9045, 2009.

83. TA Larsen, AJ Olson, and DS Goodsell. Morphology of protein-protein interfaces. *Structure*, 6(4):421–427, 1998.

84. T Steiner. The hydrogen bond in the solid state. *Angew Chem Int Ed*, 41(1):48–76, 2002.

85. V Oklejas, C Zong, GA Papoian, and PG Wolynes. Protein structure prediction: Do hydrogen bonding and water-mediated interactions suffice? *Methods*, 52(1):84–90, 2010.

86. C Hardin, MP Eastwood, M Prentiss, Z Luthey-Schulten, and PG Wolynes. Folding funnels: The key to robust protein structure prediction. *J Comput Chem*, 23(1):138–146, 2002.

87. C Hardin, Z Luthey-Schulten, and PG Wolynes. Backbone dynamics, fast folding, and secondary structure formation in helical proteins and peptides. *Proteins*, 34(3):281–294, 1999.

88. WG Noid, J-W Chu, GS Ayton, V Krishna, S Izvekov, GA Voth, A Das, and HC Andersen. The multiscale coarse-graining method. I. A rigorous bridge between atomistic and coarse-grained models. *J Chem Phys*, 128(24):244114, 2008.

89. A Savelyev and GA Papoian. Chemically accurate coarse graining of double-stranded DNA. *Proc Natl Acad Sci USA*, 107(47):20340–20345, 2010.

90. WG Noid. Perspective: Coarse-grained models for biomolecular systems. *J Chem Phys*, 139(9):090901, 2013.

91. C Clementi, H Nymeyer, and JN Onuchic. Topological and energetic factors: What determines the structural details of the transition state ensemble and "en-route" intermediates for protein folding? An investigation for small globular proteins. *J Mol Biol*, 298(5):937–953, 2000.

92. MP Eastwood, C Hardin, Z Luthey-Schulten, and PG Wolynes. Statistical mechanical refinement of protein structure prediction schemes: Cumulant expansion approach. *J Chem Phys*, 117(9):4602–4615, 2002.

93. JD Bryngelson. When is a potential accurate enough for structure prediction? Theory and application to a random heteropolymer model of protein folding. *J Chem Phys*, 100(8):6038–6045, 1994.

94. JA Hegler, J Lätzer, A Shehu, C Clementi, and PG Wolynes. Restriction versus guidance in protein structure prediction. *Proc Natl Acad Sci USA*, 106(36):15302–15307, 2009.

95. BA Reva, AV Finkelstein, and J Skolnick. What is the probability of a chance prediction of a protein structure with an RMSD of 6 Angstrom? *Fold Des*, 3(2):141–147, 1998.

96. SS Cho, Y Levy, and PG Wolynes. P versus Q: Structural reaction coordinates capture protein folding on smooth landscapes. *Proc Natl Acad Sci USA*, 103(3):586–591, 2006.

97. M Chen, X Lin, W Zheng, JN Onuchic, and PG Wolynes. Protein folding and structure prediction from the ground up: The atomistic associative memory, water mediated, structure and energy model. *J Phys Chem B*, 120(33):8557–8565, 2016.

98. S Plimpton. Fast parallel algorithms for short-range molecular dynamics. *J Comp Phys*, 117:1–19, 1995.

99. DM Hinckley, GS Freeman, JK Whitmer, and JJ de Pablo. An experimentally-informed coarse-grained 3-site-per-nucleotide model of DNA: Structure, thermodynamics, and dynamics of hybridization. *J Chem Phys*, 139(14):144903, 2013.

100. GS Freeman, DM Hinckley, JP Lequieu, JK Whitmer, and JJ de Pablo. Coarse-grained modeling of DNA curvature. *J Chem Phys*, 141(16):165103, 2014.

101. B Zhang, W Zheng, GA Papoian, and PG Wolynes. Exploring the free energy landscape of nucleosomes. *J Am Chem Soc*, 138(26):8126–8133, 2016.

102. DA Potoyan, W Zheng, DU Ferreiro, PG Wolynes, and EA Komives. Pest control of molecular stripping of nfκb from DNA transcription sites. *J Phys Chem B*, 120(33):8532–8538, 2016.

103. MY Tsai, B Zhang, W Zheng, and PG Wolynes. Molecular mechanism of facilitated dissociation of Fis protein from DNA. *J Am Chem Soc*, 138:13497–13500, 2016.

104. DA Potoyan, W Zheng, EA Komives, and PG Wolynes. Molecular stripping in the nf-κb/iκb/DNA genetic regulatory network. *Proc Natl Acad Sci USA,* 113(1):110–115, 2016.

105. H Zhao, D Winogradoff, M Bui, Y Dalal, and GA Papoian. Promiscuous histone mis-assembly is actively prevented by chaperones. *J Am Chem Soc,* 138(40):13207–13218, 2016.

106. M Chen, W Zheng, and PG Wolynes. Energy landscapes of a mechanical prion and their implications for the molecular mechanism of long-term memory. *Proc Natl Acad Sci USA,* 113(18):5006–5011, 2016.

107. ER Kandel, Y Dudai, and MR Mayford. The molecular and systems biology of memory. *Cell,* 157(1):163–186, 2014.

108. SU Heinrich and S Lindquist. Protein-only mechanism induces self-perpetuating changes in the activity of neuronal Aplysia cytoplasmic polyadenylation element binding protein (CPEB). *Proc Natl Acad Sci USA,* 108:2999–3004, 2011.

CHAPTER 5

Elastic Models of Biomolecules

Qiang Cui

CONTENTS

5.1 Introduction 191
5.2 Computational Models 192
    5.2.1 Elastic Network Models for Biomolecules 192
    5.2.2 Continuum Mechanics Models for Proteins and DNA 194
    5.2.3 Elastic Models for Membrane 196
5.3 Applications 198
    5.3.1 ENMs, Normal Modes, and Structural Transitions 198
    5.3.2 FEMs and Mechanical Responses 201
    5.3.3 Elastic Models for Membranes 208
5.4 Concluding Remarks 211
    Acknowledgments 213
    References 213

5.1 INTRODUCTION

Despite the tremendous progress of atomistic simulations due to developments of novel hardware [1,2] and sampling algorithms [3–5], many biological problems can benefit from analysis using coarse-grained (CG) models [6–11]. In addition to the obvious advantage of being able to access larger length scales and longer time scales, another motivation for conducting CG simulations is to help better understand what properties are most essential to the phenomena of interest and what can be either neglected or significantly approximated [12]. CG models can be constructed for biomolecules at either a particle or continuum level. Since particle-based CG models are reviewed by many excellent chapters in this book and in recent literature [8,9,13], we will focus on models at the continuum level, including the elastic network model (ENM) that is particle in nature but closely related to the elastic description of materials.

In the following, we first briefly summarize the technical details of the relevant models. The value and limitations of these models are then discussed with the help of several applications; we will draw mainly from our own work although some relevant studies by others will also be referred to. The goal is not to provide a comprehensive review of continuum(-like) models for biological simulations, since the application of continuum models to certain biological problems, such as change of membrane morphology, has a long history [14–16]. Rather, we aim to illustrate that by integrating insights from particle-based simulations, continuum modeling can be applied to gain unique mechanistic insights into problems. Therefore, we hope that our chapter helps to stimulate further developments that better integrate particle and continuum levels of description for biomolecular applications.

5.2 COMPUTATIONAL MODELS

5.2.1 Elastic Network Models for Biomolecules

The ENMs are particle-based CG models for biomolecules. We include them here because these models typically feature a small number of parameters and their applications mostly focus on collective motions of biomolecules, thus ENMs share many features with a continuum mechanical model.

Pioneered by the work of Tirion [17], who observed that key features of protein fluctuations can be captured with a highly CG network model, the simplest ENM [18] treats each amino acid as a single bead, and the beads interact with each other through a set of harmonic springs (Figure 5.1):

$$U_{\text{ENM}}(\mathbf{R}; \mathbf{R}^0) = \frac{1}{2} \sum_{i<j} k_{ij}(R_{ij} - R_{ij}^0)^2 \Theta(R_c - R_{ij}^0) \qquad (5.1)$$

Here, k_{ij} is the force constant, R_c is the cutoff beyond which the beads no longer interact (Θ is the Heaveside function), and $\mathbf{R}^0$ indicates the reference structure. A closely related model proposed by Bahar and coworkers [19] is the Gaussian network model (GNM), in which the potential function takes the similar form,

$$U_{\text{GNM}}(\mathbf{R}; \mathbf{R}^0) = \frac{1}{2} \sum_{i<j} k_{ij}|\mathbf{R}_{ij} - \mathbf{R}_{ij}^0|^2 \Theta\left(R_c - R_{ij}^0\right) \qquad (5.2)$$

FIGURE 5.1   An elastic network model (ENM) representation of the protein lysozyme (PDB id: 3LZT) with $R_c = 8$ Å.

With these potential functions, the equilibrium properties of a biomolecule are analyzed with normal mode analysis (NMA) [20], which involves the diagonalization of the (mass-weighted) hessian (second-derivative) matrix of the potential function. As discussed in the literature [21], GNM leads to isotropic fluctuations while ENM gives rise to more realistic anisotropic thermal fluctuations. Therefore, while GNM is well suited for analyzing the magnitude of protein fluctuations, ENM is more useful for providing insights into the dominant collective motions of biomolecules.

In the original models [17–19], the force constants k_{ij} are taken to adopt a uniform value; if only normal mode eigenvectors are of interest, the numerical value of the constant is irrelevant. Thus, the only parameter in the model is R_c, whose appropriate value depends on the resolution of the model. With a single-bead-per-residue resolution, the appropriate value is about 8–10 Å (see Section 5.3.1).

The ENM has been extended in diverse and creative ways for different purposes in recent years. For example, k_{ij}s are no longer uniform constants; they are either distance/environment dependent [22–24] or fitted by matching residual fluctuations from atomistic simulations [25,26]. By systematically perturbing the force constants and analyzing the impact on the eigenvectors, it is possible to identify connections that are particularly important to functionally relevant collective motions and thus protein allostery [27]. Periodic boundary condition has been incorporated to properly reflect the effect of crystal contacts on protein fluctuations and for a better comparison with x-ray diffraction data (see example below) [28–30]. Anharmonic potentials have also been introduced so that "springs" between residue can be broken; this revision is essential in applications to large-scale conformational transitions [31].

5.2.2 Continuum Mechanics Models for Proteins and DNA

The particles in ENM can be regarded as the nodes of a continuum mechanics model in a mesh (finite element) representation. Depending on the level of local connectivity, the effective elasticity of a protein is spatially heterogeneous. Several researchers have also developed models for biomolecules explicitly in the framework of elasticity. For example, several groups have developed elastic models for DNA and used them to describe (super)coiled forms of DNA [32–34]. Sun and coworkers developed elastic models for α helices and coiled coils [35,36]; they applied the models in the analysis of biomolecular motors and muscle filaments [37].

For more heterogeneous biomolecular structures, numerical continuum models are necessary. Bathe has explored the use of finite element model (FEM) representation of proteins and analyzed their normal modes [38]; the results are generally consistent with atomistic and ENM simulations, again highlighting the importance of protein shape in determining the nature of collective motions [39,40]. Together with the Chen group, we have explored the use of FEM for the analysis of mechanosensitive channels (MscL) [41–43]. The protein and lipid membrane are represented by tetrahedron elements, whose materials properties are, in general, heterogeneous and calibrated based on either atomistic simulations or available experimental data. The size of the elements can be determined adaptively, small for regions of interest and large for far-away areas, making the simulation framework ideal for very large systems such as an ion channel embedded in a large sheet of membrane. The interactions between elements are generally short ranged and can be parameterized based on atomistic simulations. Therefore, once parameterized, the FEM is ideally suited for studying the structural response of the biomolecule to diverse external mechanical perturbations of different form and scale [42].

Since most biomolecular systems are surrounded by solvents, we and others have attempted to integrated continuum mechanics and continuum solvent models [44–46]. For this purpose, we introduced the molecular surface FEM models [47]. Taking a short DNA as an example (Figure 5.2), the molecular surface is first calculated and triangularized with a probe water radius of 1.4 Å; the triangularized surface is then simplified to reduce the number of the surface triangles, and the volume enclosed by this simplified surface is subsequently discretized into a three-dimensional (3D) mesh consisting of tetrahedral elements with the heavy atoms of DNA. The final FEM model (Figure 5.2e) consists of 2,655 nodes and 13,728 finite elements. The mapping between heavy atoms of DNA and specific FEM nodes makes it straightforward to define "chemical nodes" (red nodes in Figure 5.2f, in

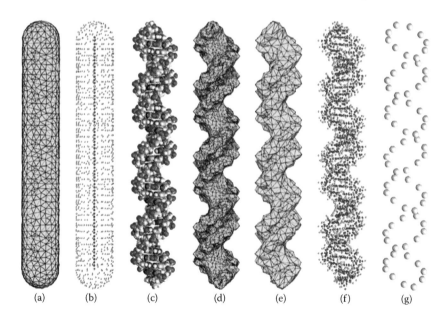

(a) (b) (c) (d) (e) (f) (g)

FIGURE 5.2 Illustration of continuum models for a short (36 base pairs [bp]) B-DNA [47]. (a and b) Continuum level poly-ion model. The triangularized surface is shown in (a) and the "chemical nodes" that bear the partial charges for continuum solvation calculations are shown in red in (b). (c–g) Various steps in constructing the molecular surface continuum model. (c) shows the atomistic model with van der Waals surface, which is then triangularized in (d) and simplified in (e). The chemical nodes are shown in (f), and the "phosphorus nodes" that bear the partial charges are shown in (g).

total 1,470), which encode the key chemical characteristics of the molecule (i.e., charge distribution and solute/solvent interface) that are required in electrostatics solvation calculations [48]. During the continuum mechanics calculations, forces due to elastic properties, inter-element interactions, and solvation are combined to solve for the response of the system to external perturbations.

5.2.3 Elastic Models for Membrane

Lipid membrane, which constitutes the boundary of cells and cell organelles, was once regarded as merely a passive medium for transmembrane proteins. It has been increasingly recognized that the coupling between protein and lipid membrane is essential to many biological processes such as signal transduction, ion channel gating, and organelle formation [49–51]. Although some of the effects are local in nature, such as the roles of annular lipids [52,53], many membrane-mediated processes implicate a rather long length scale that ranges from several nanometers [49] to submicrons [54,55]. For such problems, continuum mechanics models are appropriate. Often referred to as the Helfrich model [14–16], the simplest form of the free energy of an elastic membrane sheet is given by

$$\Delta G_{\text{bend}} = \frac{B}{2} \int_S \left(\frac{1}{R_1} + \frac{1}{R_2} - C_0 \right)^2 dS \qquad (5.3)$$

where R_1 and R_2 indicate the local principal radii of curvature of a neutral surface, C_0 the spontaneous curvature of a lipid monolayer, and B the flexural rigidity of the membrane. More sophisticated models that include additional free energy components have been proposed and employed in the literature to study biomembrane structure and dynamics [56–58].

Although Equation 5.3 is continuum in nature and therefore expected to be valid for only large length scales, our recent analysis [59] indicated that the model remains surprisingly consistent with a particle-based description even for highly curved membranes where the radius of curvature approaches the length of a lipid molecule (see Figure 5.3 for an illustration and discussion below). Indeed, the Helfrich type of model has been successfully applied to a broad range of membrane remodeling/morphology problems that implicate significant membrane curvatures [60,61].

For problems that involve both membrane and proteins, the continuum model can be revised or adapted in different ways. For membrane-centric applications, such as lipid vesicle remodeling by proteins, the effect

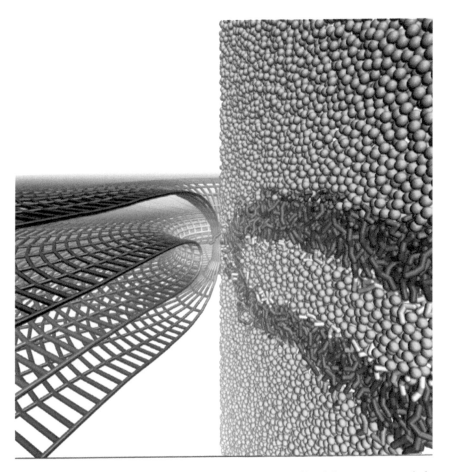

FIGURE 5.3 A schematic illustration for the comparison of lipid fusion pore morphology predicted from coarse-grained (CG) particle (MARTINI) simulations and continuum mechanics minimizations [59]. For quantitative results, see Figure 5.8.

of the protein can be taken into consideration by modifying local membrane materials properties (e.g., B and C_0) [62–64]. For applications where the structural and chemical features of the protein are essential, such as insertion of transmembrane proteins into the membrane, it is possible to integrate the elastic treatment of membrane and the continuum electrostatic model for the solvation of the protein; the key technical details concern the choice of boundary conditions [65–67] for the elasticity problem at the protein–membrane interface and dielectric boundary conditions for the continuum electrostatic modeling [44,45,68]. For example, Grabe and coworkers have implemented an efficient elasticity solver [46] in the popular continuum electrostatic program known as the adaptive

Poisson–Boltzmann solver (APBS) [48] for studying various transmembrane protein problems that implicates local membrane deformations induced by, for instance, charged residues. Alternatively, one may also represent the continuum membrane numerically with FEMs.

5.3 APPLICATIONS

In the following, we discuss several recent applications of the models discussed above to illustrate the unique value as well as limitations of these CG models.

5.3.1 ENMs, Normal Modes, and Structural Transitions

As many studies illustrated in the literature [21,40,69,70], there is often a high degree of overlap between the low-frequency normal modes and structural differences between various states of a protein. Thus by analyzing the nature of the low-frequency modes, it is possible to gain insights into structural transitions relevant to the functional cycle of large proteins. It remains difficult to quantitatively predict structural transitions based on low-frequency modes alone, but using these as a basis set together with additional experimental data (e.g., fluorescence resonance energy transfer, small angle x-ray scattering, or electron microscopy) can be very effective at predicting new structural states, especially at intermediate or low resolutions [71–73].

At these resolutions, CG models like ENM are particularly well suited because they require minimal structural information and have a very small number of parameters. Moreover, it has been well recognized that collective properties such as low-frequency normal modes tend to be insensitive to the fine details of the model [39,40,74]. For example, for a complex structure like the ribosome, the first batch of low-frequency normal modes from ENM and an atomistic model overlap strikingly well [75]. However, one has to be careful about the treatment of poorly connected or loosely packed regions (e.g., the so-called "tip effect" [76,77]). For instance, as shown in our previous work of relatively small RNA molecules [78], the degree of overlap between eigenvectors computed using ENM and those computed based on atomistic models (either harmonic [79] or quasi harmonic [80] analysis) is high for a well-packed riboswitch, but substantially reduced for the poorly packed hammerhead ribozyme.

One unique advantage of ENM is that the reference structure, by design, is an energy minimum. Therefore, ENM calculations can be done at any structure, while force field–based normal mode calculations should only be done following a stringent energy minimization. This feature of ENM has been taken advantage of in several studies that explore structural transitions, either for the purpose of low-resolution structural refinement [71] or for the exploration of low-energy transition pathways [81]. For example, we have explored the use of ENM in the analysis of a MscL of large conductance (MscL), which acts as a "safety valve" in bacteria under osmotic stress [82–84]. The closed structure of MscL was solved using x-ray crystallography [85] (see Figure 5.6a), while the open state has only been characterized with spectroscopic techniques [86] and conductance measurements [87,88]. The gating transition occurs in the millisecond time scale, making unbiased atomistic simulations difficult even with modern computational hardware [89,90]. In Section 5.3.2, we briefly review our study of the MscL gating transition using a finite element approach, but a simpler model is to treat the protein with an ENM. In our study [42,43], the potential function took the form

$$V(\mathbf{x}, \mathbf{s}) = \frac{1}{2}k^{\text{cov}} \sum_{i<j} (R_{ij} - R_{ij}^0)^2 \Theta(R_c^{\text{cov}} - R_{ij}^0)$$

$$+ \frac{1}{2}k^{\text{noncov}} \sum_{i<j} (R_{ij} - R_{ij}^0)^2 \Theta(R_c^{\text{noncov}} - R_{ij}^0)\Theta(R_{ij}^0 - R_c^{\text{cov}})$$

$$+ \frac{1}{2}k^{\text{pull}} \sum_{i \in S} (\mathbf{s}_i - \mathbf{x}_i)^2$$

Here, we treat the covalently bonded ($R_c^{\text{cov}} = 1.7$ Å) and nonbonded ($R_c^{\text{noncov}} = 5.0$ Å) connections using different force constants, whose values are determined by matching atomistic normal mode calculations. The last term represents a pulling energy that mimics the effect of membrane tension; this is implemented by attaching harmonic springs to a selected number of atoms (the "pulling set" S, see Figure 5.4 caption) and gradually moving the end positions of the springs ($\mathbf{s}_i$).

During such pulling calculations, one way to predict the protein displacement ($\delta\mathbf{x}_i$) is to minimize the energy function above after each update of $\mathbf{s}_i$. Alternatively, one may use a second-order expansion of the energy function and solve for $\delta\mathbf{x}_i$ from the following matrix equation:

$$\begin{pmatrix} \mathbf{Q}_{00} + k^{\text{pull}}\mathbf{I} & \mathbf{Q}_{01} \\ \mathbf{Q}_{10} & \mathbf{Q}_{11} \end{pmatrix} \begin{pmatrix} \delta\mathbf{x}_{1\cdots m} \\ \delta\mathbf{x}_{m+1\cdots N} \end{pmatrix} = \begin{pmatrix} k^{\text{pull}}\delta\mathbf{s} \\ 0 \end{pmatrix} \qquad (5.4)$$

where $\mathbf{Q}$ indicates the Hessian matrix for the potential function excluding the last spring term, and subscripts "0" and "1" indicate atoms in and out

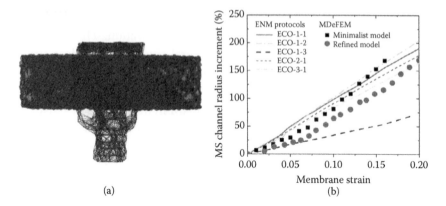

(a)   (b)

FIGURE 5.4   Analysis of the gating transition of MscL using ENM [42,43]. (a) An illustration of the ENM for a MscL embedded in a lipid bilayer. (b) Radius of the channel pore
during the pulling simulation using different pulling sets. In protocols 1-1, 1-2, and 1-3,
selected backbone atoms [1-1: TM1 (16–40) and TM2 (77–106); 1-2: $|z| < 20$ Å; 1-3: TM2
(77–106)] are pulled explicitly in a radially outward fashion. In protocols 2-1 and 3-1, an
elastic membrane bilayer is included explicitly and the outer layer (2 Å thick) of lipid atoms
are pulled in a radially outward fashion; protocol 2-1 solves Equation 5.4, while protocol 3-1
minimizes the total energy explicitly.

of the pulling set S (which contains m atoms out of the total number N),
respectively.

We explored several different pulling strategies, with an ENM that
includes only the MscL protein or the MscL embedded in a lipid bilayer
(Figure 5.4a). The details are given in Refs. [42,43] and some representative
results are shown in Figure 5.4b, which describes the pore radius increase
as a function of the strain imposed on the elastic membrane. In protocols
where a subset of the protein atoms are directly pulled to mimic membrane
tension (ECO-1-1/1-2/1-3 sets), the results are, as expected, sensitive to
the choice of the pulling set. For example, when only the transmembrane
helices 2 (TM2) are pulled (ECO-1-3), the channel opens up very slowly;
with TM1s also explicitly pulled, the channel opens up much faster, in
agreement with FEM simulations and also predictions of structural models
[88]. The differences clearly highlight the importance of TM1-membrane
interaction during the gating transition. Indeed, when the membrane is
also represented with an elastic network (ECO-2-1/3-1 sets), the results
are similar to those obtained with the FEM approach (see below) using a
minimalist model for the channel but differ from the more refined FEM calculations. This is expected considering that the elastic properties of the protein depend entirely on the density of connectivity and not on the chemical

nature of the underlying amino acid composition. This comparison highlights that while qualitative mechanical response can be captured with a simple ENM description, more quantitative results can only be obtained with a more heterogeneous treatment of the materials properties.

Before leaving this section, we mention that many ENM studies compare computed vibrational properties with x-ray crystallographic data such as the Debye–Waller factors (the "B" factors). To make such a comparison rigorously valid, it is important to include the crystalline environment in the ENM calculations. This was first done by Phillips and coworkers [91] who included the first set of neighbors of a protein in a crystal and then performed normal mode calculations using GNM. More recently, Riccardi et al. implemented the Born–von Kármán boundary condition in ENM-based NMA [29]. These and related studies [28,30] indeed indicated that including the crystalline environment may have a significant impact on the predicted motions and related properties, such as the covariance matrix that characterizes correlations among residues (Figure 5.5a through c). Including the proper boundary condition is also important to a better understanding of diffuse scattering [92] and refinement of crystal structures at medium resolution [93].

As a specific example, in Figure 5.5d, the correlation between experimental isotropic temperature factors and computed ones using ENM with different cutoffs and boundary conditions is compared. When an isolated protein molecule is used, it appears that the highest correlation with experimental temperature factors is obtained with a very large cutoff (16 Å); the best correlation is about 0.47 when only the lowest frequency (5%) modes are included and about 0.57 when all modes are included. When the crystalline environment is properly described (e.g., with P1PBC), the correlation is already about 55% when only the lowest frequency modes are included and the correlation actually decreases with the cutoff distance. Similar trends are also observed for the comparison of anisotropic displacement factors [29]. Further analysis of other properties, such as vibrational density of states and vibrational contribution to protein heat capacity, indicated that ENMs with a very large cutoff give unphysical descriptions. Thus, comparison of ENM results for isolated proteins to experimental x-ray diffraction data should be taken with care.

5.3.2 FEMs and Mechanical Responses

The application of ENM to MscL clearly illustrated that for more quantitative analysis of mechanical response of biomolecules, a more flexible

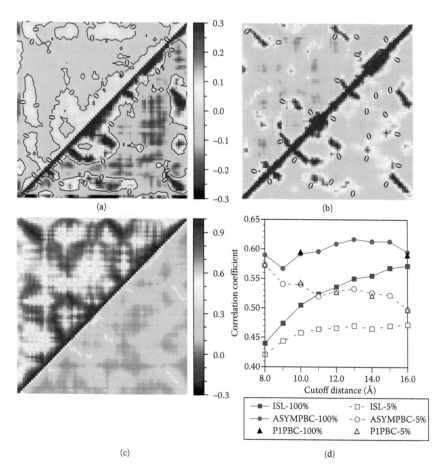

FIGURE 5.5 Normal mode analysis (NMA) with Cα-based ENMs in the crystalline environment [29]. (a–c) Comparison of residual correlations computed for the PDZ2 domain of syntenin (PDBID: 1R6J). (a) An isolated (ISL) protein calculation with ENM using a homogeneous force constant (upper triangle) versus the HCA model [23] (lower triangle). (b) ENM calculation with the HCA model using two different ways of treating the crystalline environment: ASYMPBC (upper triangle, including the hessian contributions from protein chains in the same asymmetric unit) and P1PBC (lower triangle, including the hessian contributions from all protein chains in the unit cell). (c) Born–von Kármán boundary condition calculations: correlation computed using only the three acoustic modes (upper triangle) versus using all modes (lower triangle). (d) The correlation coefficient between ENM-computed and experimental isotropic temperature factors as a function of the ENM cutoff distance (R_c) using various ways of treating the crystal environment for a set of 83 proteins; the percentile indicates the number of modes included in computing the temperature factors. The results illustrate that including the crystal environment clearly improves the agreement with experiment.

continuum mechanics framework is needed to describe their heterogeneous materials properties and potentially complex mechanical stimulations. For this purpose, in collaboration with the Chen group, we have adopted the finite element approach that has been widely applied in the (solid) mechanics community [41–43]. In particular, the materials properties of different protein components and the interactions among them can be derived based on atomistic molecular dynamics (MD) simulations; the degree of heterogeneity depends on the problem and is easily controlled. For MscL, for example, an intermediate resolution model is illustrated in Figure 5.6a. The key transmembrane helices (TM1, TM2) and three other short helices (S1, S2, and S3) are treated with similar mechanical properties (Young's moduli and Poisson ratio), with the exception that the segment containing Pro43 adopts different parameters to allow helical kinks to develop. The loops that connect the helices are treated with thin mechanical springs and the materials properties are parameterized by matching normal modes at the atomistic and continuum levels. Different continuum models for the lipid membrane can also be adopted [42].

Some representative results of our FEM studies are summarized in Figure 5.6. At this stage, the results are still semiquantitative in nature because solvation effects are not yet included (see below) and the protein model has only a limited degree of compositional heterogeneity.

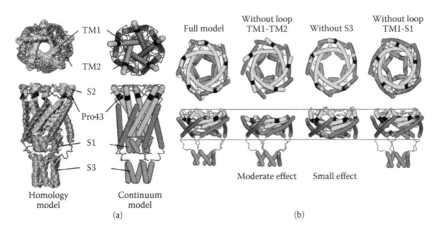

FIGURE 5.6 Representative results for the FEM simulation of the MscL gating transition under equalbiaxial tension [41–43]. (a) A comparison of the homology model for the *E. coli* MscL in the closed state and the FEM representation. (b) Predicted open state models by FEM simulations for the full protein model and several constructs with selected elements removed. *(Continued)*

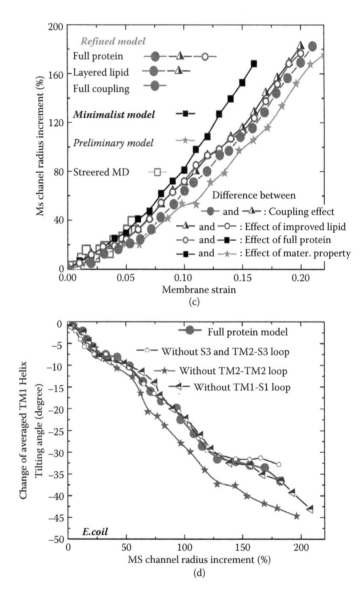

FIGURE 5.6 (*Continued*) Representative results for the FEM simulation of the MscL gating transition under equalbiaxial tension [41–43]. (c) The evolution of the channel pore as a function of the membrane strain computed with continuum models of different levels of sophistication; the steered molecular dynamics (MD) results of Ref. [89] are also included for comparison. (d) The average tilting angle of TM1 helices as a function of the membrane strain for the models shown in panel (b).

Nevertheless, the value of the FEM framework can already be demonstrated. The general trends in the structural evolution of the channel during the gating transition are consistent with expectations from experimental studies [87,94] and several computational analyses [89,95,96]; the pore opens in an iris fashion and the transmembrane helices (TM1, TM2) undergo significant tilts; the S1 helices are pulled open and become part of the pore in the open state. Notable bending in the transmembrane helices are also observed; although the magnitude of the bending is difficult to compare directly with experiments, deformations in these helices are also proposed in structural models constructed based on available experimental data [88]. Compared to atomistic simulations (steered MD [89]), the FEM simulation readily accesses more significant opening of the channel with a fraction of the computational cost. The computational efficiency of the FEM framework makes it possible to repeat calculations using continuum models of different levels of complexity. By comparing the results from those calculations, one gains insights into the relative importance of different factors. For example, as shown in Figure 5.6c, the largest impact on the degree of channel opening at a given membrane strain is observed when the full protein is included (as opposed to only the transmembrane helices in a "minimalist model") and when the protein materials properties are parameterized based on system-specific atomistic simulations (as opposed to taking generic parameters from the literature [35] in the "preliminary model"). By contrast, other simulation details such as the refined treatment of the bilayer properties (treating the bilayer as a three-component rather than a homogeneous sheet) contribute less significantly.

As another example of the analytic value of the FEM framework, we show simulation results with specific continuum mechanical elements excluded from the simulations. These were inspired by relevant mutation experiments in which loops connecting different helices are either truncated or cut [97–99]. As shown in Figure 5.6b and d, removing the S3 helices had the smallest impact on the gating transition. Removing the loops that connect TM1/TM2 or TM1/S1 leads to larger changes although the magnitude remains somewhat subtle. When the TM1/S1 loops are removed, without the "guidance" of these linkers, the pore formed by the S1 helices becomes distorted; removing the TM1/TM2 loops reduces the average TM1 tilting angle by about 10° and thus leads to a different pore size. Therefore, these simulations suggest that the inter-helical loops are not crucial to gating yet make some contributions; S3 helices, by contrast, are unlikely to be explicitly involved in gating. These insights are qualitatively consistent with experimental observations; for example, after the cleavage of the interhelical loops, MscL was observed to be still functional yet the mechanosensitivity was enhanced, suggesting that the loops act as springs that resist the opening of the channel [97]. Other unique aspects of FEM simulations not discussed here include the simulation of MscL response to different mechanical perturbations, such as to membrane bending, different laboratory experiments (patch clamp vs. nano-indentation), and the presence of another MscL channel nearby [43].

It should be pointed out that the membrane strain required to completely open the MscL channels in even the most refined FEM that we have published so far remains rather high compared to the available experimental data; moreover, the pore opening is predicted to be largely monotonic in nature rather than going through multiple substates as measured experimentally [100,101]. We understand that these limitations are due to the fact that solvation effects have not been included in the reported FEM study; solvation of the channel, which exposes a mixed set of hydrophobic and hydrophilic residues upon opening, is expected to stabilize the open state and therefore reduce the required membrane tension [101]. Therefore, ongoing efforts have focused on an efficient integration of continuum mechanics and continuum solvation models in the FEM framework.

As a proof-of-concept study, DNA bending under different salt concentrations [102–107] has been analyzed [47]. We established a molecular surface-based FEM representation of B-type DNA (Figure 5.2), which also contain "chemical nodes" that encode information required in the continuum solvation calculations. In the FEM simulations, one end of the DNA is subject to an external pulling force (Figure 5.7a), and the equilibrium shape of the DNA is solved by considering force contributions from the DNA materials properties and continuum solvation; the latter is treated with either a simple Debye–Hückel model based only the charges of the chemical nodes (as in traditional theoretical models for DNA [103,106,107]) or a more detailed nonlinear Poisson–Boltzmann model that also considers the molecular surface. The equilibrium shape change due to external force is used to extract the persistent length of the DNA under different salt concentrations.

As shown in Figure 5.7, the derived persistence length from such a "force-based" protocol depends linearly on the inverse of the salt concentration; we note that a renormalized salt concentration was derived in our work to consider the finite length of the DNA studied (the example shown here contains 300 bp). The result does not vary significantly when the

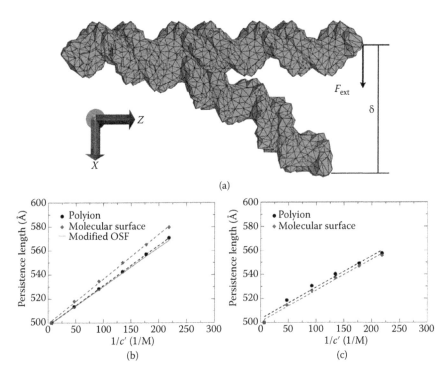

(a)

(b)

(c)

FIGURE 5.7 Representative results for the FEM simulation of B-DNA bending at different salt concentrations [47]. (a) A scheme that illustrates the bending of DNA in the $x − z$ plane by an external force $\vec{F}_{ext}$ at the right terminus. The displacement δ at the right terminus in the x-direction is exaggerated for clarity. (b) Predicted persistence length (P) versus $1/c'$ (renormalized inverse salt concentration) for the polyion and molecular surface models of a 300 bp DNA from (b) energy-based and (c) force-based simulations. The Debye–Hückel model is used for the electrostatic interactions, and the Young's modulus of DNA is parameterized such that the persistence length at 0.15 M NaCl is 500 Å.

molecular surface model is replaced by the simpler "poly-ion" model that describes the DNA simply as a rod (Figure 5.2a); this is likely because a rather simple scheme is used to represent the charge distribution of the DNA (Figure 5.2g). With more complex charge distributions, the molecular surface representation is found to be more reliable; however, we also noted that the molecular surface representation may lead to considerable numerical noise in the computed solvation force and therefore convergence issues for systems with a high degree of anisotropy. At a quantitative level, the numerical results are consistent with available experimental data. For example, the force-based protocol predicts that the persistence length increases from 50 to ~56 nm when the salt concentration is varied from 0.15 to 0.004 M; the trend compares favorably with experimental measurements [105] of

45±2 nm and 58±2 nm for the two conditions, respectively. The results in Figure 5.7 also show that the force-based protocol leads to persistence length in good agreement with an energy-based analysis that considers the free energy of DNA under uniform bending, as assumed in the Odijk–Skonick–Fixman (OSF) theory [106,107] (modified to consider the finite DNA length [47]). This is a valuable validation because the force-based protocol is more general and can treat complex mechanical loadings such as those involved in protein induced DNA looping [108].

5.3.3 Elastic Models for Membranes

An example that illustrates the complementarity of continuum mechanical (elastic) and particle models for membranes concerns our recent study of membrane fusion pores (see Figure 5.3 for a schematic illustration). Since membrane fusion pores involve bilayers of large curvature, one motivation of comparing continuum and particle models is to establish the applicability of continuum models to membrane remodeling problems. Accordingly, we compared [59] the shape of membrane fusion pores using atomistic (CHARMM36 [109]), CG (MARTINI [110]), and linear elastic models. The fusion pore, however, is difficult to converge in particle-based simulations and the shape experiences slow fluctuations even after microseconds of simulations at the MARTINI level (Figure 5.8b). Therefore, we focused on the comparison of continuum and MARTINI models. Specifically, we first simulate the fusion pore using MARTINI with the bilayers subject to different asymptotic distance separation restraints (R_b values). The neutral surfaces for the converged fusion pore are then fitted to a polynomial, which is then subject to further minimization using an elastic model described by Equation 5.3.

As shown in Figure 5.8a, the agreement between continuum and MARTINI in terms of the shape of the fusion pore is remarkable; they clearly illustrate that the fusion pore structure is neither toroidal [111–113] nor curvature free [114] as proposed in earlier studies. As the R_b value decreases, the curvature of the fusion pore increases, and the difference between continuum and MARTINI is visible only with the smallest R_b value studied (4.85 nm). With such R_b value, the maximum square mean curvature of the inner monolayer in the fusion pore is about 0.2 nm^{-2}, thus the radius of curvature approaches the thickness of a monolayer. Even for these cases, the bending free energy difference between the pore shape predicted by the MARTINI and continuum models is on the order of 20 $k_B T$ or less. Therefore, applicability of the continuum model is remarkably robust.

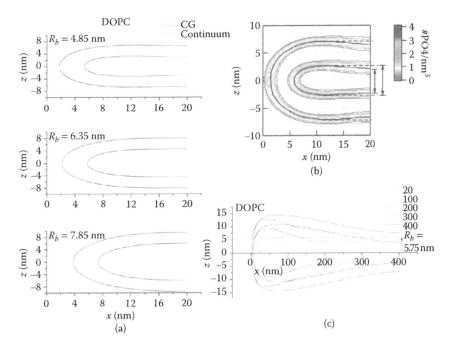

FIGURE 5.8 A comparison of membrane (DOPC) fusion pores at the continuum (Equation 5.3) and particle (MARTINI) levels [59]. (a) Neutral surfaces obtained from MARTINI simulation results are compared to those further minimized by the continuum model. (b) Phosphate density maps of the fusion pore from MARTINI simulation. The dashed horizontal line highlights the maximum distance between the monolayers on each side, and the vertical arrows to the right highlight the reduction in this distance due to bowing. (c) Continuum model fusion pores of systems of various sizes, illustrating that increasing the size allows for profound bowing during energy minimization.

Some of the MARTINI simulations with small R_b values led to an unexpected "bowing feature"; as shown in Figure 5.8b, separation of the fusing bilayers close to the central fusion pore is larger than the asymptotically flat regions. However, due to the slow relaxation of the fusion pore, it was difficult to evaluate whether the bowing feature is realistic based on the MARTINI simulations alone. Remarkably, the shape with the bowing feature was also observed in the continuum minimization as the solution with the lower free energy, once the functional form that parameterizes the continuum membrane was modified to allow nonmonotonic solutions. Clearly, by allowing bowing, curvature for regions near the fusion pore is reduced and therefore the bending free energy is substantially lowered. Moreover, as the length scale of the membrane involved in the fusion pore was further expanded to hundreds of nanometers, which remain difficult to access using even MARTINI type of models, the continuum simulations led to even more dramatic bowing features (Figure 5.8c). These results suggest that bowing can be an effective mechanism to lower the free energy cost of fusion pore formation, especially when a large patch of membrane is involved; for protein-assisted fusion, this feature might provide greater areas for the proteins to work on. This bowing feature was recently confirmed by an independent study by Cohen and coworkers [115].

Although the continuum model for membrane is applicable to a broad set of problems, we also caution that the continuum approximation can become the problematic for the protein/membrane interface, for which the discrete nature of annular lipids may become important [52,66]. As an example, we study the association of two gramicidin A (gA) dimers in bilayers that feature different degrees of hydrophobic mismatch with both MARTINI and continuum models [116]. With MARTINI calculations (Figure 5.9b), the association is strongly favored by the magnitude of hydrophobic mismatch: with a small degree of mismatch (DMPC), the association is only about 2 kcal/mol; with a high degree of mismatch (DSPC), the dimerization of two gA channels is more than 10 kcal/mol. At long range (see inset of Figure 5.9b), however, there is a soft barrier of about ~1 kcal/mol for the association in all three cases.

The qualitative g in the association PMF can be understood within the framework of continuum model for the bilayer. Indeed, association of proteins driven by the release of local membrane strain due to hydrophobic mismatch has been thoroughly studied experimentally [53,118] and theoretically [65]. Specifically, by computing the membrane strain energy,

$$E_{\text{strain}} = \int_{r_1, r_2 > 10 \text{Å}} \frac{K_A}{2} \left(\frac{u(x, y)}{h_0} \right)^2 dx dy \qquad (5.5)$$

which involves only membrane beyond the annular lipid shell, we showed that the small barrier can be captured with a continuum model, similar to previous studies [65]; the physical nature of the barrier is the higher degree of membrane compression between the gA dimers [116]. With the same type of continuum calculation, however, the magnitude of the association free energy is substantially smaller than the prediction of the MARTINI model, clearly highlighting the contribution from annular lipids. As illustrated in Figure 5.9c, the annular lipids have rather different orientational properties compared to bulk lipids; the directional cosines indicate that

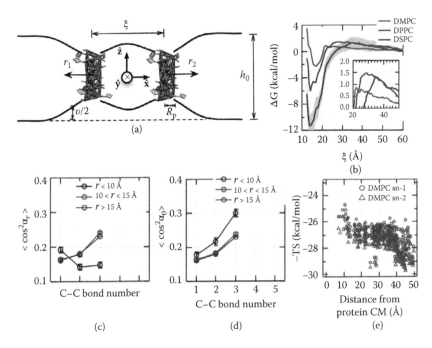

FIGURE 5.9 Analysis of gramicidin A (gA) dimer association in the presence of hydrophobic mismatch [116]. (a) Schematic illustration of the coordinate system used to characterize the association of two gA dimers and analysis of membrane stress/strain. (b) Computed potential of mean force for three degrees of hydrophobic mismatch; the inset zooms in the barrier region. (c) The orientation of the annular lipids are substantially different from the bulk, as reflected by the directional cosines for the lipid tail C–C bonds. Accordingly, the configurational entropy [117] computed for the annular lipids are also notably different. Neglect of the annular lipid properties in the continuum model is the reason that gA dimer association is difficult to describe quantitatively at the continuum level.

the annular lipid tails are largely tangential to the protein/lipid interface, unlike the umbrella ensemble sampled in the bulk. Accordingly, the configurational entropy for annular lipids are also substantially different. Therefore, releasing the annular lipids upon gA dimer association is entropically favored, and this effect is not captured by continuum models.

5.4 CONCLUDING REMARKS

In this short chapter, we have briefly reviewed continuum mechanics type of models for biomolecules, including the ENMS, finite element continuum mechanical models, and elastic sheet models for lipid membranes. The application of these models is illustrated using recent studies from our own research. In addition to highlighting the fact that continuum mechanical models can be applied to a range of problems in biophysics, we also hope to emphasize the value of examining the same problem using both particle and continuum models; by comparing results from particle and continuum models, one is able to gain new physical insights regarding trends that are robust to approximations as well as contributions that are sensitive to specific details.

This point is illustrated here with both the membrane fusion pore problem and gA dimer association. In the fusion pore study, the bowing feature was difficult to establish using either the particle (MARTINI) or the continuum (Helfrich) model alone: it was observed first in the MARTINI simulations, which motivated us to revise the functional form that parameterizes the membrane shape to allow non-monotonic solutions. On the other hand, interpretation of the MARTINI simulations was complicated by the slow relaxation of the fusion pore, thus the significance of the bowing feature was only established and its origin elucidated by analyzing the bending energy at the continuum level. In the study of gA dimer association, by studying what features of the MARTINI free energy profile can or can not be captured by a continuum analysis, new insights were gained regarding physical factors that dictate the barrier and driving force for the association; again, due to limitations in decomposing free energies, such insights would have been difficult to obtain based on MARTINI simulations alone. In turn, analysis of the MARTINI results explained limitations of standard continuum models for treating protein/lipid interfaces.

As a future development, further integration of continuum mechanics models with computational protocols at different length and temporal scales is an exciting direction to pursue. For example, the combination of continuum mechanics and continuum electrostatics attempted in our recent work only serves as a starting point and additional developments are needed to make such computations efficient and numerically robust. Integration between continuum mechanics and atomistic simulations has been reported in several studies of protein–DNA complexes [119,120]; we anticipate such models are powerful for the analysis of other protein machines involved in manipulating cellular components. Finally, including dynamical elements in the continuum modeling framework such that solvent/lipid flows can be included would help further broaden the scope of biological problems that can be addressed with multiscale computations.

ACKNOWLEDGMENTS

We appreciate the generous support from the National Science Foundation and the National Institutes of Health over the past few years for our developments in the general area of multiscale simulations (CHE-095728, CHE-1300209, R01-GM071428, R01-GM084028, and R01-GM106443). We also acknowledge all coworkers who have contributed to the work discussed here, especially Prof. G. N. Phillips, Jr. on ENM models, Prof. X. Chen on FEM modeling of biomolecules, and Prof. M. B. Jackson on the application of continuum models to membranes. Computational resources from the Extreme Science and Engineering Discovery Environment (XSEDE), which is supported by NSF grant number OCI-1053575, are greatly appreciated; computations are also supported in part by NSF through a major instrumentation grant (CHE-0840494) to the chemistry department, and the graphics processing unit (GPU) computing facility supported by the Army Research Office (W911NF-11-1-0327).

REFERENCES

1. R. O. Dror, R. M. Dirks, J. P. Grossman, H. F. Xu, and D. E. Shaw. Biomolecular simulation: A computational microscope for molecular biology. *Annu. Rev. Biophys.*, 41:429–452, 2012.
2. K. J. Kohlhoff, D. Shukla, M. Lawrenz, G. R. Bowman, D. E. Konerding, D. Belov, R. B. Altman, and V. S. Pande. Cloud-based simulations on Google Exacycle reveal ligand modulation of GPCR activation pathways. *Nat. Chem.*, 6:15–21, 2014.
3. D. M. Zuckerman. Equilibrium sampling in biomolecular simulations. *Annu. Rev. Biophys.*, 40:41–62, 2011.
4. E. Weinan and E. Vanden-Eijnden. Transition-path theory and path-finding algorithms for the study of rare events. *Annu. Rev. Phys. Chem.*, 61:391–420, 2010.
5. C. R. Schwantes, R. T. McGibbon, and V. S. Pande. Perspective: Markov models for long-timescale biomolecular dynamics. *J. Chem. Phys.*, 141:090901, 2014.
6. V. Tozzini. Coarse-grained models for proteins. *Curr. Opin. Struct. Biol.*, 15:144–150, 2005.
7. M. G. Saunders and G. A. Voth. Coarse-graining of multiprotein assemblies. *Curr. Opin. Struct. Biol.*, 22:144–150, 2012.
8. W. G. Noid. Perspective: Coarse-grained models for biomolecular systems. *J. Chem. Phys.*, 139:090901, 2013.
9. S. J. Marrink and D. P. Tieleman. Perspective on the Martini model. *Chem. Soc. Rev.*, 42:6801–6822, 2013.

10. A. Liwo, C. Czaplewski, S. Oldziej, and H. A. Scheraga. Computational techniques for efficient conformational sampling of proteins. *Curr. Opin. Struct. Biol.*, 18:134–139, 2008.

11. C. Clementi. Coarse-grained models of protein folding: Toy models or predictive tools? *Curr. Opin. Struct. Biol.*, 18:10–15, 2008.

12. Q. Cui, L. Zhang, Z. Wu, and A. Yethiraj. Generation and sensing of membrane curvature: Where materials science and biophysics meet. *Curr. Opin. Solid State Mater. Sci.*, 17:164–174, 2013.

13. D. A. Potoyan, A. Savelyev, and G. A. Papoian. Recent successes in coarse-grained modeling of DNA. *WIREs Comput. Mol. Sci.*, 3:69–83, 2013.

14. H. J. Deuling and W. Helfrich. Curvature elasticity of fluid membranes—Catalog of vesicle shapes. *J. de Phys.*, 37:1335–1345, 1976.

15. W. Helfrich. Steric interaction of fluid membranes in multilayer systems. *Z. Naturforch. A*, 33(305–315), 1978.

16. D. Boal. *Mechanics of the Cell*. Cambridge: Cambridge University Press, 2002.

17. M. M. Tirion. Low amplitude motions in proteins from a single-parameter atomic analysis. *Phys. Rev. Lett.*, 77:1905–1908, 1996.

18. A. R. Atilgan, S. R. Durell, R. L. Jernigan, M. C. Demirel, O. Keskin, and I. Bahar. Anisotropy of fluctuation dynamics of proteins with an elastic net29 work model. *Biophys. J.*, 80:505–515, 2002.

19. I. Bahar, A. R. Atilgan, and B. Erman. Direct evaluation of thermal fluctuations in proteins using a single-parameter harmonic potential. *Fold. Des.*, 2:173–181, 1997.

20. Q. Cui, and I. Bahar, editors. *Normal Mode Analysis: Theory and Applications to Biological and Chemical Systems*. New York, NY: Chapman and Hall/CRC, 2006.

21. I. Bahar and A. J. Rader. Coarse-grained normal mode analysis in structural biology. *Curr. Opin. Struct. Biol.*, 15:586–592, 2005.

22. D. A. Kondrashov, Q. Cui, and G. N. Phillips, Jr. Optimization and evaluation of a coarse-grained model of protein motion using X-ray crystal data. *Biophys. J.*, 91(8):2760–2767, 2006.

23. K. Hinsen, A. J. Petrescu, S. Dellerue, M. C. Bellissent-Funel, and G. R. Kneller. Harmonicity in slow protein dynamics. *Chem. Phys.*, 261:25–37, 2000.

24. L. Yang, G. Song, and R. L. Jernigan. Protein elastic network models and the ranges of cooperativity. *Proc. Natl. Acad. Sci. U S A*, 106:12347–12352, 2009.

25. J. W. Chu and G. A. Voth. Coarse-grained modeling of the actin filament derived from atomistic-scale simulations. *Biophys. J.*, 90:1572–1582, 2006.

26. E. Lyman, J. Pfaendtner, and G. A. Voth. Systematic multiscale parameterization of heterogeneous elastic network models of proteins. *Biophys. J.*, 95:4183–4192, 2008.

27. W. J. Zheng and B. R. Brooks. Identification of dynamical correlations within the myosin motor domain by the normal mode analysis of an elastic network model. *J. Mol. Biol.*, 346:745–759, 2005.

28. K. Hinsen. Structural flexibility in proteins: Impact of the crystal environment. *Bioinformatics*, 24(4):521–528, 2008.
29. D. Riccardi, Q. Cui, and G. N. Phillips, Jr. Application of elastic network models to proteins in the crystalline state. *Biophys. J.*, 96:464–475, 2009.
30. M. Y. Lu and J. P. Ma. Pim: Phase integrated method for normal mode analysis of biomolecules in a crystalline environment. *J. Mol. Biol.*, 425:1082–1098, 2013.
31. W. J. Zheng. All-atom and coarse-grained simulations of the forced unfolding pathways of the SNARE complex. *Proteins Struct. Funct. Bioinform.*, 82:1376–1386, 2014.
32. M. G. Munteanu, K. Vlahovicek, S. Parthasarathy, I. Simon, and S. Pongor. Rod models of DNA: Sequence-dependent anisotropic elastic modelling of local bending phenomena. *Trends Biochem. Sci.*, 23:341–347, 1998.
33. W. K. Olson and V. B. Zhurkin. Modeling DNA deformations. *Curr. Opin. Struct. Biol.*, 10:286–297, 2000.
34. R. Phillips, M. Dittrich, and K. Schulten. Quasicontinuum representations of atomic-scale mechanics: From proteins to dislocations. *Annu. Rev. Mater. Res.*, 32:219–233, 2002.
35. S. Choe and S. X. Sun. The elasticity of a-helices. *J. Chem. Phys.*, 122:244912, 2005.
36. O. N. Yogurtcu, C. W. Wolgemuth, and S. X. Sun. Mechanical response and conformational amplification in alpha-helical coiled coils. *Biophys. J.*, 99:3895–3904, 2010.
37. G. H. Lan and S. X. Sun. Dynamics of myosin-V processivity. *Biophys. J.*, 88:999–1008, 2005.
38. M. Bathe. A finite element framework for computation of protein normal modes and mechanical response. *Proteins Struct. Funct. Bioinform.*, 70:1595–1609, 2008.
39. M. Y. Lu and J. P. Ma. The role of shape in determining molecular motions. *Biophys. J.*, 89:2395–2401, 2005.
40. F. Tama, and C. L. Brooks III. Symmetry, form, and shape: Guiding principles for robustness in macromolecular machines. *Annu. Rev. Biophys. Biomol. Struct.*, 35:115–134, 2006.
41. Y. Tang, G. Cao, X. Chen, J. Yoo, A. Yethiraj, and Q. Cui. A finite element framework for studying mechanical response of macromolecules: Application to the gating of the mechanosensitive channel. *Biophys. J.*, 91:1248–1263, 2006.
42. X. Chen, Q. Cui, Y. Y. Tang, J. Yoo, and A. Yethiraj. Gating mechanisms of mechanosensitive channels of large conductance, I: A continuum mechanics-based hierarchical framework. *Biophys. J.*, 95:563–580, 2008.
43. Y. Y. Tang, J. Yoo, A. Yethiraj, Q. Cui, and X. Chen. Gating mechanisms of mechanosensitive channels of large conductance, II: Systematic study of conformational transitions. *Biophys. J.*, 95:581–596, 2008.

44. Y. C. Zhou, M. Holst, and J. A. McCammon. A nonlinear elasticity model of macromolecular conformational change induced by electrostatic forces. *J. Math. Anal. Appl.*, 340:135–164, 2008.

45. N. R. Latorraca, K. M. Callenberg, J. P. Boyle, and M. Grabe. Continuum approaches to understanding ion and peptide interactions with the membrane. *J. Memb. Biol.*, 247:395–408, 2014.

46. K. M. Callenberg, O. P. Choudhary, G. L. de Forest, D. W. Gohara, N. A. Baker, and M. Grabe. APBSmem: A graphical interface for electrostatic calculations at the membrane. *PLOS ONE*, 5:e12722, 2010.

47. L. Ma, Y. Tang, A. Yethiraj, X. Chen, and Q. Cui. A combined continuum mechanics and continuum electrostatics (CM/CE) computational framework for macromolecules: Application to the salt concentration dependence of DNA bendability. *Biophys. J.*, 96:3543–3554, 2009.

48. N. A. Baker, D. Sept, S. Joseph, M. J. Holst, and J. A. McCammon. Electrostatics of nanosystems: Application to microtubules and the ribosome. *Proc. Acad. Natl. Sci. U.S.A.*, 98:10037–10041, 2001.

49. R. Phillips, T. Ursell, P. Wiggins, and P. Sens. Emerging roles for lipids in shaping membrane-protein function. *Nature*, 459:379–385, 2009.

50. J. T. Groves and J. Kuriyan. Molecular mechanisms in signal transduction at the membrane. *Nat. Struct. Mol. Biol.*, 17:659–665, 2010.

51. D. Schmidt and R. MacKinnon. Voltage-dependent K+ channel gating and voltage sensor toxin sensitivity depend on the mechanical state of the lipid membrane. *Proc. Natl. Acad. Sci. U S A*, 105:19276–19281, 2008.

52. A. G. Lee. How lipids affect the activities of integral membrane proteins. *Biochim. Biophys. Acta Biomem.*, 1666:62–87, 2004.

53. O. S. Andersen and R. E. Koeppe. Bilayer thickness and membrane protein function: An energetic perspective. *Annu. Rev. Biophys. Biomol. Struct.*, 36:107–130, 2007.

54. C. Kung. A possible unifying principle for mechanosensation. *Nature*, 436:647–654, 2005.

55. Y. Shibata, J. Hu, M. M. Kozlov, and T. A. Rapoport. Mechanisms shaping the membranes of cellular organelles. *Annu. Rev. Cell Dev. Biol.*, 25:329–354, 2009.

56. L. V. Chernomordik and M. M. Kozlov. Protein-lipid interplay in fusion and fission of biological membranes. *Annu. Rev. Biochem.*, 72:175–207, 2003.

57. F. L. Brown. Elastic modeling of biomembranes and lipid bilayers. *Annu. Rev. Phys. Chem.*, 59:685–712, 2008.

58. F. L. H. Brown. Continuum simulations of biomembrane dynamics and the importance of hydrodynamic effects. *Q. Rev. Biophys.*, 44:391–432, 2011.

59. J. Yoo, M. B. Jackson, and Q. Cui. A comparison of coarse-grained and continuum models for membrane bending in lipid bilayer fusion pores. *Biophys. J.*, 104:841–852, 2013.

60. M. M. Kozlov, F. Campelo, N. Liska, L. V. Chernomordik, S. J. Marrink, and H. T. McMahon. Mechanisms shaping cell membranes. *Curr. Opin. Cell Biol.*, 29:53–60, 2014.

61. M. Terasaki, T. Shemesh, N. Kasthuri, R. W. Klemm, R. Schalek, K. J. Hayworth, A. R. Hand, M. Tankova, G. Huber, J. W. Lichtman, T. A. Rapoport, and M. M. Kozlov. Stacked endoplasmic reticulum sheets are connected by helicoidal membrane motifs. *Cell*, 154:285–296, 2013.

62. G. S. Ayton and G. A. Voth. Multiscale simulation of protein mediated membrane remodeling. *Semi. Cell Dev. Biol.*, 21:357–362, 2010.

63. E. Lyman, H. S. Cui, and G. A. Voth. Reconstructing protein remodeled membranes in molecular detail from mesoscopic models. *Phys. Chem. Chem. Phys.*, 13:10430–10436, 2011.

64. J. Yoo. Computational and Theoretical Studies of Lipid Membrane and Protein-Membrane Interactions. Ph.D. thesis. University of Wisconsin-Madison, 2010.

65. T. Ursell, K. C. Huang, E. Peterson, and R. Phillips. Cooperative gating and spatial organization of membrane proteins through elastic interactions. *PLoS Comput. Biol.*, 3(5):e81, 2007.

66. T. Kim, K. Lee, P. Morris, R. W. Pastor, O. S. Andersen, and W. Im. Influence of hydrophobic mismatch on structures and dynamics of gramicidin A and lipid bilayers. *Biophys. J.*, 102:1551–1560, 2012.

67. J. Yoo and Q. Cui. Three-dimensional stress field around a membrane protein: Atomistic and coarse-grained simulation analysis of gramicidin A. *Biophys. J.*, 104:117–127, 2013.

68. A. Panahi and M. Feig. Dynamic heterogeneous dielectric generalized Born (DHDGB): An implicit membrane model with a dynamically varying bilayer thickness. *J. Chem. Theory Comp.*, 9:1709–1719, 2013.

69. J. P. Ma. Usefulness and limitations of normal mode analysis in modeling dynamics of biomolecular complexes. *Structure*, 13:373–380, 2005.

70. A. W. van Wynsberghe and Q. Cui. Interpretating correlated motions using normal mode analysis. *Structure*, 14:1647–1653, 2006.

71. F. Tama, O. Miyashita, and C. L. Brooks. Normal mode based flexible fitting of high-resolution structure into low-resolution experimental data from cryoem. *J. Struct. Biol.*, 147:315–326, 2004.

72. W. J. Zheng and B. R. Brooks. Modeling protein conformational changes by iterative fitting of distance constraints using reoriented normal modes. *Biophy. J.*, 90:4327–4336, 2006.

73. W. J. Zheng and M. Tekpinar. Accurate flexible fitting of high-resolution protein structures to small-angle x-ray scattering data using a coarse-grained model with implicit hydration shell. *Biophys. J.*, 101:2981–2991, 2011.

74. P. Doruker, R. L. Jernigan, and I. Bahar. Dynamics of large proteins through hierarchical levels of coarse-grained structures. *J. Comput. Chem.*, 23:119–127, 2002.

75. G. Li, A. Van Wynsberghe, O. N. A. Demerdash, and Q. Cui. *Normal Mode Analysis: Theory and Application to Biological and Chemical Systems.* New York: Chapman and Hall/CRC Press, 2006.

76. M. Y. Lu, B. Poon, and J. P. Ma. A new method for coarse-grained elastic normal-mode analysis. *J. Chem. Theory Comput.*, 2:464–471, 2006.

77. F. Xia, D. D. Tong, L. F. Yang, D. Y. Wang, S. C. H. Hoi, P. Koehl, and L. Y. Lu. Identifying essential pairwise interactions in elastic network model using the alpha shape theory. *J. Comput. Chem.*, 35:1111–1121, 2014.

78. A. W. van Wynsberghe and Q. Cui. Comparisons of mode analyses at different resolutions applied to nucleic acid systems. *Biophys. J.*, 89:2939–2949, 2005.

79. G. Li and Q. Cui. A coarse-grained normal mode approach for macromolecules: An efficient implementation and application to Ca^{2+}-ATPase. *Biophys. J.*, 83:2457–2474, 2002.

80. R. M. Levy, M. Karplus, J. Kushick, and D. Perahia. Evaluation of the configurational entropy for proteins—Application to molecular-dynamics simulations of an alpha-helix. *Macromolecules*, 17:1370–1374, 1984.

81. O. Miyashita, J. N. Onuchic, and P. G. Wolynes. Nonlinear elasticity, protein quakes, and the energy landscapes of functional transitions in proteins. *Proc. Natl. Acad. Sci. U S A*, 100:12570–12575, 2003.

82. A. Anishkin and C. Kung. Microbial mechanosensation. *Curr. Opin. Neuro.*, 15:397–405, 2005.

83. E. Perozo and D. Rees. Structure and mechanism in prokaryotic mechanosensitive channels. *Curr. Opin. Struct. Biol.*, 13:432–442, 2003.

84. E. S. Haswell, R. Phillips, and D. C. Rees. Mechanosensitive channels: What can they do and how do they do it? *Structure*, 19:1356–1369, 2011.

85. G. Chang, R. H. Spencer, A. T. Lee, M. T. Barclay, and D. Rees. Structure of the MscL homolog from Mycobacterium tuberculosis: A gated mechanosensitive ion channel. *Science*, 282:2220–2226, 1998.

86. E. Perozo, D. M. Cortes, P. Sompornpisut, A. Kloda, and B. Martinac. Open channel structure of MscL and the gating mechanism of mechanosensitive channels. *Nature*, 418:942–948, 2002.

87. E. Perozo. Gating prokaryotic mechanosensitive channels. *Nat. Rev. Mol. Cell Biol.*, 7:109–119, 2006.

88. S. I. Sukharev, S. R. Durell, and H. R. Guy. Structural models of the MscL gating mechanism. *Biophys. J.*, 61:917–936, 2001.

89. J. Gullingsrud and K. Schulten. Gating of MscL studied by steered molecular dynamics. *Biophys. J.*, 85:2087–2099, 2003.

90. S. Yefimov, E. van der Giessen, P. R. Onck, and S. J. Marrink. Mechanosensitive membrane channels in action. *Biophys. J.*, 94:2994–3002, 2008.

91. S. Kundu, J. S. Melton, D. C. Sorensen, and G. N. Phillips, Jr. Dynamics of proteins in crystals: Comparison of experiment with simple models. *Biophys. J.*, 83(2):723–732, 2002.

92. D. Riccardi, Q. Cui, and G. N. Phillips, Jr. Evaluating elastic network models of crystalline biological molecules with temperature factors, cor-

related motions, and diffuse x-ray scattering. *Biophys. J.*, 99:2616–2625, 2010.

93. B. K. Poon, X. R. Chen, M. Y. Lu, N. K. Vyas, F. A. Quiocho, Q. H. Wang, and J. P. Ma. Normal mode refinement of anisotropic thermal parameters for a supramolecular complex at 3.42-a crystallographic resolution. *Proc. Natl. Acad. Sci. U S A*, 104:7869–7874, 2007.

94. E. Perozo, A. Kloda, D. M. Cortes, and B. Martinac. Physical principles underlying the transduction of bilayer deformation forces during mechanosensitive channel gating. *Nat. Struct. Biol.*, 9:696–703, 2002.

95. Y. F. Kong, Y. F. Shen, T. E. Warth, and J. P. Ma. Conformational pathways in the gating of escherichia coli mechanosensitive channel. *Proc. Natl. Acad. Sci. U S A*, 99(9):5999–6004, 2002.

96. M. S. Turner and P. Sens. Gating-by-tilt of mechanically sensitive membrane channels. *Phys. Rev. Lett.*, 93:118103, 2004.

97. B. Ajouz, C. Berrier, M. Besnard, B. Martinac, and A. Ghazi. Contributions of the different extramembranous domains of the mechanosensitive ion channel MscL to its response to membrane tension. *J. Biol. Chem.*, 275:1015–1022, 2000.

98. C. C. Hase, A. C. Le Dain, and B. Martinac. Molecular dissection of the large mechanosensitive ion channel (MscL) of *E. coli*: Mutants with altered channel gating and pressure sensitivity. *J. Memb. Biol.*, 157:17–25, 1997.

99. B. Martinac. Mechanosensitive ion channels: Molecules of mechanotransduction. *J. Cell Sci.*, 117:2449–2460, 2004.

100. S. I. Sukharev, W. J. Sigurdson, C. Kung, and F. Sachs. Energetic and spatial parameters for gating of the bacterial large conductance mechanosensitive channel, MscL. *J. Gen. Physiol.*, 113:529–539, 1999.

101. S. Sukharev and A. Anishkin. Mechanosensitive channels: What can we learn from 'simple' model systems? *Trends Neurosci.*, 27:345–351, 2004.

102. M. Tricot. Comparison of experimental and theoretical persistence length of some polyelectrolytes at various ion strengths. *Macromolecules*, 17:1698–1704, 1984.

103. M. O. Fenley, G. S. Manning, and W. K. Olson. Electrostatic persistence length of a smoothly bending polyion computed by numerical counterion condensation theory. *J. Phys. Chem.*, 96:3963–3969, 1992.

104. G. S. Manning. The persistence length of DNA is reached from the persistence length of its null isomer through an internal electrostatic stretching force. *Biophys. J.*, 91:3607–3616, 2006.

105. J. R. Wenner, M. C. Williams, I. Rouzina, and V. A. Bloomfield. Salt dependence of the elasticity and overstretching transition of single DNA molecules. *Biophys. J.*, 82:3160–3169, 2002.

106. T. Odijk. Polyelectrolytes near rod limit. *J. Poly. Sci. Part B-Poly. Phys.*, 15:477–483, 1977.

107. J. Skolnick and M. Fixman. Electrostatic persistence length of a wormlike polyelectrolyte. *Macromolecules*, 10:944–948, 1977.

108. H. G. Garcia, P. Grayson, L. Han, M. Inamdar, J. Kondev, P. C. Nelson, R. Phillips, J. Widom, and P. A. Wiggins. Biological consequences of tightly bent DNA: The other life of a macromolecular celebrity. *Biopolymers*, 85:115–130, 2007.

109. J. B. Klauda, R. M. Venable, J. A. Freites, J. W. O'Connor, D. J. Tobias, C. Mondragon-Ramirez, I. Vorobyov, A. D. MacKerrel Jr., and R. W. Pastor. Update of the CHARMM all-atom additive force field for lipids: Validation on six lipid types. *J. Phys. Chem. B*, 114:7830–7843, 2010.

110. S. J. Marrink, H. J. Risselada, S. Yefimov, D. P. Tieleman, and A. H. de Vries. The MARTINI force field: Coarse grained model for biomolecular simulations. *J. Phys. Chem B*, 111:7812–7824, 2007.

111. M. M. Kozlov, S. L. Leikin, L. V. Chernomordik, V. S. Markin, and Y. A. Chizmadzhev. Stalk mechanism of vesicle fusion. Intermixing of aqueous contents. *Eur. Biophys. J.*, 17(3):121–9, 1989.

112. Y. A. Chizmadzhev, F. S. Cohen, A. Shcherbakov, and J. Zimmerberg. Membrane mechanics can account for fusion pore dilation in stages. *Biophys. J.*, 69(6):2489–500, 1995.

113. V. S. Markin, M. M. Kozlov, and V. L. Borovjagin. On the theory of membrane fusion. The stalk mechanism. *Gen. Physiol. Biophys.*, 3(5):361–377, 1984.

114. V. S. Markin and J. P. Albanesi. Membrane fusion: Stalk model revisited. *Biophys. J.*, 82:693–712, 2002.

115. R. J. Ryham, M. A. Ward, and F. S. Cohen. Teardrop shapes minimize bending energy of fusion pores connecting planar bilayers. *Phys. Rev. E*, 88:062701, 2013.

116. J. Yoo and Q. Cui. Membrane-mediated protein-protein interactions and connection to elastic models: A coarse-grained simulation analysis of gramicidin A association. *Biophys. J.*, 104:128–138, 2013.

117. J. Schlitter. Estimation of absolute and relative entropies of macromolecules using the covariance-matrix. *Chem. Phys. Lett.*, 215:617–621, 1993.

118. J. A. Lundbæk, S. A. Collingwood, H. I. Ingólfsson, R. Kapoor, and O. S. Andersen. Lipid bilayer regulation of membrane protein function: Gramicidin channels as molecular force probes. *J. R. Soc. Interface*, 7:373–395, 2010.

119. E. Villa, A. Balaeff, and K. Schulten. Structural dynamics of the lac repressor-DNA complex revealed by a multiscale simulation. *Proc. Acad. Natl. Sci. U S A*, 102:6783–6788, 2005.

120. T. D. Lillian, M. Taranova, J. Wereszczynski, I. Andricioaei, and N. C. Perkins. A multiscale dynamic model of DNA supercoil relaxation by topoisomerase IB. *Biophys. J.*, 100:2016–2023, 2011.

CHAPTER 6

Knowledge-Based Models

RNA

Raúl Méndez, Andrey Krokhotin, Marino Convertino, Jhuma Das, Arpit Tandon, and Nikolay V. Dokholyan

CONTENTS

6.1	Introduction	222
6.2	RNA Molecule Description	225
6.3	Coarse-Grained Modeling of RNA Molecules	229
	6.3.1 Backbone Models	230
	6.3.1.1 One-Bead Model: NAST	230
	6.3.1.2 Two-Bead Model	231
	6.3.2 Nucleobase Models (Inspired by Rosetta)	233
	6.3.3 Backbone–Nucleobase Hybrid Models	236
	6.3.3.1 Three-Bead Model	236
	6.3.3.2 Five-Bead Models	238
	6.3.3.3 Seven-Bead Models	240
6.4	Sampling Methods	242
	6.4.1 MD Simulations	242
	6.4.2 MC Simulations	243
	6.4.3 DMD Simulations	244
6.5	Guiding Conformational Sampling by Means of Experimental Restraints	245
	6.5.1 Most Common Experimental Methods to Derive Distance Restraints	245
	6.5.2 Secondary Structure from SHAPE in Addition to Long-Range Restraints	247
	6.5.3 Incorporating HPR Data as Restraints into DMD Simulations	251

6.6 Comparison of the Performance of Different Methods: RNA-Puzzles 253

6.7 Conclusions 261

References 262

6.1 INTRODUCTION

Nucleic acids were discovered early in 1869 by young Swiss physician Friederich Miescher, as a consequence of his studies on the molecules of life using leucocyte cultures (Dahm 2005). Miescher isolated a novel molecule from the cell nuclei with distinct biochemical properties from proteins. This was the first reported deoxyribonucleic acid (DNA) extraction. He termed the newly discovered molecule "nuclein" because it was isolated from cell nuclei. Several other experiments yielded the ensuing discovery of ribonucleic acid (RNA) in prokaryote organisms, which do not have nuclei. In the beginning of the twentieth century, the difference between DNA and RNA was still unclear. The two molecules were referred to as "pancreas nucleic acid" and "yeast nucleic acid," respectively, referring to the source for purification. Only in the decade between 1930 and 1940 were the chemical compositions of DNA and RNA determined, resulting in the current distinction between sugar molecules (ribose and 2-deoxyribose for RNA and DNA, respectively), and one specific nucleobase (uracil and thymine in RNA and DNA, respectively) in the two nucleic acids (Worthington Allen 1941).

The next landmark discovery concerning RNA was its central role in protein biosynthesis. In 1956, Francis Crick stated his famous "central dogma" of molecular biology (Crick 1970), which claims that *DNA makes RNA and RNA makes proteins*. The "dogma" emphasized the role of RNA as messenger molecule. The discovery of the lactose (lac) operon, a cluster of metabolic genes transcribed together under the same promoter, in *Escherichia coli* allowed the study of the messenger RNA (mRNA) and the genetic code (i.e., the equivalence between nucleotide sequence in mRNA and actual amino acids in proteins). Later, radiolabeling experiments in the 1950s showed that amino acids bind ribosomes, those large complexes made of three long RNA strands and several proteins known to interact with mRNA. In the 1960s, the first ribosomes were isolated as they move along a single mRNA molecule during protein biosynthesis, catalyzing the new peptide bond formation (Geiduschek and Haselkorn 1969). During the same period of time, the transfer RNA (tRNA) was shown to covalently bind amino acids at the $3'$-terminus and to recognize a codon sequence that codes for a specific amino acid. The genetic code was fully elucidated using *in vitro* experiments, establishing a consistent equivalence between codons and amino acids.

The first tRNA molecule was sequenced in 1965 (Holley et al. 1965) and later in 1975, its three-dimensional structure was determined (Ladner et al. 1975), revealing secondary structure conservation between homologous sequences. Also the first RNA virus, a bacteriophage, was sequenced, pointing to the existence of the enzyme reverse transcriptase, which is able to synthesize DNA from RNA within the host organism (note that this observation introduces an important change in the "central dogma," i.e., *RNA can also make DNA*).

Beyond roles in protein synthesis as mRNA, tRNA, and ribosomal RNA (rRNA), most notably the 1980s yielded a new round of discoveries pertaining to a new type of RNA molecule, namely the noncoding RNA (ncRNA). Many ncRNAs have catalytic activity (a property that besides the rRNA was previously believed to be the sole domain of proteins) such as Ribonuclease P, which catalyzes its tRNA preprocessing mechanism (Frank and Pace 1998). At the time, another breakthrough discovery was that eukaryotic genes are not continuous pieces of DNA, but discontinuous fragments that code for protein (exons) intercalated by noncoding fragments (introns) which are removed after transcription (pre-mRNA) by splicing (Green 1986). Moreover, intron removal is catalyzed by a complex of proteins and small nuclear ribonucleic particles (snRNPs) (Kramer 1996). In the last two decades, multitudes of ncRNAs have been found that are involved in several mechanisms such as DNA genome editing, gene regulation, interference or silencing, etc. For a detailed review, please refer to Cech and Steitz (2014).

The increasing importance of RNA in biology raised the "RNA world" hypothesis, which advances the view that RNA is the precursor molecule to life because of its ability to self-replicate, catalyze reactions, and store genetic information (Copley, Smith, and Morowitz 2007). Interestingly, from the evolutionary point of view, RNA replication machinery (or "replicons") evolves quicker than DNA, since reverse transcriptases and RNA replicases do not have exonuclease activity and there is no equivalent RNA repairing system as for damaged DNA molecules (Hansen, Long, and Schultz 1997).

To understand the dynamics, the physicochemical properties, and the functions of RNA molecules, we need to determine the atoms' precise spatial arrangement. Similarly to proteins, there are two main techniques to elucidate RNA three-dimensional (3D) structures: x-ray crystallography (Thomsen and Berger 2012) and nuclear magnetic resonance (NMR) (Scott and Hennig 2008). What makes RNA structure determination a more challenging task is the higher number of possible molecular conformations at physiological temperature. Despite the lower number of building blocks (i.e., only 4 nucleotides (nt), as compared to 20 amino acids in proteins), the number of conformations available to RNA polymers is much greater than what is accessible to proteins. In proteins, the conformations of amino acids' side chains impose serious steric hindrance to the backbone, so that in practice the angles of the rotatable bonds of the backbone are restricted to certain range of values (Ramachandran, Ramakrishnan, and Sasisekharan 1963) thus making the protein conformational space somewhat limited. RNA molecules, in contrast, have significantly large number of conformations accessible to the ribose subunits. Although some rotameric states can be found for RNA backbone conformations, they are less accurately defined than for proteins (Murray et al. 2003).

X-ray crystallography has allowed for the elucidation of a number of structures, notably the 3D structure of the ribosome was found in 2000 (Schluenzen et al. 2000; Ban et al. 2000; Wimberly et al. 2000). While the power of this technique yields exact structures at high resolution, care must be taken in sample preparation to avoid aggregates or misfolded structures (Reyes, Garst, and Batey 2009). NMR, while it is exceptionally useful and does not require crystallization, presents its own set of challenges. Solving an NMR spectrum for an RNA molecule is in general much harder than for proteins mostly because (1) ^{1}H isotope proportion is much lower in RNA than in proteins; (2) chemical shifts from different active species such as ^{1}H, ^{15}N, or ^{13}C are difficult to distinguish because they are in very similar chemical environments; (3) the 4 nts, adenosine monophosphate (A), uridine monophosphate (U), guanosine monophosphate (G), and cytosine monophosphate (C), are more chemically similar *per se* compared to the 20 possible amino acids; and (4) RNA molecules display helical structures that are rather linear, and not globular like proteins, making 2D correlation spectrums more difficult to resolve (Addess and Feigon 1996). In addition, backbone proton signal that is characteristic to polypeptides and can be easily identified, is not as clear in RNA, because of the intercalated phosphate groups.

The limitations of experimental methodologies require that accurate computational techniques be applied to RNA structure prediction to gather knowledge about the different functional roles of RNA molecules. Any computational modeling of a biomolecule consists of two important components: a description of the system, which involves an energy function that accounts for intra- and interatomic interactions, along with a sampling engine to explore the energy landscape derived from that energy function and, ideally, be able to identify the native conformation of the system as the absolute energy minimum. The description of the system can be as realistic as to include all atoms, typically coming with a molecular mechanics force field made of the sum of pairwise bonded and nonbonded interactions. There is always a tradeoff between the accuracy of the model description and the appropriate sampling of the energy landscape. Detailed atomistic models require efficient sampling of the conformational landscape (involving computationally extensive simulations); otherwise the optimal predicted structure may just be a conformation trapped in a local energy minimum that is not the absolute energy minimum. Coarse-grained methods simplify the system description and the interactions of their components. They provide a less rugged energy landscape that is easier to sample and consequently lightens the computational load as well. Coarse-grained methods are the common choice in RNA modeling because they provide reasonably accurate results as compared to all-atom models at a lower computational expense.

In this chapter, we summarize the most common coarse-grained models used to predict RNA 3D structure and provide a qualitative description of their energy functions. This classification is based on the study published by Xia and Ren (2013). This chapter is structured as follows. In Section 6.2, the RNA molecule is revisited. Section 6.3 introduces the most common coarse-grained methods in RNA modeling. Section 6.4 summarizes the principles for the typical sampling algorithms used in combination with the previously described models. Section 6.5 illustrates how experimental information can be used to improve sampling and hence the final quality of the models. Section 6.6 is the comparison of the state-of-the-art RNA structure prediction methods as reported in the international blind contest *RNA-Puzzles*. Not all the methods cited in that section are coarse-grained strictly speaking, but they best represent what it is possible to achieve using in the RNA structure-modeling field. Finally, Section 6.7 draws some conclusions.

6.2 RNA MOLECULE DESCRIPTION

Nucleic acids play a central role in almost all cellular processes. RNA molecules, like DNA, form polymeric chains, but unlike the latter, they are single-stranded (except in special cases of RNA–RNA interaction) polymers. Each RNA monomer is made of a ribose sugar (containing a hydroxyl group at position 2′, see Figure 6.1), a phosphate group in position 5′, and like DNA, a purine (adenine or guanine) or pyrimidine (cytosine and uracil) nucleobase at position 1′ (Figure 6.1). At physiological conditions, phosphate groups are negatively charged, while nucleobases can

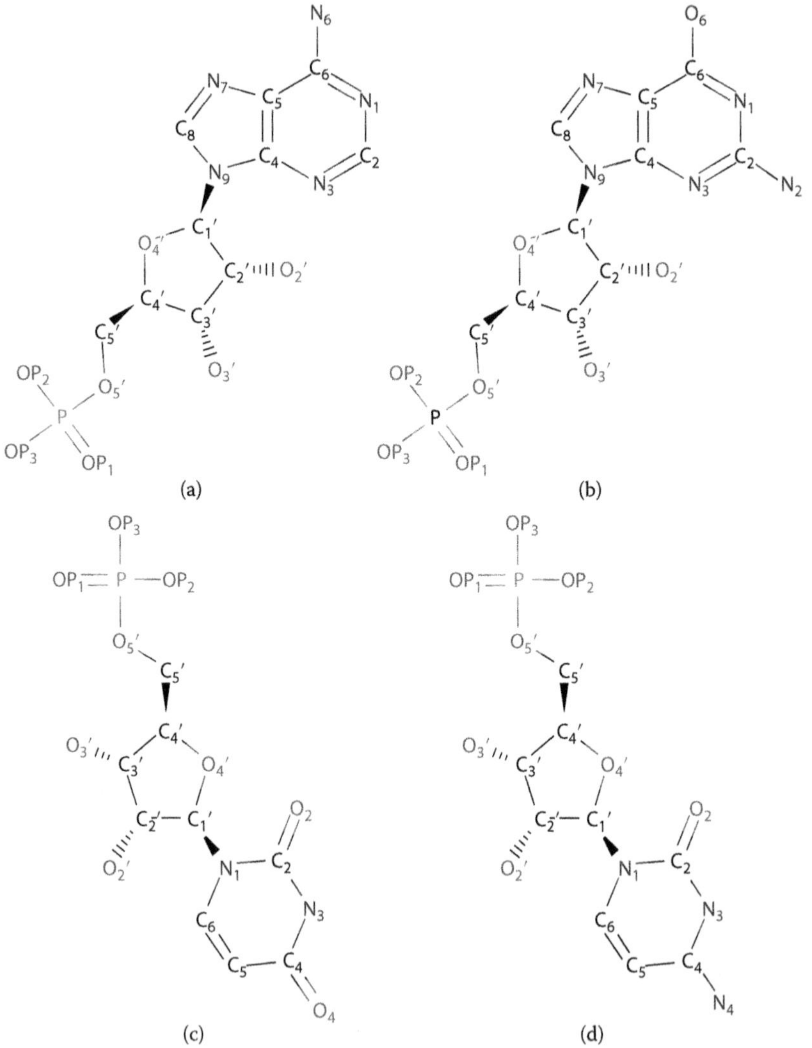

FIGURE 6.1 2D representations of the four nucleotides found in RNA: (a) adenosine 5′-monophosphate, (b) guanosine 5′-monophosphate, (c) uridine 5′-monophosphate, and (d) cytidine 5′-monophosphate.

make hydrogen bond interactions with other bases of the same strand or from different RNA molecules. Such base pairing is the major determinant of the 3D structure of RNA.

There are several possible base pairings in RNA molecules. In the canonical Watson–Crick pairing, adenine interacts via hydrogen bonds to uracil, while guanine interacts with cytosine (Figure 6.2). Hoogsteen base pairing also takes place between adenine (A) and uracil (U) and between guanine (G) and a positively charged cytosine (C^+). A–U pairing involves atom N_6 as a hydrogen bond donor and N_7 as a hydrogen bond acceptor, which bind the same N_3–O_4 Watson–Crick side of the pyrimidine base. G–C^+ pairing is made between O_6 (G, acceptor) and N_4 (C^+, donor), and N_7 (G, acceptor) and N_3^+ (C^+, donor). Note the base flipping in the pyrimidine base as compared to Watson–Crick pairing (see Figures 6.1 and 6.2 for details about atom nomenclature).

Both Watson–Crick and Hoogsteen pairing yield hydrogen bond interactions that result in characteristic secondary structure elements, such as

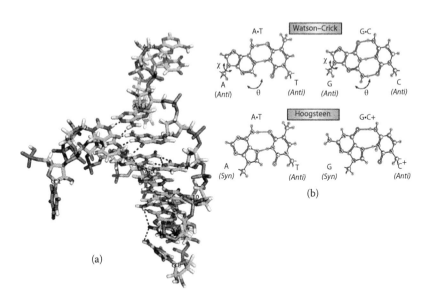

FIGURE 6.2 (a) Watson–Crick base pairing from the RNA structure of a small interfering RNA (siRNA)-like, a noncoding RNA that interferes with the expression of certain genes, duplex from *Archaeoglobus fulgidus* (PDB-ID: 2BGG). Hydrogen bonds are highlighted as dashed black lines. (b) Hoogsteen base pairing scheme. Note that instead of the thymine as a pyrimidine, we should have uracil for the RNA, but the hydrogen bond pattern remains the same. (Reprinted by permission from MacMillan Publishers Ltd. *Nature*, Nikolova et al., 470: 498–502, copyright 2011.)

stem loops (Watson et al. 2013) (Figure 6.3a), tetraloops (Woese, Winker, and Gutell 1990) (Figure 6.3b), and pseudoknots (Staple and Butcher 2005) (Figure 6.3c and d). The tertiary structure of RNA can be in the form of a triplex, where either a third nucleotide is Watson–Crick paired to a Hoogsteen duplex in the major groove (Toor et al. 2008) (Figure 6.4a) or in the minor groove, involving the addition of ribose 2′-hydroxyl interactions. It can also form a quadruplex, made of four G nucleobases in a "Hoogsteen ring" (Figure 6.4b) or by G–C or A–U pairs using a combination of Watson–Crick and noncanonical base pairing in the minor groove (Su et al. 1999). Other tertiary structure motifs include coaxial stacking (Quigley and Rich 1976), in which two RNA duplex form a contiguous helix; an A-minor motif

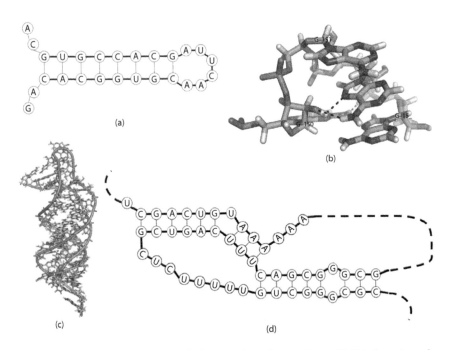

FIGURE 6.3 (a) Schematic view of a base pairing of a stem loop. (b) Tetraloop is made of sequence GAAA, which is a hairpin loop of the P4–P6 domain of the *Tetrahymena thermophila* intron (PDB-ID: 1GID). This RNA molecule is a class I self-splicing intron, which is able to catalyze its own excision from an RNA precursor (Cate et al. 1996). (c) Stick representation of the 3D structure of an RNA pseudoknot, from the human telomerase (P2b-P3 fragment, PDB-ID: 1YMO). (d) Schematic view of a base pairing of pseudoknot from a human telomerase, showing the typical pattern made of two stem loops, in which half of one stem is intercalated between the two halves of the second stem. (From Chen, J-L and C W Greider, *Proceedings of the National Academy of Sciences of the United States of America*, 102, 8080–8085.)

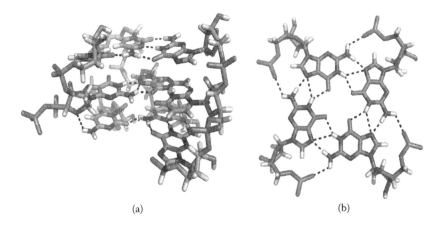

(a)   (b)

FIGURE 6.4 (a) Major groove triplex in the group II introns from *Oceanobacillus Iheyensis* (PDB-ID: 3BWP). Note the three layers of triplexes made of C377-C360-G383, C288-G359-U384, and C289-C358-G385. (b) Structure of a Hogsteen paired G-quadruplex from a synthetic tetraplex construct (UGGGGU, PDB-ID: 1RAU).

(Doherty et al. 2001), made of an unpaired nucleotide inserted into a minor groove of an RNA duplex; and a ribose zipper that involves two RNA chains interacting through hydrogen bonds between the ribose 2′-hydroxyl atoms (Cate et al. 1996), which can act simultaneously as a hydrogen bond acceptor and donor.

Due to the many combinations made possible through these structural motifs, RNA can serve many important biological functions. In a further challenge to the canonical role of RNA as the messenger between DNA and protein, high-throughput genomic analysis demonstrated that about 97% of the transcriptional output corresponds to ncRNA sequences (Cech and Steitz 2014), which can either come from introns or RNA genes. The behavior and role of these molecules have not yet been fully elucidated, indicating a significant role for computational modeling, which is the subject of the next section.

6.3 COARSE-GRAINED MODELING OF RNA MOLECULES

In an attempt to reduce the computational cost associated with multiple degrees of freedom in RNA modeling and to enhance sampling efficiency, several methods represent RNA molecules using different degrees of detail/coarse-graining. Typically, most of the computational methods to predict RNA 3D structure can be divided into three families, according to the degree of "coarse-graining": (1) methods that focus on the coarse-grain representation of the backbone, (2) those that focus on the nucleobase, and (3) others that consider both backbone and nucleobase. The selection of a particular type of coarse-grained method also requires choosing a potential energy model that describes interactions between particles (i.e., pseudoatoms or beads). In this section, we provide a qualitative review of the most important coarse-grained methods for RNA structure modeling along with their energy functions.

6.3.1 Backbone Models

When considering only the backbone to model RNA 3D structure, the simplest choice is to model each nucleotide with one pseudoatom (or bead) and pseudobonds connecting them. This yields the so-called one-bead model. Additionally, a second bead can be considered along the backbone in the two-bead models. One of the most well-known one-bead models is the Nucleic Acid Simulation Tool (NAST) (Jonikas et al. 2009), while Vfold applies a two-bead model for RNA structure prediction (Cao and Chen 2005, 2009).

6.3.1.1 One-Bead Model: NAST

The NAST, developed by Altman and collaborators in 2009 (Jonikas et al. 2009), represents RNA nucleotides as single bead centered in the C_3' atom, with a Van der Waals radius of 4.0 Å (smaller radius would allow residues to intertwine within helical regions). A molecular dynamics (MD) engine (Alder and Wainwright 1959; Rahman 1964) is used to sample Cartesian degrees of freedom (see next section for more details). The potential energy function penalizes harmonic deviation from the ideal pseudobond length, angles, and tertiary contacts. In addition, there is a Fourier summation that accounts for the pseudorotation bonds and a repulsive Van der Waals term. Optimal parameters for the different energy terms are obtained in a two-step process. Probability distributions of distances, angles, and dihedrals, derived from a curated set of RNA structures, are fitted to ideal normal distributions (Figure 6.5). These ideal distributions are then converted to energies through the Boltzmann formula and fitted to the corresponding term in the potential energy, for which parameters are derived.

In addition to the primary nucleotide sequence as input, NAST requires the secondary structure information (base pairing) as well as available data about tertiary contacts (the latter increases performance significantly, since it reduces conformational space considerably). No information about hydrogen bonds is used, yet this method is able to find native-like structures

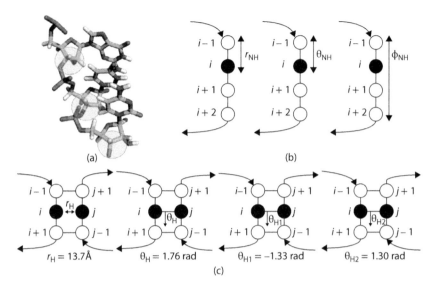

FIGURE 6.5 The NAST model and energy function. (a) Coarse-grained representation of an RNA molecule, using one bead per nucleotide centered at the C3′ atom. (b) Geometric restraints for nonhelical regions in the NAST energy function. (c) Geometric restraints used in helical regions.

that preserve global molecular shape. Typically, several clusters of structures are generated and rank ordered according to the experimental and predicted data on solvent accessible area, small-angle x-ray scattering (SAXS) data, and NAST energy (Jonikas et al. 2009).

The method is able to handle large RNA molecules, >100 nt. Predictions made for a 76-nt yeast phenylalanine tRNA and the 158-nt P4–P6 domain of *Tetrahymena thermophila* showed that NAST can attain structures as close as 8 Å to x-ray structure for the 76-nt tRNA, and up to 16 Å for the P4–P6 domain. This method can be used as a baseline or preliminary tool to produce a low-resolution model that can be further refined using more realistic energy functions.

6.3.1.2 Two-Bead Model

Still within the backbone coarse-grained representation, the resolution of the RNA model can be improved by using two beads to represent each nucleotide. The first two-bead model, VFold, was developed by Cao and Chen (2005). VFold utilizes two pseudoatoms per nucleotide along the RNA backbone, placed on the phosphate and the C_4 sugar atoms, respectively (see Figure 6.6); since the bonds $P-O_5-C_5-C_4$ and the bonds $C_4-C_3-C_3-P$ are approximately planar (Olson 1975; Olson and Flory 1972), they

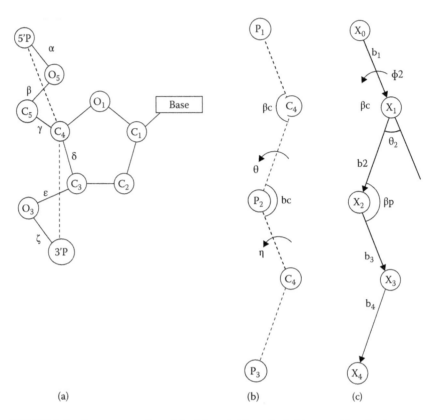

FIGURE 6.6 Coarse-grained model of the VFold method. (a) Virtual bond scheme for nucleotide backbone. (b) The bond angles (βC, βP) and the torsional angle (η). (c) Vector model used to determine the atomic coordinates from the torsional angles.

can be considered rotameric (Murray et al. 2003). This allows us to use a limited number of degrees of freedom: two pseudobonds connecting P–C_4 of nucleotide i and C_4–P connecting nucleotide $i+1$, respectively, two pseudobond angles for the P–C–P and C_4–P–C_4 triads, and two dihedral angles through the P–C_4 and C_4–P bonds in nucleotide i and $i + 1$, respectively (Figure 6.6). An analysis of known RNA structure revealed a distance of 3.9 Å for pseudobonds (Rich et al. 1961). The modeling of the different secondary structure elements utilizes distinct methods for each type of characteristic. Within helices, the pseudobond angles can be considered $105 \pm 5°$ and $95 \pm 5°$ (RNA-A helix (Biswas, Mitra, and Sundaralingam 1998), while for the torsion angles, θ, η, Duarte et al. (Duarte and Pyle 1998; Duarte, Wadley, and Pyle 2003) reported values close to $170°, 210°$).

The Vfold method performs an exhaustive enumeration of all the possible conformations (torsion angles θ, η) compatible with a given nucleotide sequence using the rotameric property of RNA. In contrast, the more flexible loops are modeled by self-avoiding walks on a diamond lattice, in which the bond length of the lattice equals that of the pseudobond (Flory 1969; Mattice and Suter 1994; Rapold and Mattice 1995). Mapping helix atoms onto the closest diamond lattice site makes joining between loops and off lattice helix nodes. In both cases, helices and loops, a translation from the previous pseudobond vector and composite rotation on the first pseudorotation and bond angle, generate different conformations (Cao and Chen 2005). Some initial coordinates from reference RNA structure are used for the $P-C_4-P$ and C_4-P-C_4 pseudobonds.

By employing this pseudobond/diamond lattice, this method can enumerate all the possible conformations of a given sequence to compute the minimized partition function of the system and obtain the lowest free energy conformation. Only base-stacking interactions are considered. They can be a canonical base stack if both base pairs are any of the A–U, G–C, or G–U, or a base mismatch if only one of the base pairs is from the previous list. The base stack an has effect on the enthalpy (including base pair mismatches) and entropy, while the nonstacked nucleotide (at loops) contributes only to the entropy.

The method can accurately predict thermodynamic properties such as heat capacity for different 58–59 nt RNA fragments (Cao and Chen 2005). In general, the VFold method is suited to predict secondary structure rather than tertiary structure, although the formalism in which it is based would allow the latter as well. VFold was later modified to accept an extra pseudobond (and an extra pseudoatom) between C_4 and N_1 (pyrimidines) or N_9 (purines) (Figure 6.7). This extra pseudobond accounts for the base orientation, so that more sophisticated secondary structure element such as pseudoknots up to 65 nt can be predicted accurately (Cao and Chen 2009).

6.3.2 Nucleobase Models (Inspired by Rosetta)

The Baker group has adapted their protocol to model protein structure *ab initio* (Rohl et al. 2004) and applied it to RNA. The fragment assembly RNA (FARNA) uses a library of 3-nt fragments and a low-resolution energy function to bias the Monte Carlo (MC) simulation process toward native RNA structures (Das and Baker 2007).

To avoid inclusion of fragments belonging to evolution related molecules, the library of fragments is taken from a single crystal structure, the large ribosomal subunit from *Haloarcula marismortui*, Protein Data

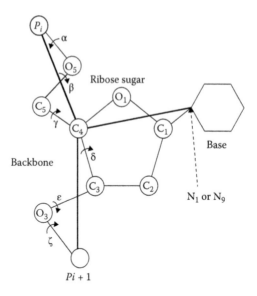

FIGURE 6.7   In addition to the two pseudobonds between P–C_4–P atoms, a third pseu-dobond is added to the VFold model to better describe the base orientation between the C_4–N_1 (pyrimidine) and C_4–N_9 (purine) atoms.

Bank identifier (PDB-ID) (Bernstein et al. 1977) 1FFK. Considering a two letter alphabet (purines/pyrimidines) yields more than 300 3-nt fragments.

The energy function includes several terms: (1) a term to ensure RNA compactness proportional to the radius of gyration, (2) a term to avoid steric clashes, (3) a term to reward Watson–Crick base pairing as a knowledge-based potential, (4) a term to enforce coplanarity between base pairs also as a knowledge-based potential of an orientation between two base pair planes, and a distance among the latter, and (5) an additional term to reward stacking of $-1k_BT$, where k_B is the Boltzmann constant. This energy function is derived from and applied to a coarse-grained rep-resentation of the RNA molecule, in which there is a Cartesian reference system placed on the geometric center of a given base pair. The x-axis goes through the N1 atom for the purines or the N3 atom for the pyrimidines; the z-axis is perpendicular to the base plane (Figure 6.8). Noncanonical base pairing is not included in the energy function, however; it is found to be recapitulated (Das and Baker 2007).

When applied to a set of 20 RNA structures up to 46 nt, most of them had Watson–Crick interactions predicted, and 11 of them showed root-mean-square distance (RMSD) <4.0 Å with respect to the native structure. In general, poor predictions in FARNA are the results of inaccuracies in the

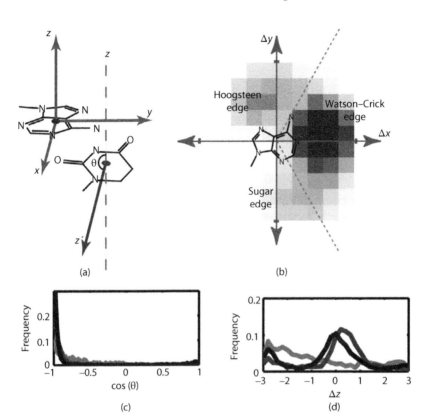

FIGURE 6.8 (a) The FARNA coordinate reference system. (b) Distribution of the Δx, Δy coordinates for a uridine residue in the vicinity of adenosine residues in the ribosome crystal structure. The logarithm of this distribution provides the knowledge-based potential. (c) Distribution of angles between base planes. (d) Distribution of relative stagger of base planes. Black line corresponds to the observed values, light gray, the models built by FARNA without coplanarity, and dark gray, values considering coplanarity. (Reprinted from Das, R and D Baker. 2007. Automated de novo prediction of native-like RNA tertiary structures. *Proceedings of the National Academy of Sciences of the United States of America* 104: 14664–14669. With permission from PNAS, Copyright [2007] National Academy of Sciences, U.S.A.)

energy function rather than a lack of sampling. In order to improve RNA models, the Baker group introduced a second refinement step using the traditional all-atom Rosetta energy function (Rohl et al. 2004), which includes pairwise terms to account for van der Waals forces, hydrogen bonds, and packing of hydrophobic groups interactions, and a penalty for burying hydrophilic groups. The entire protocol was named FARFAR (Full Atom Refinement FARNA). In addition to the previous terms, the latest updates from the Rosetta development community included a term to account for the carbon–hydrogen bonds, an alternative orientation-dependent model for desolvation, and a term to describe the screened electrostatics interactions between phosphate groups (Das, Karanicolas, and Baker 2010).

The two-step procedure is able to model RNA structure up to an RMSD < 2.0 Å from native for 14 out of 32 RNA fragments of short lengths (6–23 residues). Failures in RNA structure prediction may be due to lack of proper sampling (Das, Karanicolas, and Baker 2010). Moreover, the same authors tested FARFAR on the reverse design problem, by redesigning the sequence (nucleobases) on 15 x-ray high-resolution RNA template structures with good recapitulation of Watson–Crick and non-Watson–Crick base pairing at 40% and 62%, respectively. Some of the divergent designed sequences were more stable than native; however, the fact that they were not selected by natural evolution suggests functional restrictions (interaction with other macromolecules, like proteins) that cannot be properly modeled by force fields.

6.3.3 Backbone–Nucleobase Hybrid Models

These models use several beads to describe atoms in both backbone and nucleobases, to improve the accuracy of the results. Naturally, a rise in computational expense is the cost of using more sophisticated models. Here, we describe two of the most common hybrid models: the three-, five- and seven-bead models.

6.3.3.1 Three-Bead Model

In the three-bead model, introduced by Dokholyan and collaborators (Ding et al. 2008), the geometry of an RNA molecule is described by three beads representing phosphate, sugar, and nucleobase (Figure 6.9). The beads are placed in the center of mass of the corresponding chemical moieties. To describe forces acting between the atoms of RNA, two types of interactions are introduced: bonded and nonbonded. Bonded interactions are used to preserve the connectivity and local geometry of RNA chain. The nonbonded interactions include base pairing (A–U, G–C, and U–G), base stacking, short-range phosphate–phosphate repulsion, and hydrophobic interactions. All the interactions are approximated by square-well potentials in order to make the model compatible with a discrete molecular dynamics (DMD) sampling engine, as described in the next section. By choosing an appropriate discretization potential step, the computer simulations can be accelerated. Due to the specific nature of DMD, all the potentials are expressed as a function of the interatomic distances. For example,

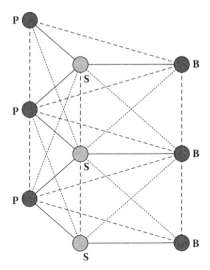

FIGURE 6.9   Three-bead model of RNA. The beads represent phosphate (P), sugar (S), and nucleobase (B). Covalent interactions are shown by thick lines, dashed lines show bond angular restraints, and dashed-dotted lines show dihedral restraints.

to place a restraint on a bond angle in the RNA backbone, three pairwise terms are required to restrict the atoms related by a certain bond angle. For bonded potentials, infinite square wells are used to preserve connectivity of a chain. The parameters of bonded interactions were derived from an RNA database with high spatial resolution (Murray et al. 2003). Base pairing interactions are modeled using a "reaction" algorithm (Ding et al. 2003), which allows formation of direction-dependent hydrogen bonds. Phosphate–phosphate electrostatic repulsion is modeled by the Debye–Hückel potential (Debye and Hückel 1923). Parameters for the stacking and hydrophobic interactions were determined by sequence-dependent free energy decomposition for an individual nearest-neighbor hydrogen bond (INN-HB) model (Mathews et al. 1999). An essential part of the simulation algorithm is explicit modeling of loop entropy. Due to the reduction of the degrees of freedom as a consequence of a coarse-grained nature of the model, the entropy of a system tends to be underestimated in the simulations, which leads to formation of long unnatural loops that trap RNA in nonnative conformations. To solve this problem, the authors use experimentally determined free energies of the loops, which were tabulated for loops of different sizes and types (i.e., hairpin, bulge, or internal loops) (Mathews et al. 1999). These values are used to guide formation of base pairs during the course of a simulation. The model was validated on the set of 153 RNAs, ranging in length from 10 to 100 nt. The accuracy of the predictions was evaluated using two criteria: (1) RMSD between predicted and experimentally determined structure from the PDB and (2) percentage of native contacts formed in predicted structure. It was shown that for RNA molecules with length less than 50 nt, predicted RMSD is less than 6 Å, and for 85% of all predicted structures, the RMSD is less than 4 Å. The average percentage of native contacts predicted in the simulations was 94%. The performance of this model can be greatly enhanced with the use of experimental restraints, as described in Section 6.5.

6.3.3.2 Five-Bead Models

In the model developed by Ren and collaborators (Xia et al. 2010), every nucleotide is represented by five pseudoatoms: two pseudoatoms for the phosphate and the sugar and three pseudoatoms for the nucleobase (Figure 6.10). Three pseudoatoms describe well every nucleobase, each

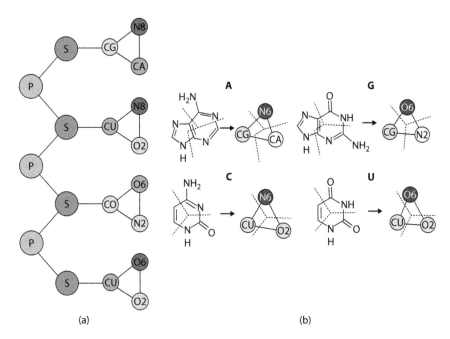

(a) (b)

FIGURE 6.10 (a) Coarse-grained five-bead model of RNA. Phosphate (P) and sugar (S) are represented by one bead. Three beads are used to represent nucleobase. (b) Composition of the beads. The atoms of nucleobase are grouped into different pseudoatoms by lines to coarse-grained pseudoatoms. (Reprinted with permission from Xia et al., 13497–13506. Copyright 2010 American Chemical Society.)

representing a certain group of real atoms in the nucleobase. It was found that five different types of pseudoatoms are enough to describe all four nucleobases, since the same groups of atoms are present in different nucleobases. Potential energy of the coarse-grained field consists of two terms, corresponding to bonded and nonbonded interactions. Bonded interactions are described by the sum of bond stretching, bond angle bending, and dihedral angle rotation terms:

$$E_{\text{bonded}} = K_{\text{bond}} (b - b_0)^2 + K_a (\theta - \theta_0)^2 + \sum_{n=1}^{3} K_n \left[1 + \cos(n\phi - \delta_n)\right]$$

Here, b is a bond length, θ is a bond angle, and ϕ is a dihedral angle. In total, there are 14 different types of bonds, 25 types of bond angles, and 28 types of dihedral angles, each corresponding to their own set of parameters. These parameters were extracted from statistical distributions, fitted to a normal distribution, of bond lengths and bond and dihedral angles obtained from a set of 668 RNA structures from the PDB. The Buckingham potential describes the nonbonded interactions as

$$E_{\text{nonbonded}} = \varepsilon_{ij} \left[-2.25 \left(\sigma_{ij}/r_{ij}\right)^6 + 1.84 \times 10^5 e^{-12.00(r_{ij}/\sigma_{ij})} \right]$$

Here, ε_{ij} is the depth of the potential well, σ_{ij} is the radius, and r_{ij} is the distance between the atoms. There are 19 pairs of (σ_{ij}, r_{ij}) parameters describing nonbonded interactions between different types of pseudoatoms. The values of these parameters were derived upon least square fit of the Buckingham potential to the following potential of mean force:

$$E_{\text{nonbonded}}(r) = -k_B T \ln g(r)$$

Here, $g(r)$ is a radial distribution function (RDF) defined as

$$g_{ij}(r) = \frac{1}{N_i d_j} \frac{n_{ij}(r)}{4\pi r^2 \delta r}$$

Here, n_{ij} is a number of pairs at the distance from r to $r + dr$, N_i is the total number of type i particles in the system, and d_j is the mean density of type j particles. All 19 RDFs were calculated using the same data set of RNA structures as was used to determine the parameters for bonded interactions. The nonbonded interactions were further refined on the set of seven structurally diverse RNA by minimizing the difference between predicted and experimental structures. This method was tested on a set of 15 RNA molecules with length 12–27 nt. MD simulations with simulated annealing led 75% of tested RNA molecules to native-like structures at least once during the course of the simulation. With imposed restraints on secondary structure content, all 15 RNAs were folded to within 6.5 Å relatively to native structures.

In an alternative five-bead model by Levitt (Bernauer et al. 2011), the beads are placed at the positions of P and C_4' backbone atoms, and at the positions of C_2, C_4, and C_6 nucleobase atoms. The authors define statistical potential as

$$E = -kT \sum_{ij} \ln \left(\frac{P_{\text{obs}}(d_{ij})}{P_{\text{ref}}(d_{ij})} \right)$$

Here, $P_{\text{obs}}(d_{ij})$ is a probability to find pseudoatoms of types i and j at the distance d, calculated from the set of RNA structures from the PDB. On total, 77 RNA structures were used, chosen to have high experimental resolution, and to be nonredundant. $P_{\text{ref}}(d_{ij})$ is a reference probability calculated from overall mole fractions of atom types i and j (Lu and Skolnick 2001). The authors used their model to identify native-like RNA structure from a set of near-native models (decoys). Three sets of decoys were generated using position-restrained MD, normal modes, and assembly from RNA-like fragments, which had no native base pairing enforced. The performance of the model was evaluated by counting the number of decoys with energy lower than the energy of the native structure. Out of five RNAs used to generate decoys with position-restrained MD, for one of the RNA structures, only one decoy was found to have lower energy than native structure. Out of 15 RNAs used to produce decoys with normal modes, the native structures had the lowest energy for 14 RNAs, and only for one RNA 11 decoys scored better than the native structure. The most difficult task was to distinguish native structure from the set of decoys produced from RNA fragments. Out of 19 tested RNAs, only in nine of them the native structure was scored as having the lowest energy. In each of other 10 RNAs, the number of decoys scored better than native structure was close to 500.

6.3.3.3 Seven-Bead Models

HiRE-RNA (Pasquali and Derreumaux 2010) is a coarse-grained model that relies on seven-bead representation of a single nucleotide: one bead for phosphate, four beads for sugar, O_5', C_5', C_4', C_1', one bead for the pyrimidine bases (C, U), and two beads for the purine bases (G, A). The beads representing pyrimidine and purine bases are placed in the centers of mass of nonhydrogen atoms of the corresponding aromatic rings (Figure 6.11). The HiRE-RNA force field is represented by a sum of local, nonbonded, and hydrogen-bonding terms. The local interaction term is expressed as

$$E_{\text{local}} = \sum_b K_b \left(r - r_{\text{eq}} \right)^2 + \sum_\alpha K_\alpha \left(\alpha - \alpha_{\text{eq}} \right)^2$$
$$+ \sum_d K_d \left[1 + \cos(\tau - \gamma) \right]$$

Nonbonded and hydrogen bond interactions are expressed as

$$E_{\text{nonbonded}}(r_{ij}) = \varepsilon_{ij} \left[\left(\frac{G\left(\sigma_{ij}\right)}{r_{ij}} \right)^6 e^{-2r_{ij}} + \frac{3}{5}\tanh\left[2\left(r_{ij} - \sigma_{ij} - \frac{1}{2} \right) - 1 \right] \right]$$

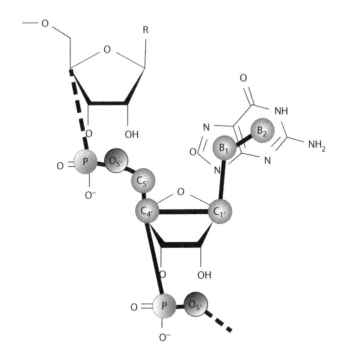

FIGURE 6.11   Coarse-grained representation of seven-bead model. Phosphate (P), O5′, C5′, C4′, and C1′ are represented by single beads. Beads for purine (two beads) and pyrimidine (one bead) are placed in the centers of corresponding aromatic rings. (Reprinted with permission from Pasquali, Samuela, and Philippe, 11957–11966. Copyright 2010 American Chemical Society.)

Here, r_{ij} and σ_{ij} are instantaneous and equilibrium distances between atoms, ε_{ij} is a coupling constant, and $G(\sigma_{ij})$ is a numerical parameter depending on equilibrium distance between i and j. Hydrogen-bonding interactions are described by a sum of two-body, three-body, and four-body terms. Recently, the authors of HiRE-RNA revised their force field (Cragnolini et al. 2014), introducing new features, in particular noncanonical base pairing and the possibility for a base to form multiple contacts. The performance of the improved force field was evaluated on a set of seven topologically different RNAs ranging in length between 22 and 79 nt. For three modeled structures, including two hairpins (PDB-IDs 1F9L and 1N8X) and a duplex (PDB-ID 433D), the reached precision was 2–3 Å as compared to the x-ray structure. For the triple helix (PDB-ID: 2K96) of 49 nt, the precision was 4.3 Å. For two pseudoknots (PDB-IDs: 2G1W and 1L2X) of 22 and 28 nt, respectively, the experimental structure was predicted as the most stable structure; however, the authors have not reported obtained RMSD. The last predicted structure was the riboswitch (PDB-ID: 1Y26) of 79 nt. This prediction was done using secondary structure restraints, resulting in a precision of 7–8 Å for the best models.

6.4 SAMPLING METHODS

Modeling of RNA, in order to predict its native conformation, ideally involves the exploration of the full conformational landscape. This is almost unreachable using the current state-of-the-art computers. Most common sampling algorithms aim at generating a representative number of conformations to allow a reasonable estimation of thermodynamic and structural properties. In this section, we summarize the basic working principles behind those sampling algorithms used for modeling RNA structures, i.e., MD, MC, and DMD simulations.

Note that, to enhance the sampling efficiency and rapid exploration of the conformational space overcoming big energy barriers, these sampling methodologies often use a variety of different enhancement strategies, such as umbrella sampling (Torrie and Valleau 1977) and steered MD (Patel et al. 2014).

6.4.1 MD Simulations

MD simulations intend to compute the real dynamics of a given system. Regardless of the model used to account for interatomic particle interactions of the system, whether all-atom or coarse-grained, the positions and velocities of each particle in the system are calculated by integrating Newton's equations of motion numerically (discretized into small time steps of the order of 1 fs). At each time step, the forces acting on each particle are computed from the gradient of the potential energy. The corresponding accelerations derived from those forces are then used to update positions and velocities of each particle at the current step. New forces are then computed and the process iterates for the entire simulation trajectory. Forces are considered constant within an integration step. At the beginning of the simulation, the velocities are randomly assigned to each particle according to a Maxwell–Boltzmann distribution that depends on the simulation temperature. The potential energy between two interacting atoms typically contains bonded and nonbonded terms (Laurendeau 2005; McCammon, Gelin, and Karplus 1977; Gelin and Karplus 1975). The bonded terms usually include three harmonic potential terms: bond, angle, and torsional angles. The harmonic potential is applied to each atom with respect to the equilibrium values of bond distances and bend angles, and a Fourier summation for the torsional angles. Nonbonded interactions in all-atom simulations are typically the sum of the electrostatic and van der Waals interaction functions. Notably, analogous interactions are developed in case of more coarse-grained models. Several methods reviewed in Section 6.3 adopted an MD-based sampling method combined with a coarse-grained potential for modeling RNA 3D structure: NAST (Jonikas et al. 2009), a five-bead model (Xia et al. 2010), and HiRe-RNA (Pasquali and Derreumaux 2010).

6.4.2 MC Simulations

In MC simulations, a new conformation is generated by randomly changing one or several degrees of freedom of the system. At this point, the potential energy is calculated and compared with the energy of the conformation at the previous time step. If the energy of the new conformation is lower than the previous one, the current conformation is accepted. If the new energy is higher than its predecessor, then a random number between 0 and 1 is generated and compared to the computed Boltzmann factor $\exp\left[-\left(E_{i+1} - E_i\right)/k_B T\right]$. If the random number is lower than the Boltzmann factor, the new conformation is accepted; otherwise it is rejected (Metropolis acceptance criterion, Metropolis and Ulam 1949). Generating system trajectories using an MC method is computationally less demanding than using MD, requiring only the energies of the system and not the forces.

However, MD provides time dependence of the variables of the system, while in MC, each move attempt depends exclusively on its predecessor.

Das and collaborators (Das and Baker 2007; Das, Karanicolas, and Baker 2010) developed an MC-based sampling approach within the Rosetta family of methods to sample the RNA conformational space and find the native conformation as described in Section 6.3.2.

6.4.3 DMD Simulations

DMD simulation is a special case of a generalized MD simulation, in which the particles move with constant velocity upon no interaction, and change velocities upon interaction and/or elastic collision so that total energy and momenta are preserved (Alder and Wainwright 1959). Elastic collisions between two interacting atoms take place when the increase in the potential energy cannot be compensated by a decrease in the kinetic energy (Figure 6.12). After a collision, the velocities and positions are updated for the colliding atoms, keeping the potential energy constant and generating a new list of putative collisions with the neighbor atoms. The consecutive repetition of these events yields a complete trajectory.

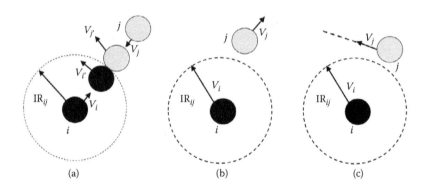

(a)   (b)   (c)

FIGURE 6.12 DMD interactions and collisions. (a) When two particles approach each other within the interaction range, IR_{ij}, higher than the minimum distance for hard-core repulsion and lower than the maximum interacting distance, they interact. Whenever within the interaction range, if two atoms are approaching, they will experience hard-core repulsion (collision) if the decrease in the kinetic energy does not compensate for the increase in the potential energy. Upon collision, two particles with divergent velocities will not interact (b). Another example of noninteraction is the case in which the particles are approaching but not intersecting the interaction range (c). (With kind permission from Springer Science+Business Media: *Computational Modeling of Biological Systems: From Molecules to Pathways*, Discrete molecular dynamics simulation of biomolecules, 2012, 55–73, Ding, F and N V Dokholyan. Copyright 2012b.)

Besides the "event-driven" nature of DMD, it has been successfully parallelized so that it can be used in combination with replica exchange algorithm (Earl and Deem 2005), where different simulations are run in parallel at different temperatures. In order to overcome potential energy barriers, a given pair of replicas can exchange temperature according to a Metropolis criterion. The Dokholyan group has developed a tool to model RNA native structures that integrates replica exchange DMD sampling with a three-bead model representation of the RNA molecule and a discrete stepwise interaction potential. This tool has proven to be computationally efficient and generates reliable native-like structures of RNA (see previous section for details) (Ding et al. 2008; Sharma, Ding, and Dokholyan 2008).

What makes DMD particularly interesting is the fact that it does not require the computation of the gradients of the potentials to obtain forces. Thus, it does not have to integrate any differential equation, allowing longer time steps and simulation runs as compared to conventional MD, at lower computational cost. Also the discrete nature of the potential energy function makes calculations much faster than a traditional MD simulation of the same time length.

6.5 GUIDING CONFORMATIONAL SAMPLING BY MEANS OF EXPERIMENTAL RESTRAINTS

Even with efficient sampling algorithms, the conformational space is so large that it becomes almost impossible to exhaustively sample it all. An efficient way to improve the quality of the predictions is by imposing restraints (typically distance restraints) between certain atom pairs, according to experimental data. By doing so, simulations become more reliable in overcoming force field limitations and are more likely to converge to the native state.

There are several experimental techniques that can provide restraints between atom pairs in the form of a pairwise potential to improve RNA structure prediction methods. In this section, we describe the most common techniques. It is not the purpose to deeply review them all here, but rather provide an introduction and appropriate reference to the original works to the interested reader.

6.5.1 Most Common Experimental Methods to Derive Distance Restraints

As for protein structure determination, the two most well-developed techniques to determine RNA 3D structures are x-ray crystallography and NMR. Since x-ray crystallography builds the entire 3D model for the RNA molecule under consideration upon analysis of the diffraction pattern (Kendrew et al. 1958), it will not be considered here as a mean to provide experimental restraints. However, one can derive knowledge-based potentials on the deposited structures in the PDB or use a particular solved structure as a template in homology modeling to build the structure of a homologous RNA piece (Flores et al. 2010). Despite the intrinsic limitations to solve NMR spectra (see Section 6.1), computer simulation is needed to build up the final model with the torsion angles derived from J-coupling constants and the distances from the Nuclear Overhauser Effect SpectroscopY (NOESY) cross peaks (Addess and Feigon 1996), taking into account that the data to parameter ratio is much poorer than in x-ray crystallography. In particular, it is not necessary to resolve the entire NOESY spectrum in order to model RNA structure; using a few distance restraints is enough for DMD to attain conformations close to the native.

In the recent years, a set of chemical probing methods have been developed to extract structural features on RNA by exploiting the chemical properties of its nucleotides. The most used chemical reagents comprise a mixture of H_2O_2, ascorbic acid, and Fe(II)-ethylenediaminetetraacetic acid (EDTA) to generate hydroxyl radicals; dimethyl sulfate (DMS), 1-cyclohexyl-(2-morpholinoethyl) carbodiimide metho-p-toluenesulfonate (CMCT), kethoxal; and the selective 2'-hydroxyl acylation analyzed by primer extension (SHAPE) method. We will treat more deeply the hydroxyl radical probing (HRP) and SHAPE methods since they have been able to produce reliable results in conjunction with a DMD sampling and three-bead model combined method (Ding et al. 2012; Gherghe et al. 2009).

DMS modifies nucleobases by methylating N_1 atom in the adenosine and N_3 at cytosine. Those chemical modifications can only take place in single-stranded RNA, base paired at the end of a helix or a base pair next to a G–U. The modified RNA is then analyzed through reverse transcriptase polymerase chain reaction (RT-PCR) (Freeman, Walker, and Vrana 1999), and run on a gel. Since modified sites cannot be base paired, they produce different size bands, as the polymerase fails to add a nucleotide in this particular position. Protected positions can be due to base pairing, tertiary contacts, or interactions with other macromolecules. The band pattern generated by DMS treated can thus be used to infer RNA secondary structure (Tijerina, Mohr, and Russell 2007; Krokhotin et al. 2017).

Similarly, CMCT also reacts to exposed N_3 atoms in uridine or N_1 atoms in guanine, preventing base pairing, so that the unprotected sites correspond to single-stranded RNA, and the protected to double stranded. The RT-PCR and gel allow also assignment of a secondary structure (Fritz et al. 2002). The kethoxal (1,1-dihydroxy-3-ethoxy-2-butanone) reagent forms a covalent adduct with adenosine upon reaction on N_1 and N_2 atoms that prevents base pairing, so that when the reverse transcriptase reaches the modified position, it falls off yielding a band in the gel (Gopinath 2009). By combining the three chemical probing, DMS, CMT, and kethoxal, it is possible to gather data about base pairing and/or tertiary contacts.

Hydroxyl radical probing (HRP) consists of the reaction of hydroxyl radicals with RNA, by nucleophilic attack on the ribose, breaking the phosphate backbone. Hydroxyl radicals are chemically unstable, but can be generated using a mixture of H_2O_2, ascorbic acid, and Fe(II)-EDTA. Upon digestion, RNA molecules are cleaved and then amplified by RT-PCR. Since the last nucleotide remains intact upon cleavage, the recovered band length on a gel is informative about the last nucleotide cleaved. The different protection strength against HRP does not depend on base pairing (secondary structure) since the backbone is not involved, but rather on tertiary contacts (Karaduman et al. 2006). In Section 6.5.3, we explain how to incorporate the data on HRP accessibility to bias RNA structure modeling, in the context of DMD.

The SHAPE method uses some reactants such as N-methylisotoic anhydride (NMIA) and 1-methyl-7-nitroisatoic anhydride (1M7) that make covalent adducts with RNA upon reaction with a flexible 2′ hydroxyl group. Adduct formation at 2′OH position is quantified by extension of a complimentary DNA using reverse transcriptase and two primers flanking the RNA fragment upon consideration. The 2′-O modified nucleotide stops reverse transcriptase one nucleotide before. By comparing the band profile upon reaction to either NMIA or 1M7 with respect to a nonreactive control and to a sequencing control for nucleotide identification in the same gel, one can get precise data about the secondary structure (Merino et al. 2005). In Section 6.5.2, we discuss the use of SHAPE data as restraints in DMD simulation to model 3D RNA structure.

6.5.2 Secondary Structure from SHAPE in Addition to Long-Range Restraints

SHAPE chemistry can be used to determine RNA base pairing very accurately, since 2′-OH reactivity depends on whether bases are paired and/or making tertiary contacts. Secondary structure is typically assigned using SHAPE reactivity data in combination with secondary structure prediction software. In brief, the reactivity data (Figure 6.13) can be converted into a free energy profile (Deigan et al. 2009) that can be input to dynamic programming algorithms to predict RNA secondary structure (Mathews et al. 2004). In the case of the $^{t}RNA^{Asp}$, SHAPE data in combination with dynamic programing produced the correct secondary structure, as displayed in Figure 6.14. Base pairing is implemented as restraints in a multibody energy potential that accounts for both distance and orientation restraints between base–base, base–sugar, and base–phosphate pseudoatoms (Figure 6.15), in addition to a soft attractive potential up to 50 Å term between bases. Distances are taken from statistics on known RNA structures.

A particular derivation of the HRP chemistry, named tethered hydroxyl radical probing (t-HRP), can be used to infer long-range contact information and derive pairwise distance restraint potential energy (Gherghe et al. 2009). Pairwise interactions are generated between a Fe(II)-EDTA moiety covalently bound between GpC steps adjacent to a single nucleotide bulge upon reaction to methydiumpropyl-EDTA (MPE) (Figure 6.16). In the context of the DMD, a pairwise step potential can be set up with a minimum energy in the range of 25 Å to the tethered position, whose depth is assigned from the normalized t-HRP reactivity relative intensity (as compared to nontethered RNA molecule), interpreted as the probability of cleavage within the cutoff distance. Some soft penalties can extend the interaction range up to 35 Å (Figure 6.17). Long-range restraints are incorporated into DMD simulations in a second step after base pairing (see Figure 6.18 for the DMD protocol). The last stage involves clustering and analysis of the most populated cluster. The method has been applied to the refinement of $^{t}RNA^{Asp}$ structure up to 4 Å, when one or several sets of MPE (differing in the nucleotide tethered) are used (Gherghe et al. 2009). However, in the absence of long-range restraints, the closest model attained is at 11 Å from the crystal structure.

In a more general context, the combined use of DMD with restraints from tertiary contact data available from different experiments plus secondary structure (base pairing) data yielded high-resolution models for four RNAs: domain III of the cricket paralysis virus internal ribosome entry site (CrPV) (49 nt), a full-length hammerhead ribozyme from *Schistosoma mansoni* (HHR) (67 nt), *Saccharomyces cerevisiae* $^{t}RNA^{Asp}$ (75 nt), and the P546 domain of the *T. thermophila* group I intron (P546) (158 nt), at 3.6, 5.4, 6.4, and 11.3 Å, respectively, from the crystal structure (Lavender et al. 2010). The restraint potentials were set generically as 2 kcal/mol well depth and 15 Å width.

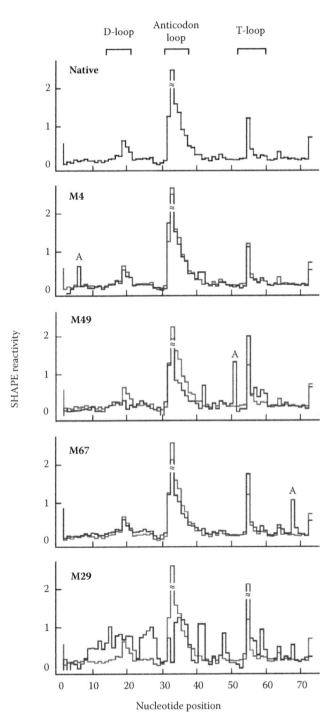

FIGURE 6.13 SHAPE reactivity profile for ᵗRNAᴬˢᵖ. The SHAPE profile tells us the compatibility of the MPE binding sites with ᵗRNAᴬˢᵖ tertiary structure. Here the profile for each mutant is shown as the dark bar plot, superimposed to the profile on the wild type in gray. The position where the bulged adenosine is inserted is indicated. (Reprinted with permission from Gherghe et al., 2541–2546. Copyright 2009 American Chemical Society.)

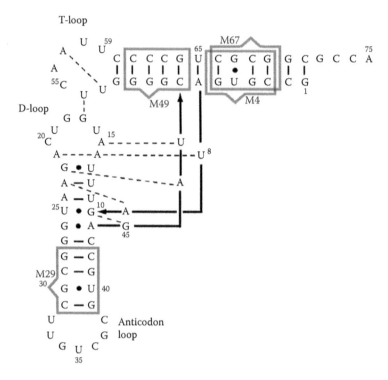

FIGURE 6.14 Secondary structure for ${}^{t}RNA^{Asp}$, showing the different mutants, which are numbered according to the closest residue to the tethered Fe(II)-EDTA group. (Reprinted with permission from Gherghe et al., 2541–2546. Copyright 2009 American Chemical Society.)

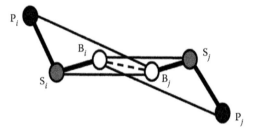

FIGURE 6.15 Three-bead model indicating the different pairwise interactions between: bases, B_i–B_j or base pairing; nucleotide orientation, defined by the base–sugar interactions, B_i–S_j and B_j–S_i; and the base–phosphate interaction B_i–P_j and B_j–P_i, which define RNA rigidity. (Reprinted with permission from Gherghe et al., 2541–2546. Copyright 2009 American Chemical Society.)

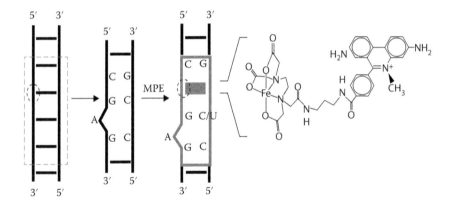

FIGURE 6.16 Intercalation of PME in a bulged helix to replace four canonical base pair. MPE has its Fe(II)-EDTA moiety oriented toward the bulged adenosine nucleotide. (Reprinted with permission from Gherghe et al., 2541–2546. Copyright 2009 American Chemical Society.)

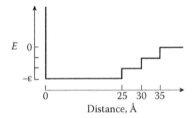

FIGURE 6.17 Discrete energy potential applied to the long-range interaction derived from the site-directed hydroxyl chemistry. (Reprinted with permission from Gherghe et al., 2541–2546. Copyright 2009 American Chemical Society.)

6.5.3 Incorporating HPR Data as Restraints into DMD Simulations

The traditional HRP chemistry reactivity profile, on the other hand, provides information about solvent accessibility for the backbone atoms, meaning that low accessible backbone atoms are making a high number of interatomic contacts (Figure 6.19a). For a particular nucleotide, there is thus a negative correlation between the number of atoms in contact at a given distance cutoff and its HRP reactivity (Figure 6.19b). The Dokholyan group implemented a procedure, based on the three-bead model, which maximizes the number of contacts between sugar beads according to the data derived from HRP, in DMD simulations. The biasing potential is made of two terms, one generically attractive to most of the nucleotide pairs, ensuring RNA collapse and packing, and a second term that is repulsive once a given nucleotide has reached its maximum number of assigned

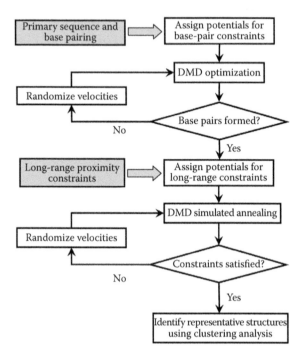

FIGURE 6.18 DMD protocol for RNA structure refinement, including the secondary structure data from SHAPE and long-range restraints from site-directed hydroxyl radical cleavage. It starts from the RNA sequences and the base pairing data. Once the base pairing is satisfied, the weighted potentials to account for the long-range interactions are added. The system is then cooled down. Upon fulfillment of the latter restraints, there is a final equilibration run. (Reprinted with permission from Gherghe et al., 2541–2546. Copyright 2009 American Chemical Society.)

contacts as derived from the HRP reactivity profile. If a given nucleotide achieved the maximum number of contacts, it will only accept a new one if the kinetic energy during DMD simulation overcomes the penalty energy barrier of overpacking (Ding et al. 2012) (Figure 6.20a).

The entire DMD protocol is made up of three steps. Initially, conventional DMD simulations are performed to ensure base pairing. A second round involves replica exchange DMD simulations and HRP restraints as explained above. The second step expects to have an enriched population of structures in agreement with the experimental data (Figure 6.20b). At the end, the limited number of trajectory snapshots is selected based on their energy and correlation between number of contacts and HRP reactivities. These structures are clustered according to the RMSD. Representative structures of the most populated clusters tend to reproduce native RNA

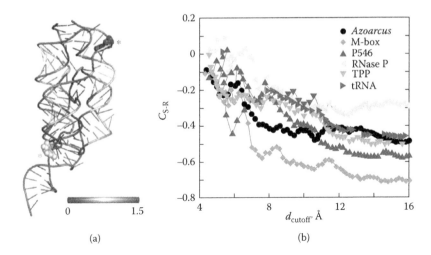

(a) (b)

FIGURE 6.19 (a) Structure of the M-box riboswitch shaded according to HRP reactivity. Dark zones correspond to low HRP reactivity and are buried, while light areas are rather solvent exposed. However, two residues are displayed as spheres that correspond to a solvent accessible residue with high HRP reactivity (light spheres) and a buried residue with low HRP reactivity (dark spheres). (b) Pearson correlation coefficient between the HRP reactivity and the number of atomic contacts C_{S-R} as a function of the distance cutoff. The best correlation in absolute values appears to be in the range of 14–16 Å. (Reprinted by permission from MacMillan Publishers Ltd. *Nature Methods*, Ding et al., 9 (6): 603–8, copyright 2012.)

structures, even for large RNA fragments up to 230 nucleotides (Ding et al. 2012; Krokhotin et al. 2012).

6.6 COMPARISON OF THE PERFORMANCE OF DIFFERENT METHODS: RNA-PUZZLES

In recent years, several computational techniques have been developed for the *de novo* 3D structure prediction of middle-sized RNA sequences (i.e., up to a few hundred). Therefore, it has become essential to benchmark the quality and the performance of the current methods and tools for structural predictions. In order to meet this need of the scientific community, Dr. Westhof and colleagues have established RNA-Puzzles (http://ahsoka.u-strasbg.fr/rnapuzzlesv2), a collective experiment for blind RNA 3D structure prediction. The main aim of RNA-Puzzles is the comparison of strengths and limits of the existent methodologies for modeling RNA structures in order to assess the status of research in the field and promote its further development (Cruz et al. 2012).

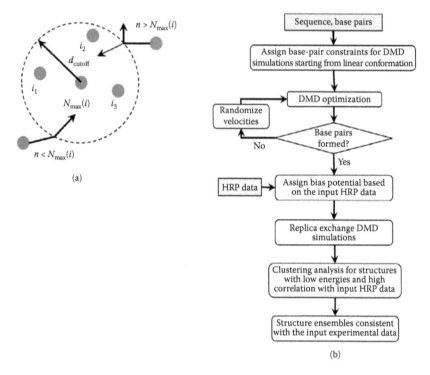

(a)

(b)

FIGURE 6.20 Incorporating HRP reactivities into energy potentials in DMD simulations. (a) A number of maximum (N_{max}) contacts are assigned to each nucleotide according to the cutoff distance. A nucleotide approaching a given i can form new contact if the number of neighbors (denoted by $i_1, i_2, ...$) is smaller than N_{max}. If it is bigger, a new contact can form only if the total kinetic energy is sufficient to overcome the overpacking energy penalty. (b) Schematic representation of the flow chart for the HRP-directed DMD simulation protocol. (Reprinted by permission from MacMillan Publishers Ltd. *Nature Methods*, Ding et al., 9 (6): 603–8, copyright 2012.)

In RNA-Puzzles, crystallographic RNA structures are provided to the structural biology community in order to challenge several research groups to generate accurate 3D structural models. Starting from 2011, eight exercises (i.e., puzzles) have been proposed (Figure 6.21), two of which were open at the time this chapter was written.

The quality of each submitted model is evaluated using the RMSD and the deformation index (DI). RMSD accounts for the average distance between the model and the target structure (from x-ray crystallography), upon optimal superposition of the former onto the latter:

$$RMSD(A, B) = \sqrt{\frac{\sum_{i=1}^{N} d_i^2}{N}}$$

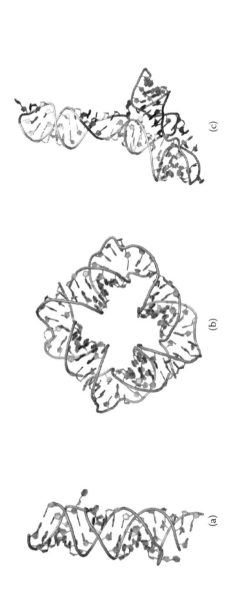

FIGURE 6.21 RNA-Puzzles crystallographic molecular structures and challenges. (a) Challenge 1: *"What is the structure of the following sequence: 5′-CCGCCGCGGCCAUGCCUGUGGCGG-3′ knowing that the crystal structure shows a homodimer that contains two strands of the sequence? The strands hybridize with blunt ends (C–G closing base pairs)."* Crystal structure provided by T. Hermann (From Dibrov, S et al., *Acta Crystallographica Section D*, 67, 97–104, 2011), 46 nt. (b) Challenge 2: *"The crystal structure shows a 100 nucleotide square that assembles from four inner and four outer strands. 3D coordinates of the nucleotides in the inner strands were provided. What are the structures of the outer strands?"* Crystal structure provided by T. Hermann (From Dibrov, S et al., *Acta Crystallographica Section D*, 67, 97–104, 2011). (c) Challenge: 3 *"A domain of a riboswitch was crystallized. The sequence is the following: 5′-CUCUGGAGAGAACCGUUUAAUCGGUCGCCGAAGGAGGAGCACUCUGCCAUAUGCAGAGUGAAACUCUCAGGCAAAGGACAGAGAG-3′. The crystallized sequence was slightly different (an apical loop was replaced by a GAAA loop) but it was not mentioned to protect the work of crystallographers."* Crystal structure provided by D. Patel (From Huang, L et al., *Molecular Cell*, 40, 774–786, 2010), 84 nt.   *(Continued)*

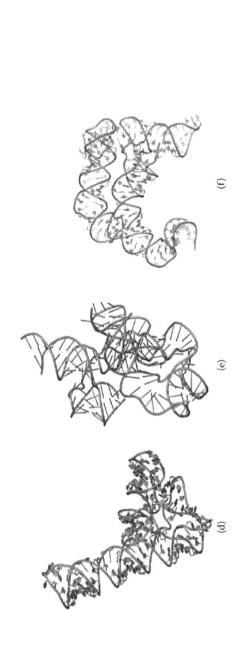

(d)

(e)

(f)

FIGURE 6.21 (*Continued*) RNA-Puzzles crystallographic molecular structures and challenges. (d) Challenge 4: *"What is the structure of the following sequence:* 5'-GGCUUAUCAAGAGAGGUGGAGGGACUGGCCCGAUGAAACCCGGCAACCACUAGUCUAGUCUAGCGUUGGCUAGCUUCGGCUGACGCUAGGCUA GUGGUGCCAAUUCCUGCAGCGGAAACGUUGAAACGAUGAGCCA-3?" Crystal structure provided by A. Ferré-D'amaré (From Baird, N J et al., *RNA*, 18, 759–770, 2012), 126 nt. (e) Challenge 6: *"This problem consists in predicting the structure of an RNA molecule with the following sequence:* 5'-CGGCAGGUGCUCCCGACCCUGCGGUCGGAGUUAAAAGGGAAGCCGGUGCAAGUCCGGCACGUCCGCCACGUGACGGGGAGUCGCC CCUCGGGAUGUGUGCCACUGGCCCGAAGGCCGGGAAGGCGGAGGGCGGGGAAGGGCGGAUCCGAGUCAGGAAACCUGCCUGCCG-3'." Crystal struct-ure provided by A. Serganov (From Peselis, A and A Serganov, *Nature Structural & Molecular Biology*, 19, 1182–1184, 2012), 168 nt. (f) Challenge 10: *"This problem consists in predicting the structure of a T-box riboswitch and tRNA 1:1 complex with the following sequence:* 5'UGCGAUGAGAAGAAGA GUAUUAAGGAUUUACUAGGAUUAGCGACUCUAGGAUAGUGAAAGCUAGGAUAGUAACCUUAAGAAGGCACUUCGAGCA-3' (*T-box*). 5'-GC GGAGUAGUUCAGUCGGUUAGAACACCACCUUGCCAAGGUGGGGGUCGCGGGUUCGAAUCCCGUCUUCCGCUCCA-3' (*tRNA*)." Crystal structure provided by A. Ferré-D'amaré (From Zhang, J and A R Ferré-D'amaré, *Nature*, 500, 363–366, 2013), 171 nt. 3D structural models and challenges are reported from the RNA-Puzzles website (http://ahsoka.u-strasbg.fr/rnapuzzlesv2).

where d_i is the distance between the i atom in the modeled (A) and experimental (B) structures, and N is the number of considered atoms. *DI* corrects the RMSD between a model and the corresponding target structure by considering the interaction network fidelity (INF) between the two, i.e., how similar the base pairing and base stacking are:

$$DI(A, B) = \frac{RMSD(A, B)}{MCC(A, B)}$$

$$INF(A, B) = MCC(A, B) = \sqrt{PPV \times STY}$$

$$PPV \left(\text{specificity}\right) = \frac{|TP|}{|TP| + |FP|}$$

$$STY \left(\text{sensitivity}\right) = \frac{|TP|}{|TP| + |FN|}$$

where $MCC(A, B)$ is the Matthews correlation coefficient calculated as the geometric mean of the specificity and sensitivity for the base–base interaction predictions (Matthews 1975; Parisien et al. 2009). Stereochemistry accuracy of RNA structural models, however, is estimated with MolProbity (Davis et al. 2007), a web service that performs a number of structure quality checks on macromolecules, such as torsional angles and atomic clashes. The *clash score* is chosen as a measure of stereochemical correctness. Since the RMSD metric depends largely on the size of the RNA molecules considered, it requires an assessment of the statistical significance. The groups of Weeks and Dokholyan (Hajdin et al. 2010) have shown that all-against-all RMSD between pairs of computationally random generated RNA conformations (decoys) is normally distributed, and that the RMSD expected by chance correlates with the length of the molecule via a power law:

$$\langle RMSD \rangle = a \times N^{0.41} - b$$

where N is the number of nucleotides and a and b are constants whose values depend on the use of secondary structure information as restraints when sampling the RNA conformational space. Hence, for a given comparison RMSD, its associated p-value can be determined from the cumulative density function of the normal distribution (assuming a standard deviation of 1.8):

$$p = \left[1 + \text{erf}\left(\frac{(RMSD - \langle RMSD \rangle)/1.8}{\sqrt{2}}\right)\right] / 2$$

Generally, RNA structural models with a p-value lower than .01 represent a statistically significant prediction of a global RNA fold.

RNA-Puzzles participants adopt a variety of sampling methods and computer-assisted strategies to model the RNA 3D structure provided in each single puzzle (Hajdin et al. 2010; Kosinski, Cymerman, and Feder 2003; Chen 2008; Cao and Chen 2011; Sripakdeevong, Kladwang, and Das 2011; Das and Baker 2008; Das, Karanicolas, and Baker 2010; Ding and Dokholyan 2012a; Parisien and Major 2008; Rother et al. 2011, 2012; Cao and Chen 2005, 2006a,b; Gherghe et al. 2009; Lavender et al. 2010; Cruz et al. 2012; Sijenyi et al. 2012). For example, the Bujnicki group used a hybrid strategy, based on homology modeling, previously developed for protein modeling in the course of the Critical Assessment of protein Structure Prediction (CASP) experiment (Kosinski, Cymerman, and Feder 2003). The Chen lab used Vfold, a multiscale, free energy landscape-based RNA folding model (Chen 2008; Cao and Chen 2011). The Das group used the stepwise assembly method (Sripakdeevong, Kladwang, and Das 2011) as implemented in Rosetta (Das and Baker 2008; Das, Karanicolas, and Baker 2010). The Dokholyan group adopted a combined three-bead model RNA molecule representation with a DMD sampling approach (Ding and Dokholyan 2012a; Ding et al. 2008). The Major group applied their in-house developed MC-Fold and MC-Sym pipeline (Parisien and Major 2008). The results of their contributions to the past six challenges are summarized in Table 6.1. Generally, all submitted models correctly reproduced the base-pair interactions and accurately predicted the structure of helical regions in RNA. Loops remained the most difficult elements to model. Similarly, non-Watson–Crick base pairs were not consistently reproduced, clearly indicating the need for a better description of alternative interactions between RNA base pairs in the current computational techniques. Furthermore, the high values of *clash scores* in the submitted models can be improved by a more precise parameterization of geometric features like atomic bond, distances, and angles. The human and computer times required to produce the submitted models vary according to the approach adopted by individual research groups, and the computational resources that they employed. Therefore, it is difficult to have a direct comparison of the time-efficiency of each specific approach. Overall, the results of this first blind exercise for *de novo* RNA 3D structure prediction reflect the state of the art of the field.

TABLE 6.1 Comparison of RNA 3D Structural Models Submitted for RNA Puzzles Challenges

Challenge 1 (November 2011)					
Lab	RMSD	DI	Clash Score	*p*-Value	References
Das	3.413	3.657	0	2.220e−16	(Sripakdeevong, Kladwang, and Das 2011; Das and Baker 2008; Das, Karanicolas, and Baker 2010)
Major	4.064	4.567	66.4	3.830e−15	(Parisien and Major 2008)
Chen	4.113	5.008	0.68	4.774e−15	(Chen 2008; Cao and Chen 2005, 2006a,b, 2011)
Bujnicki	4.60	5.754	54.73	4.985e−14	(Kosinski, Cymerman, and Feder 2003; Rother et al. 2011, 2012; Boniecki et al. 2003; Surles et al. 1994)
Santalucia	5.688	6.750	28.40	3.203e−12	(Sijenyi et al. 2012)
Dokholyan	6.939	8.552	31.74	3.306e−10	(Hajdin et al. 2010; Ding and Dokholyan 2012a; Ding et al. 2008; Gherghe et al. 2009; Lavender et al. 2010)

Challenge 2 (November 2011)					
Lab	RMSD	DI	Clash Score	*p*-Value	References
Bujnicki	2.3	2.827	14.54	0.000e+00	(Kosinski, Cymerman, and Feder 2003; Rother et al. 2011, 2012; Boniecki et al. 2003; Surles et al. 1994)
Das	2.496	2.899	19.25	0.000e+00	(Sripakdeevong, Kladwang, and Das 2011; Das and Baker 2008; Das, Karanicolas, and Baker 2010)
Dokholyan	2.543	3.089	9.77	0.000e+00	(Hajdin et al. 2010; Ding and Dokholyan 2012a; Ding et al. 2008; Gherghe et al. 2009; Lavender et al. 2010)
Chen	2.83	3.739	18.66	1.110e−16	(Chen 2008; Cao and Chen 2005, 2006a,b, 2011)
Major	2.98	3.819	134.26	2.220e−16	(Parisien and Major 2008)
Wildauer	3.479	4.403	165.57	1.988e−15	NA
Santalucia	3.650	4.537	26.4	4.274e−15	(Sijenyi et al. 2012)

(Continued)

TABLE 6.1 (*Continued*) Comparison of RNA 3D Structural Models Submitted for RNA Puzzles Challenges

Challenge 3 (November 2011)				
Lab	RMSD	DI	Clash Score	p-Value	References
Chen	7.241	9.842	1.1	2.010e−05	(Chen 2008; Cao and Chen 2005, 2006a,b, 2011)
Dokholyan	11.460	16.104	41.21	3.902e−02	(Hajdin et al. 2010; Ding and Dokholyan 2012a; Ding et al. 2008; Gherghe et al. 2009; Lavender et al. 2010)
Das	11.965	16.419	1.1	6.932e−02	(Sripakdeevong, Kladwang, and Das 2011; Das and Baker 2008; Das, Karanicolas, and Baker 2010)
Bujnicki	12.186	17.494	14.72	8.711e−02	(Kosinski, Cymerman, and Feder 2003; Rother et al. 2011, 2012; Boniecki et al. 2003; Surles et al. 1994)
Major	13.701	23.331	93.52	3.026e−02	(Parisien and Major 2008)
Lab	RMSD	DI	Clash Score	p-Value	References
Chen	3.347	5.178	4.43	0.000e+00	(Chen 2008; Cao and Chen 2005, 2006a,b, 2011)
Santalucia	4.117	6.651	52.07	0.000e+00	(Sijenyi et al. 2012)
Bujnicki	4.191	6.803	34.92	0.000e+00	(Kosinski, Cymerman, and Feder 2003; Rother et al. 2011, 2012; Boniecki et al. 2003; Surles et al. 1994)
Adamiak	4.342	7.54	29.27	0.000e+00	NA
Das	4.498	7.183	14.28	0.000e+00	(Sripakdeevong, Kladwang, and Das 2011; Das and Baker 2008; Das, Karanicolas, and Baker 2010)
Major	4.515	7.229	1.23	0.000e+00	(Parisien and Major 2008)
Dokholyan	5.366	8.516	19.42	0.000e+00	(Hajdin et al. 2010; Ding and Dokholyan 2012a; Ding et al. 2008; Gherghe et al. 2009; Lavender et al. 2010)
Mikolajczak	12.801	25.341	40.11	1.362e−06	NA

(*Continued*)

TABLE 6.1 (*Continued*)   Comparison of RNA 3D Structural Models Submitted for RNA Puzzles Challenges

Lab	RMSD	DI	Clash Score	*p*-Value	References
Das	11.699	16.426	32.84	9.015e−14	(Sripakdeevong, Kladwang, and Das 2011; Das and Baker 2008; Das, Karanicolas, and Baker 2010)
Blanchet	21.379	30.813	2.2	2.359e−02	NA
Chen	22.152	34.442	3.49	5.996e−02	(Chen 2008; Cao and Chen 2005, 2006a,b, 2011)
Dokholyan	22.769	34.744	15.23	1.126e−01	(Hajdin et al. 2010; Ding and Dokholyan 2012a; Ding et al. 2008; Gherghe et al. 2009, Lavender et al. 2010)
Bujnicki	30.968	46.451	2.75	9.996e−01	(Kosinski, Cymerman, and Feder 2003; Rother et al. 2011, 2012; Boniecki et al. 2003; Surles et al. 1994)

Lab	RMSD	DI	Clash Score	*p*-Value	References
Das	6.803	12.687	11.09	0.000e+00	(Sripakdeevong, Kladwang, and Das 2011; Das and Baker 2008; Das, Karanicolas, and Baker 2010)
Bujnicki	9.339	18.149	0.91	0.000e+00	(Kosinski, Cymerman, and Feder 2003; Rother et al. 2011, 2012; Boniecki et al. 2003; Surles et al. 1994)
Chen	11.34	22.981	7.81	4.852e−14	(Chen 2008; Cao and Chen 2005, 2006a,b, 2011)
Ding	12.496	26.208	7.8	5.143e−12	(Lavender et al. 2010; Ding and Dokholyan 2012a)

Note: Data are retrieved from the RNA-puzzles website (http://ahsoka.u-strasbg.fr/rnapuzzlesv2). Among the several models submitted by each research group per single challenge, only the ones with the lowest RMSD with respect to the experimental structure are reported herein. RMSD and DI values are given in Å.

6.7 CONCLUSIONS

In this chapter, we have reviewed some of the most employed methodologies to model the 3D structure of RNA molecules, in particular those based on knowledge-based energy potentials. It is very difficult to establish a quantitative comparison about their performance, even if they are contrasted to the same problem, like in the RNA-Puzzles initiative (Cruz et al. 2012), because neither the amount of human expertise versus computer power utilized, nor the way they handle the experimental information is the same for each method. However, at this point there are relevant questions we should address. Does the degree of coarse-graining have a clear impact on the results? What is the relevance of using experimental data to restrain the sampling? To answer those questions we can have a look at the results of the RNA-Puzzles. Interestingly, the degree of coarse-graining does not seem to always be critical. The group of Chen used Vfold, a two-bead model combined with MD sampling (Cao and Chen 2005, 2009), and are still able to produce significantly good models (RMSD with respect the x-ray structure close to 4 Å and low *clash score* values) in three out the six Puzzles experiments reviewed in Table 6.1, producing the lowest RMSD model on two occasions (on the first below 4 Å and second at 7.2 Å). Notwithstanding, the Das group, using a stepwise assembly procedure and an all-atom model (Sripakdeevong, Kladwang, and Das 2011), produced models lower than 4 Å in just two experiments, although it produced the absolute lowest RMSD model on three occasions. The hybrid approach between the two is the one employed by Bujnicki, which combines homology modeling with posterior MC refinement on a three-bead model to incorporate secondary structure restraints (Rother et al. 2011; Magnus et al. 2014); it was able to generate good models in two of the analyzed Puzzles rounds, and was the absolute lowest RMSD prediction in one of them. A similar approach by the Santalucia group also yielded good models in two rounds. The use of experimental information in the past two Puzzles experiments makes it possible to obtain models below 10 Å, for large RNA fragments of 171 and 187 nt, respectively (Challenges 10 and 5, data not shown).

REFERENCES

Addess, K J, and J Feigon. 1996. Introduction to 1H NMR spectroscopy of DNA. In *Bioorganic Chemistry: Nucleic Acids*, edited by S M Hecht, 500. New York, NY: Oxford University Press.

Alder, B J, and T E Wainwright. 1959. Studies in molecular dynamics. I. General method. *Journal of Chemical Physics* 31 (2): 459.

Baird, N J, J Zhang, T Hamma, and A R Ferré-D'Amaré. 2012. YbxF and YlxQ are bacterial homologs of L7Ae and bind K-turns but not K-loops. *RNA* 18 (4): 759–70.

Ban, N, P Nissen, J Hanssen, P B Moore, and T A Steitz. 2000. The complete atomic structure of the large ribosomal subunit at 2.4 Å resolution. *Science* 289 (5481): 905–20.

Bernauer, J, X Huang, A Y L Sim, and M Levitt. 2011. Fully differentiable coarse-grained and all-atom knowledge-based potentials for RNA structure evaluation. *RNA* 17 (6): 1066–75.

Bernstein, F C, T F Koetzle, G J Williams, E F Meyer, M D Brice, J R Rodgers, O Kennard, T Shimanouchi, and M Tasumi. 1977. The protein data bank: A computer-based archival file for macromolecular structures. *Journal of Molecular Biology* 112 (3): 535–42.

Biswas, R, S N Mitra, and M Sundaralingam. 1998. 1.76 Å structure of a pyrimidine start alternating A-RNA hexamer r(CGUAC)dG. *Acta Crystallographica. Section D, Biological Crystallography* 54 (Pt 4): 570–6.

Cao, S, and S-J Chen. 2005. Predicting RNA folding thermodynamics with a reduced chain representation model. *RNA* 11 (12): 1884–97.

Cao, S, and S-J Chen. 2006a. Predicting RNA pseudoknot folding thermodynamics. *Nucleic Acids Research* 34 (9): 2634–52.

Cao, S, and S-J Chen. 2006b. Free energy landscapes of RNA/RNA complexes: With applications to snRNA complexes in spliceosomes. *Journal of Molecular Biology* 357 (1): 292–312.

Cao, S, and S-J Chen. 2009. Predicting structures and stabilities for H-type pseudoknots with interhelix loops. *RNA* 15 (4): 696–706.

Cao, S, and S-J Chen. 2011. Physics-based de novo prediction of RNA 3D structures. *Journal of Physical Chemistry B* 115 (14): 4216–26.

Cate, J H, A R Gooding, E Podell, K Zhou, B L Golden, C E Kundrot, T R Cech, and J A Doudna. 1996. Crystal structure of a group I ribozyme domain: Principles of RNA packing. *Science* 273 (5282): 1678–85.

Cech, T R, and J A Steitz. 2014. The noncoding RNA revolution-trashing old rules to forge new ones. *Cell* 157 (1): 77–94.

Chen, J-L, and C W Greider. 2005. Functional analysis of the pseudoknot structure in human telomerase RNA. *Proceedings of the National Academy of Sciences of the United States of America* 102 (23): 8080–5; discussion 8077–9.

Chen, S-J. 2008. RNA folding: Conformational statistics, folding kinetics, and ion electrostatics. *Annual Review of Biophysics* 37: 197–214.

Copley, S D, E Smith, and H J Morowitz. 2007. The origin of the RNA world: Co-evolution of genes and metabolism. *Bioorganic Chemistry* 35 (6): 430–43.

Cragnolini, T., Laurin, Y., Derreumaux, P., & Pasquali, S. (2014). Predicting complex 3D RNA structures with a high resolution coarse-grained model. *arXiv preprint arXiv:1404.0568*.

Crick, F. 1970. Central dogma of molecular biology. *Nature* 227: 561–3.

Cruz, J A, M-F Blanchet, M Boniecki, J M Bujnicki, S-J Chen, S Cao, R Das et al. 2012. RNA-puzzles: A CASP-like evaluation of RNA three-dimensional structure prediction. *RNA* 18 (4): 610–25.

Dahm, R. 2005. Friedrich Miescher and the discovery of DNA. *Developmental Biology* 278 (2): 274–88.

Das, R, and D Baker. 2007. Automated de novo prediction of native-like RNA tertiary structures. *Proceedings of the National Academy of Sciences of the United States of America* 104 (37): 14664–9.

Das, R, and D Baker. 2008. Macromolecular modeling with rosetta. *Annual Review of Biochemistry* 77: 363–82.

Das, R, J Karanicolas, and D Baker. 2010. Atomic accuracy in predicting and designing noncanonical RNA structure. *Nature Methods* 7 (4): 291–4.

Davis, I W, A Leaver-Fay, V B Chen, J N Block, G J Kapral, X Wang, L W Murray et al. 2007. MolProbity: All-atom contacts and structure validation for proteins and nucleic acids. *Nucleic Acids Research* 35 (Web Server issue): W375–83.

Debye, P, and E Hückel. 1923. The theory of electrolytes. I. Lowering of freezing point and related phenomena. *Physikalische Zeitschrift* 24: 185–206.

Deigan, K E, T W Li, D H Mathews, K M Weeks. 2009. Accurate SHAPE-directed RNA structure determination. *Proceedings of National Academy of Sciences* 106 (1): 97–102.

Dibrov, S, J McLean, and T Hermann. 2011. Structure of an RNA dimer of a regulatory element from human thymidylate synthase mRNA. *Acta Crystallographica Section D* 67 (Pt 2): 97–104.

Dibrov, S M, J McLean, J Parsons, and T Hermann. 2011. Self-assembling RNA square. *Proceedings of the National Academy of Sciences of the United States of America* 108 (16): 6405–8.

Ding, F, J M Borreguero, S V Buldyrey, H E Stanley, and N V Dokholyan. 2003. Mechanism for the alpha-helix to beta-hairpin transition. *Proteins: Structure, Function and Genetics* 53 (2): 220–8.

Ding, F., and N V Dokholyan. 2012a. Multiscale modeling of RNA structure and dynamics. In N Leontis, E Westhof (Eds.), *RNA 3D Structure Analysis and Prediction* (pp. 167–184). Berlin Heidelberg: Springer.

Ding, F, and N V Dokholyan. 2012b. Discrete molecular dynamics simulation of biomolecules. In N V Dokholyan (Ed.), *Computational Modeling of Biological Systems: From Molecules to Pathways*, 55–73. Springer.

Ding, F, C A Lavender, K M Weeks, and N V Dokholyan. 2012. Three-dimensional RNA structure refinement by hydroxyl radical probing. *Nature Methods* 9 (6): 603–8.

Ding, F, S Sharma, P Chalasani, V V Demidov, N E Broude, and N V Dokholyan. 2008. Ab initio RNA folding by discrete molecular dynamics: From structure prediction to folding mechanisms. *RNA* 14 (6): 1164–73.

Doherty, E A, R T Batey, B Masquida, and J A Doudna. 2001. A universal mode of helix packing in RNA. *Nature Structural Biology* 8 (4): 339–43.

Duarte, C M, and A M Pyle. 1998. Stepping through an RNA structure: A novel approach to conformational analysis. *Journal of Molecular Biology* 284 (5): 1465–78.

Duarte, C M, L M Wadley, and A M Pyle. 2003. RNA structure comparison, motif search and discovery using a reduced representation of RNA conformational space. *Nucleic Acids Research* 31 (16): 4755–61.

Earl, D J, and M W Deem. 2005. Parallel tempering: Theory, applications, and new perspectives. *Physical Chemistry Chemical Physics* 7 (23): 3910.

Flores, S C, Y Wan, R Russell, and R B Altman. 2010. Predicting RNA structure by multiple template homology modeling. *Pacific Symposium on Biocomputing* (10): 216–27.

Flory, Paul. J., and Volkenstein, M. (1969), Statistical mechanics of chain molecules. *Biopolymers* 8: 699–700.

Frank, D N, and N R Pace. 1998. Ribonuclease P: Unity and diversity in a tRNA processing ribozyme. *Annual Review of Biochemistry* 67: 153–80.

Freeman, W M, S J Walker, and K E Vrana. 1999. Quantitative RT-PCR: Pitfalls and potential. *BioTechniques* 26 (1): 112–25.

Fritz, J J, A Lewin, W Hauswirth, A Agarwal, M Grant, and L Shaw. 2002. Development of hammerhead ribozymes to modulate endogenous gene expression for functional studies. *Methods* 28 (2): 276–85.

Geiduschek, B E P, and R Haselkorn. 1969. Messenger RNA. *Annual Review of Biochemistry* 38 (1): 647–76.

Gelin, B R, and M Karplus. 1975. Sidechain torsional potentials and motion of amino acids in proteins: Bovine pancreatic trypsin inhibitor. *Proceedings of the National Academy of Sciences of the United States of America* 72 (6): 2002–6.

Gherghe, C M, C W Leonard, F Ding, N V Dokholyan, and K M Weeks. 2009. Native-like RNA tertiary structures using a sequence-encoded cleavage agent and refinement by discrete molecular dynamics. *Journal of the American Chemical Society* 131 (7): 2541–6.

Gopinath, S C B. 2009. Mapping of RNA-protein interactions. *Analytica Chimica Acta* 636 (2): 117–28.

Green, M R. 1986. Pre-mRNA splicing. *Annual Review of Genetics* 20: 671–708.

Hajdin, C E, F Ding, N V Dokholyan, and K M Weeks. 2010. On the significance of an RNA tertiary structure prediction. *RNA* 16 (7): 1340–9.

Hansen, J L, A M Long, and S C Schultz. 1997. Structure of the RNA-dependent RNA polymerase of poliovirus. *Structure* 5 (8): 1109–22.

Holley, R W, J Apgar, G A Everett, J T Madison, M Marquisee, S H Merrill, J R Penswick, and A Zamir. 1965. Structure of a ribonucleic acid. *Science* 147 (3664): 1462–5.

Huang, L, A Serganov, and D J Patel. 2010. Structural insights into ligand recognition by a sensing domain of the cooperative glycine riboswitch. *Molecular Cell* 40 (5): 774–86.

Jonikas, M A, R J Radmer, A Laederach, R Das, S Pearlman, D Herschlag, and R B Altman. 2009. Coarse-grained modeling of large RNA molecules with knowledge-based potentials and structural filters. *RNA* 15 (2): 189–99.

Karaduman, R, P Fabrizio, K Hartmuth, H Urlaub, and R Lührmann. 2006. RNA structure and RNA-protein interactions in purified yeast U6 snRNPs. *Journal of Molecular Biology* 356 (5): 1248–62.

Kendrew, J C, G Bodo, H M Dintzis, G Parrish, H Wickoff, and D C Phillips. 1958. A three-dimensional model of the myoglobin molecule obtained by X-ray analysis. *Nature* 181: 662–6.

Kosinski, J, I A Cymerman, and M Feder. 2003. A 'FRankenstein's monster' approach to comparative modeling: Merging the finest fragments of fold-recognition models and iterative model refinement aided by 3D structure evaluation. *Proteins: Structure* 6(S6): 369–379.

Kramer, A. 1996. The structure and function of proteins pre-mRNA splicing. *Annual Review of Biochemistry* 65: 367–409.

Krokhotin, A, A M Mustoe, K M Weeks, and N V Dokholyan. 2017. Direct identification of base-paired RNA nucleotides by correlated chemical probing. *RNA* 23: 6–13.

Krokhotin, A., K Houlihan, and N V Dokholyan. 2012. iFoldRNA v2: folding RNA with constraints. *Bioinformatics* 31 (17): 2891–2893.

Ladner, J E, A Jack, J D Robertus, R S Brown, D Rhodes, B F Clark, and A Klug. 1975. Structure of yeast phenylalanine transfer RNA at 2.5 A resolution. *Proceedings of the National Academy of Sciences* 72 (11): 4414–8.

Laurendeau, N M. 2005. *Statistical Thermodynamics: Fundamentals and Applications.* New York, NY: Cambridge University Press.

Lavender, C A, F Ding, N V Dokholyan, and K M Weeks. 2010. Robust and generic RNA modeling using inferred constraints: A structure for the hepatitis C virus IRES pseudoknot domain. *Biochemistry* 49 (24): 4931–3.

Lu, H, and J Skolnick. 2001. A distance-dependent atomic knowledge-based potential for improved protein structure selection. *Proteins* 44 (3): 223–32.

Magnus, M, D Matelska, G Lach, G Chojnowski, M J Boniecki, E Purta, W Dawson, S Dunin-Horkawicz, and J M Bujnicki. 2014. Computational modeling of RNA 3D structures, with the aid of experimental restraints. *RNA Biology* 11 (5): 1–15.

Mathews, D H, M D Disney, J L Childs, S J Schroeder, M Zuker, and D H Turner. 2004. Incorporating chemical modification constraints into a dynamic programming algorithm for prediction of RNA secondary structure. *Proceedings of the National Academy of Sciences of the United States of America* 101 (19): 7287–92.

Mathews, D H, J Sabina, M Zuker, and D H Turner. 1999. Expanded sequence dependence of thermodynamic parameters improves prediction of RNA secondary structure. *Journal of Molecular Biology* 288 (5): 911–40.

Matthews, B W. 1975. Comparison of the predicted and observed secondary structure of T4 phage lysozyme. *Biochimica et Biophysica Acta* 405 (2): 442–51.

Mattice, W L, and U W Suter. 1994. Conformational theory of large molecules: The rotational isomeric state model in macromolecular systems. *AlChE Journal* 42 (4): 1199–1200.

McCammon, J A, B R Gelin, and M Karplus. 1977. Dynamics of folded proteins. *Nature* 267 (5612): 585–90.

Merino, E J, K A Wilkinson, J L Coughlan, and K M Weeks. 2005. RNA structure analysis at single nucleotide resolution by selective 2'-hydroxyl acylation and primer extension (SHAPE). *Journal of the American Chemical Society* 127 (12): 4223–31.

Metropolis, N, and S Ulam. 1949. The Monte Carlo method. *Journal of the American Statistical Association* 44 (247): 335–41.

Murray, L J W, W B Arendall, D C Richardson, and J S Richardson. 2003. RNA backbone is rotameric. *Proceedings of the National Academy of Sciences of the United States of America* 100 (24): 13904–9.

Nikolova, E N et al. 2011. Transient Hoogsteen base pairs in canonical duplex DNA. *Nature.* 470: 498–502.

Olson, W K. 1975. Configurational statistics of polynucleotide chains. A single virtual bond treatment. *Macromolecules* 8 (3): 272–5.

Olson, W K, and P J Flory. 1972. Spatial configurations of polynucleotide chains. I. Steric interactions in polyribonucleotides: A virtual bond model. *Biopolymers* 11 (1): 1–23.

Parisien, M, J A Cruz, E Westhof, and F Major. 2009. New metrics for comparing and assessing discrepancies between RNA 3D structures and models. *RNA* 15 (10): 1875–85.

Parisien, M, and F Major. 2008. The MC-fold and MC-sym pipeline infers RNA structure from sequence data. *Nature* 452 (7183): 51–5.

Pasquali, S, and P Derreumaux. 2010. HiRE-RNA: A high resolution coarse-grained energy model for RNA. *Journal of Physical Chemistry B* 114 (37): 11957–66.

Patel, J S, A Berteotti, S Ronsisvalle, W Rocchia, and A Cavalli. 2014. Steered molecular dynamics simulations for studying protein-ligand interaction in cyclin-dependent kinase 5. *Journal of Chemical Information and Modeling* 54 (2): 470–80.

Peselis, A, and A Serganov. 2012. Structural insights into ligand binding and gene expression control by an adenosylcobalamin riboswitch. *Nature Structural & Molecular Biology* 19 (11): 1182–4.

Quigley, G J, and A Rich. 1976. Structural domains of transfer RNA molecules. *Science* 194 (4267): 796–806.

Rahman, A. 1964. Correlations in the motion of atoms in liquid argon. *Physical Review* 136 (2A): A405–A411.

Ramachandran, G N, C Ramakrishnan, and V Sasisekharan. 1963. Stereochemistry of polypeptide chain configurations. *Journal of Molecular Biology* 7 (1): 95–9.

Rapold, R F, and W L Mattice. 1995. New high-coordination lattice model for rotational isomeric state polymer chains. *Journal of the Chemical Society, Faraday Transactions* 91 (16): 2435.

Reyes, F E, A D Garst, and R T Batey. 2009. Strategies in RNA crystallography. *Methods in Enzymology* 469: 119–139.

Rich, A, D R Davies, F H C Crick, and J D Watson. 1961. The molecular structure of polyadenylic acid. *Journal of Molecular Biology* 3 (1): 71–86.

Rohl, C A, C E M Strauss, K M S Misura, and D Baker. 2004. Protein structure prediction using Rosetta. *Methods in Enzymology* 383: 66–93.

Rother, K, M Rother, M Boniecki, and T Puton. 2012. Template-based and template-free modeling of RNA 3D structure: Inspirations from protein structure modeling. In N Leontis, E Westhof (Eds.), *RNA 3D Structure Analysis and Prediction* (pp. 67–90). Berlin Heidelberg: Springer.

Rother, M, K Rother, T Puton, and J M Bujnicki. 2011. ModeRNA: A tool for comparative modeling of RNA 3D structure. *Nucleic Acids Research* 39 (10): 4007–22.

Schluenzen, F, A Tocilj, R Zarivach, J Harms, M Gluehmann, D Janell, A Bashan et al. 2000. Structure of functionally activated small ribosomal subunit at 3.3 Å resolution. *Cell* 102: 615–23.

Scott, L G, and M Hennig. 2008. RNA structure determination by NMR. In *Bioinformatics. Volume I: Structure, Sequence Analysis and Evolution*, edited by J M Keith, Vol. 452, 29–61. *Methods in Molecular Biology*. Totowa, NJ: Humana Press.

Sharma, S, F Ding, and N V Dokholyan. 2008. iFoldRNA: Three-dimensional RNA structure prediction and folding. *Bioinformatics* 24 (17): 1951–2.

Sijenyi, F, P Saro, Z Ouyang, K Damm-Ganamet, M Wood, J Jiang, and J Santalucia. 2012. The RNA folding problems: Different levels of sRNA structure prediction. In *RNA 3D Structure Analysis and Prediction*, Vol. 27, 91–117. Berlin, Heidelberg: Springer Berlin Heidelberg.

Sripakdeevong, P, W Kladwang, and R Das. 2011. An enumerative stepwise ansatz enables atomic-accuracy RNA loop modeling. *PNAS* 108 (51): 20573–8.

Staple, D W, and S E Butcher. 2005. Pseudoknots: RNA structures with diverse functions. *PLoS Biology* 3 (6): e213.

Su, L, L Chen, M Egli, J M Berger, and A Rich. 1999. Minor groove RNA triplex in the crystal structure of a ribosomal frameshifting viral pseudoknot. *Nature Structural Biology* 6 (3): 285–92.

Thomsen, N D, and J M Berger. 2012. Crystallization and x-ray structure determination of an RNA-dependent hexameric helicase. *Methods in Enzymology* 511: 171–190.

Tijerina, P, S Mohr, and R Russell. 2007. DMS footprinting of structured RNAs and RNA-protein complexes. *Nature Protocols* 2 (10): 2608–23.

Toor, N, K S Keating, S D Taylor, and A M Pyle. 2008. Crystal structure of a self-spliced group II intron. *Science* 320 (5872): 77–82.

Torrie, G M, and J P Valleau. 1977. Nonphysical sampling distributions in Monte Carlo free-energy estimation: Umbrella sampling. *Journal of Computational Physics* 23 (2): 187–99.

Watson, J D, T A Baker, S P Bell, A Gann, M Levine, and R Losick. 2013. *Molecular Biology of the Gene*. 7th ed. Menlo Park, CA: Benjamin Cummings.

Wimberly, B T, D E Brodersen, W M Clemons Jr, R J Morgan-Warren, A P Carter, C Vonrhein, T Hartsch, and V Ramakrishnan. 2000. Structure of the 30S ribosomal subunit. *Nature* 407: 327–39.

Woese, C R, S Winker, and R R Gutell. 1990. Architecture of ribosomal RNA: Constraints on the sequence of 'tetra-loops'. *Proceedings of the National Academy of Sciences* 87 (21): 8467–71.

Worthington Allen, F. 1941. The biochemistry of the nucleic acids, purines and pyrimidines. *Annual Review of Biochemistry* 10: 221–45.

Xia, Z, D P Gardner, R R Gutell, and P Ren. 2010. Coarse-grained model for simulation of RNA three-dimensional structures. *Journal of Physical Chemistry. B* 114 (42): 13497–506.

Xia, Z, and P Ren. 2013. Prediction and coarse-grained modeling of RNA structures. In *Biophysics of RNA Folding, Biophysics for the Life Sciences*, edited by R Russell, Vol. 3, 53–68. New York, NY: Springer.

Zhang, J, and A R Ferré-D'Amaré. 2013. Co-crystal structure of a T-box riboswitch stem I domain in complex with its cognate tRNA. *Nature* 500 (7462): 363–6.

CHAPTER 7

The Need for Computational Speed

State of the Art in DNA Coarse Graining

Davit Potoyan and Garegin A. Papoian

CONTENTS

7.1 Introduction 271
7.2 Coarse Graining the Molecule of DNA 272
7.3 Computational Models 274
    7.3.1 A Bottom-Up One-Bead-per-Nucleotide DNA Force Field Based on the Renormalization Group Coarse-Graining of Atomistic Simulations 274
    7.3.2 3SPN.2c: An Experimentally Inspired Model by de Pablo and Coworkers 279
    7.3.3 oxDNA: A Nearest Neighbor Anisotropic Top-Down Model by Ouldridge, Louis, and Coworkers 282
7.4 Coarse Graining of Nucleosomes and Their Assemblies: Toward Genome-Scale Simulations of Chromatin 286
7.5 Concluding Remarks 289
    Acknowledgments 290
    References 290

7.1 INTRODUCTION

The recent boom in the postgenomic race to understand chromatin organization, and to develop corresponding bio-nanotechnology applications, has created a surge of interest in studying various DNA-based mesoscale systems. Among computational approaches, classical molecular dynamics simulations [1,2] are a powerful way of treating and visualizing various phases of DNA and their transformations. The fully atomistic description has provided many insights into the biophysics of DNA *in vitro* [3–6]. Such atomically detailed approaches, while chemically more rigorous, are not always realistic when aiming for the scales of problems encountered in the description of chromatin, viral packaging, and DNA-based nanomaterials. Fortunately, DNA coarse-grained (CG) models have been consistently improving and expanding their predictive power, starting to reproduce key structural and dynamic properties of double- (ds) and single-stranded (ss) DNA, while still providing orders of magnitude faster sampling [3,7–18]. In this chapter, we take a closer look at a selection of CG DNA models [10–18], that have established a continuous line of research and are part of an ongoing effort to go beyond simple characteristics of the double helix and into the realm of protein–DNA interactions, chromatin packaging, and nanomaterial assembly.

7.2 COARSE GRAINING THE MOLECULE OF DNA

Double-stranded DNA is the carrier of hereditary information that is physically involved in template-based synthesis and regulation of vast array of biomolecules in the cell [19]. These processes rely on unique mechanical, stereochemical, and thermodynamic properties of DNA, which allow efficient read-out of its informational content while the chain may be globally highly condensed [20]. A roughly meter-long stretch of the entire human genome fits within the micrometer length-scale confines of a cell's nucleus. This compression, achieved under physiological conditions, is remarkable knowing that DNA is a highly charged and rigid biopolymer, resisting thermal undulations below the length scale of ∼50 nm [21–23]. Despite the apparent rigidity of DNA, one frequently encounters various scenarios *in vivo*, where DNA chains fit inside viral capsids of <100 nm, condense into chromatin fibers with diameters between ∼10–100 nm, and assume various highly kinked and twisted geometric configurations when bound to various transcription factors [20,24]. Understanding the mechanical and thermodynamic behaviors of DNA is of primary interest not only for biology but also for material science. Due to the programmable nature of DNA sequences, DNA molecules have become attractive materials for designing various predetermined nano-objects [17,25]. In the last decade, our understanding of the physics of DNA on shorter length scales has improved considerably, thanks to various single molecule techniques that allow mechanical manipulation of individual molecules [26–30]. In turn, newly developed CG molecular models of DNA chains proved to be instrumental in supplying microscopic explanations for the behaviors observed in the single molecule experiments [3,7–9]. However, such computational models can be used not only for explaining and complementing experiments but also for probing and visualizing the length and timescales that are not accessible to experimental techniques. In particular, one of the ongoing challenges for CG DNA models is to achieve enough reliability to allow accurate modeling of entire nucleosomes and eventually chromatin fibers, where the structures of the latter are yet to be resolved experimentally [31–34]

The task of devising accurate and transferable CG models for DNA faces many unique challenges, such as the need to treat inherently strong electrostatic interactions, DNA chains' coupling with the surrounding ionic environment, the impact of sequence on molecules' local curvature and structural rigidity, and the need to balance canonical structures with the ability to also access highly deformed and melted forms [3,7–9]. To address these challenges, many CG DNA models have been proposed, with various levels of structural resolution and adhering to different coarse-graining philosophies [3,7–9]. It would be ideal but nearly impossible to create a universal and fully transferable CG DNA model that well describes DNA's structural, mechanical, and thermodynamic properties. On the other hand, several well-rounded models have been developed in recent years, which perform consistently well at addressing various structural and thermodynamic aspects of the ds and ss DNA chains under different conditions. Those models can be roughly classified into two broad classes: (1) top-down, where experimental data is used to parametrize the skillfully invented Hamiltonian [16,18] or (2) bottom-up, where all parameters as well as even potentially the functional form of the Hamiltonian can, in principle, be extracted from more microscopic, all-atom simulations, carried out on smaller fragments [12,35]. Both types of models exhibit distinct strengths and weaknesses, and hence, we expect them to work in synergy in resolving the outstanding challenges of developing robust and reliable DNA force fields capable of capturing the physics of millions to billions of atoms with only small fraction of particles compared with fully atomistic simulations.

In a recent review [8], we covered several CG DNA models, discussing them in great detail. We follow a similar format for the present chapter and highlight the latest developments that have happened since then. Also, to reduce the material overlap with our previous review, we cover here the CG methodologies in a more descriptive manner, putting more emphasis on the recent developments and applications. We review three CG DNA models that have by now established a continuous line of research and spawned many applications in biophysics and material science. These models are the rigorous bottom-up CG model of ds DNA by Savelyev and Papoian [10–12], the three sites per nucleotide (3SPN) family of models by de Pablo and coworkers [13–16], and the oxDNA model developed by Ouldridge, Louis, and coworkers [17,18]. Readers interested in additional background and technical details of these models are referred to the original references as well as to our previous review [8].

7.3 COMPUTATIONAL MODELS

7.3.1 A Bottom-Up One-Bead-per-Nucleotide DNA Force Field Based on the Renormalization Group Coarse-Graining of Atomistic Simulations

The Molecular Renormalization Coarse-Graining (MRG-CG) technique arose from the inverse Monte Carlo scheme proposed by Swendsen [36], then further developed for pairwise atomistic Hamiltonians by Laaksonen and Lyburatsev [37], and subsequently generalized to many-body molecular Hamiltonians by Savelyev and Papoian [10–12]. The MRG-CG is based on a powerful idea of inferring pairwise and many-body interaction potentials between CG sites by matching the partition functions of microscopic and CG models. In its current form, the method provides a systematic way of deriving CG Hamiltonians for DNA and, in fact, for an arbitrary polymer in the presence of mobile ions. In the previous works [10–12], this approach was used to derive a one-bead-per-nucleotide CG Hamiltonian for ds DNA, where extensive comparisons demonstrated a high level of fidelity of both the large-scale and finer-scale mechanical and electrostatic properties of DNA obtained from the CG simulations compared with their atomistic counterparts [12]. A noteworthy aspect of the MRG-CG is its rigorous connection to equilibrium statistical physics and critical phenomena, namely the renormalization group (RG) theory, which, in turn, provides an avenue for systematic improvements of the CG force field. In principle, these improvements can be pursued to a desired degree of precision, depending on the nature of the problem and the available resources for development. This aspect of MRG-CG distinguishes it favorably from many bottom-up coarse-graining methodologies that are not deeply rooted in statistical mechanics (e.g., matching arbitrarily chosen correlation functions lead to ad-hoc, hard-to-improve schemes) and also from typical top-down coarse-graining approaches, where one has to resort to skillful guessing of which experimental database to use for fitting, and choose somewhat arbitrarily the physical observables that need to be reproduced, often resulting in reinventing the force fields over and over again.

The force field used in the works [10–12] consists of intra-DNA, inter-ionic, and ion–DNA interactions that accurately capture the coupling of DNA's molecular mechanics with the ionic environment. The Hamiltonian of Savelyev and Papoian for the ds DNA model has the following form:

$$\mathcal{H} = \mathcal{U}_{\text{bond}} + \mathcal{U}_{\text{ang}} + \mathcal{U}_{\text{fan}} + \mathcal{U}_{\text{el}} \tag{7.1}$$

where the first two terms account for bond and bending angle fluctuations (*intra*strand interactions), and the last two terms for *inter*strand ("*fan*" interactions represented by dashed lines in Figure 7.1) and electrostatic interactions, respectively. The functional form of the individual contributions to the bond, *fan*, and angular potentials is based on quartic polynomials, to account for the asymmetric shape of DNA's structural fluctuations; hence, going beyond the usual worm-like-chain regime by taking

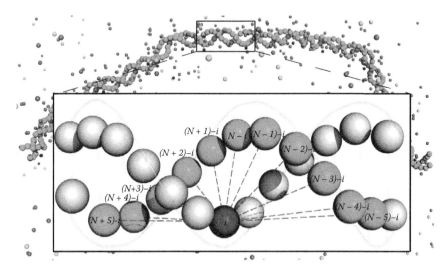

FIGURE 7.1 A chemically accurate coarse-grained (CG) model of the double-stranded (ds) DNA with explicit mobile ions [12]. Each DNA base pair is represented by two beads, each placed in the geometric center of the corresponding all-atom nucleotide. Dashed lines indicate effective interactions, which represent a superposition of stacking and base-pairing among two polynucleotides.

into account the anharmonicities of DNA chain motions,

$$U_{\text{bond, fan}} = \sum_{\alpha=2}^{4} K_\alpha (l - l_0)^\alpha, \quad U_{\text{ang}} = \sum_{\alpha=2}^{4} K_\alpha (\theta - \theta_0)^\alpha \qquad (7.2)$$

The parameters l_0 and θ_0 are equilibrium interparticle separations for bond and *fan* interactions, and the equilibrium angle for bending angle potential, respectively. The equilibrium values l_0 and θ_0, as well as the trial set of coefficients $\{K_\alpha^{(0)}\}$, were obtained by fitting these polynomials to the corresponding potentials of mean force (PMF), extracted from all-atom molecular dynamics (AA MD) simulations. However, this is only the first step, followed subsequently by optimization of all trial interaction parameters using the iterative scheme of MRG-CG. The following are the interionic interaction potentials ($\mathcal{H}_{\text{ion-ion}}$) and the potentials among beads of DNA and the ions ($\mathcal{H}_{\text{DNA-ion}}$):

$$U_{\text{ion-ion}} = \sum_{i>j} \left[\frac{A}{r_{ij}^{12}} + \sum_{k=1}^{5} B^{(k)} e^{-C^{(k)} [r_{ij} - R^{(k)}]^2} + \frac{q_i q_j}{4\pi\varepsilon_0 \varepsilon r_{ij}} \right] \qquad (7.3)$$

$$U_{\text{DNA-ion}} = \sum_{i>j} \left[\frac{A}{r_{ij}^{6}} + \sum_{k=1}^{3} B^{(k)} e^{-C^{(k)} [r_{ij} - R^{(k)}]^2} + \frac{q_i q_j}{4\pi\varepsilon_0 \varepsilon r_{ij}} \right] \qquad (7.4)$$

These are defined by the set of parameters, $\{A, B^{(k)}, C^{(k)}\}$, and the positions of Gaussian peaks and minima, $\{R^{(k)}\}$, that were used in the latest CG DNA model.

In the CG representation of the model (see Figure 7.1), each nucleotide is mapped into one effective bead, which results in an over 30-fold reduction of a DNA chain's degrees of freedom, paving the way for truly large-scale modeling [12]. Another notable feature is the explicit treatment of the electrostatic interactions via modeling ions with explicit beads. This eliminates many artifacts due to the phenomenological treatment of electrostatics in most CG DNA models, where important many-body correlations among mobile ions near highly charged DNA surfaces are largely ignored. Rigorous treatment of electrostatics allowed the Savelyev–Papoian model to achieve remarkable quantitative agreement between the calculated and experimentally obtained dependence of DNA's persistence length on the concentration of monovalent salts [12]. Additionally, the model was shown to predict subtle structural transitions in DNA nanocircles induced by variation of the solution ionic strength [12]. In subsequent work, this model along with atomistic simulations were used to shed light on the origin of intrinsic rigidity of DNA molecules, showing that the electrostatic repulsion among the bases contributes significantly more to DNA's persistence length compared to the common view in the prior literature [38].

The MRG-CG scheme, following the earlier tradition [36,37], relies on representing the desired effective Hamiltonian, $\mathcal{H} = \sum_{\alpha=1}^{N} K_\alpha S_\alpha$, as a linear combination of N relevant collective coordinates, S_α, with weights proportional to coefficients, K_α. The S_α variables can be formally treated as dynamical observables, whose (various order) correlation functions, $\langle S_\alpha \dots S_\beta \rangle$, can be computed both from atomistic and CG simulations. It turns out that the partition function matching condition (which is a one-step renormalization) requires matching of all first-order cumulants $\langle S_\alpha \rangle$ that enter the CG Hamiltonian [8,10]. Hence, the force constants, K_α, associated with each observable, S_α, are formally treated as "conjugate fields," whose numerical value has to be adjusted appropriately to generate the desired system dynamics. Because of Hamiltonian's linearity with respect to K_αs (but allowing S_αs to be nonlinear functions of CG bead coordinates), it is possible to establish a mathematical connection between these conjugate fields and the expectation values of the associated dynamical observables in terms of the covariance matrix of *all* observables [36]:

$$\Delta\langle S_\alpha \rangle = -1/(k_B T) \sum_\gamma [\langle S_\alpha S_\gamma \rangle - \langle S_\alpha \rangle \langle S_\gamma \rangle] \Delta K_\gamma \qquad (7.5)$$

where $\Delta\langle S_\alpha \rangle \equiv \langle S_\alpha \rangle_{CG} - \langle S_\alpha \rangle_{AA}$ is the difference between the expectation values of an observable, S_α, averaged over CG and AA systems, and the ΔK_γs are the sought-after corrections to the CG Hamiltonian trial parameters, $\{K_\alpha^{(0)}\}$. A set of linear equations, Equation 7.5, is iteratively solved until the convergence is reached for all observables, $\Delta\langle S_\alpha \rangle \approx 0$, $\alpha = 1 \dots N$. In this way, the process of parameter adjustment explicitly accounts for cross-correlations among various CG degrees of freedom. For example, as the Hamiltonian parameters for DNA bending angle potential are iteratively adjusted, the distributions of other CG structural degrees of freedom (e.g., bond or stacking variables) are perturbed as well, which is taken into account by the MRG-CG scheme.

In the recent contributions of Naome et al [35,39], the *fan* potential from Savelyev and Papoian was generalized in such a way that each CG site i has 6–16 intramolecular and five intermolecular pairwise interactions. Thus, the angular terms described above are accounted for in an implicit manner. Furthermore, all trained potentials are pairwise, represented as a large collection of approximate positional Dirac functions, $V_p(r_p)\delta(r_p - |r_i - r_j|)$, where $V_p(r_p)$ is tabulated in 0.05 Å increments. This somewhat unorthodox choice of the Hamiltonian turns out to have a number of attractive features. The nonlinear pairwise and many-body potentials, in the works of Savelyev and Papoian, have the advantage of representing the Hamiltonian in a highly compact functional form, which, in turn, results in a small number of parameters, and hence, a small covariance matrix and lower number of iterative steps needed for quantitative optimization. However, the specific choice of a nonlinear functions in the Hamiltonian may, in certain cases, prove to be too constraining, which should manifest in the inability to match various correlation functions obtained from atomistic and CG simulations (this, however, was not the case for DNA coarse-graining via MRG-CG in reference [12]). Furthermore, Naome et al. used a two-phase iterative procedure for obtaining the tabulated force field coefficients [35,39]. In the first stage, one obtains a first-order estimate of the interaction potentials via the Boltzmann inversion method, by setting $K_\alpha^0 = -kT \ln S_\alpha$ as the first guess and iteratively adjusting the potentials $K_\alpha^1 = K^0 + \Delta K_\alpha^0$ until reasonable convergence is reached. Afterwards, matching of $\langle S_\alpha \rangle$s is used via Equation 7.5 to further refine the potentials. The number of force field parameters is approximately 100-fold higher in this scheme compared with the MRG-CG approach.

Naome et al. validated their force field by reproducing bond, angle, and dihedral distributions without having some of these terms explicitly present in their DNA model. The force field also quantitatively captures the radial distributions among the mobile ions, mobile ions, and DNA beads, as well as provides reasonable estimates of DNA's persistent length and its dependence on the salt concentration, demonstrating agreement with single molecule experiments. The model was also further applied to a wider range of DNA conformations [39] to study structural energetic and dynamical properties of linear and cyclic DNA fragments of various length ranging from 90 to 500 bp. Authors computed strain, relaxation energies, and cyclization probabilities (j-factors) of minicircles showing that various ways of estimating DNA flexibility from their simulations are all consistent with each other. A particularly interesting observation from their study was the identification of the artifacts of using a simple cutoff treatment for the electrostatics. The artifacts are (1) higher flexibility of smaller fragments versus the larger ones at low concentrations; (2) unexpected and dramatic collapse of longer fragments at high concentrations; and (3) anomalous dependence of persistence length at lower concentrations. All of these artifacts disappear when electrostatics is treated fully via a particle-mesh Ewald scheme [1,2] without a cutoff. This observation provides a strong argument for the importance of long-ranged electrostatic interactions when modeling poly-electrolytes and explicit ions.

7.3.2 3SPN.2c: An Experimentally Inspired Model by de Pablo and Coworkers

One of the most widely used CG DNA models has been designed in the top-down fashion by the group of de Pablo [13], utilizing thermodynamic properties of the DNA solutions. Their CG representation maps the DNA nucleotides into three distinct spherical beads (Figure 7.2)[40], corresponding to sugar, phosphate, and base groups, hence the name 3SPN. Over the years, the model has gone through several iterations [13–16], which refined and extended the reach of its predictive power to a multitude of structural, kinetic, and thermodynamic properties, while at the same time reducing the externally imposed structural constraints. The latest development, termed 3SPN.2c (where c indicates curvature and 2 is the version), stands out as a well-rounded model capable of reproducing diverse facets of DNA behaviors such as (1) maintaining stability of major and minor grooves close to crystallographic values; (2) quantitatively capturing persistence lengths of both ss and ds DNA and their dependence on the solution ionic strength, (3) reproducing effects of sequence, ionic strength, and temperature on hybridization and melting; and (4) capturing kinetic effects of the dependence of the sDNA oligomer hybridization rate coefficients on the DNA sequence and the solution ionic strength.

The Hamiltonian of the 3SPN model accounts for bonded interactions, represented by V_{bond}, V_{angle}, and V_{dihed} terms, which stand for bonded, angular, and dihedral terms, respectively. The nonbonded interactions are modeled by stacking V_{stack}, base-pairing V_{bp}, excluded volume V_{excl}, solvent mediated V_{solve}, and charge–charge potential terms,

$$V_{total} = V_{bond} + V_{angle} + V_{dihed} + V_{stack} + V_{bp} + V_{excl} + V_{qq} + V_{solv} \quad (7.6)$$

A key addition to the latest 3SPN model [16] is the replacement of the Gō-like structure-based terms with angle-dependent potentials, which capture the directional aspects of stacking and base-pairing interactions. The inclusion of anisotropic potentials made the structure-based terms unnecessary for maintaining stable helical configuration, which were acting as structural constraints limiting the range of sampled conformations around the canonical structure. The new angularly modulated potentials

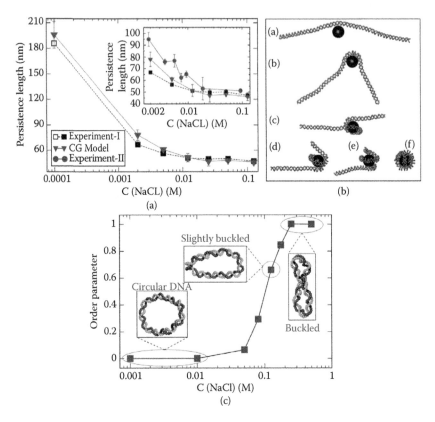

FIGURE 7.2 Illustrations of some of the applications of the MRG-CG model of DNA by Papoian–Savleyev. (a) Dependence of persistence length on the salt concentration extracted from the CG simulations matches with the experimental data with quantitative accuracy (Reproduced from from A Savelyev and GA Papoian. *Proc Natl Acad Sci U S A*, 107, 20340–20345, 2010.) (b) Cao et al. have used the CG DNA model to simulate DNA nanosphere aggregation, which mimics the formations of nucleosomal arrays [40] (Reproduced from Cao et al, *Soft Matter*, 7, 506–514, 2011.) (c) Model predicts the phase transition from the circular form of DNA to the buckled conformation upon increasing the salt concentration. (Reprinted with permission from A Savelyev and GA Papoian. *Proc Natl Acad Sci U S A*, 107, 20340–20345, 2010. Copyright 2011 American Chemical Society.)

make for more physically consistent representation of ds DNA, enabling exploration of important structures that arise in DNA melting, bubble formation, or due to bending by proteins, which is relevant for transcription factor binding and nucleosome formation. The CG beads are spherical and, hence, excluded volume interactions remain isotropic, which is not ideal but seems to be a reasonable approximation. The electrostatics is modeled with a Debye–Huckel potential [41], where the partial charges were chosen to account for counterion condensation [42,43]. The model performs better at low salt concentrations but is less satisfactory in the high concentration of ions. The earlier version of the model, 3SPN1, was modified to include explicit ions [15]; however, this latter approach is still rather experimental, hence the explicitly parametrized ions are not part of the newer model of 3SPN2. For the parametrization of 3SPN2, extensive experimental data on hybridization were used for quantifying interstrand nonbonded interactions, while the electrophoretic mobility of DNA strands were used for determining intrastrand stacking interactions. The enhanced sampling metadynamics simulations were used to match the equilibrium probabilities with experimentally measured free energies using the Boltzmann inversion approach. The bonded terms were adjusted manually and the whole cycle was repeated in iterative fashion until sufficient agreement with experiments was reached. As a consequence of using thermal data for parameterization, the melting curves and melting temperatures obtained for duplexes and hairpins were in near quantitative agreements with the experimental data. The major and minor grooves are well defined and stable throughout typical simulations, with important measures of structural properties, such as duplex widths and groove widths, reproduced with high accuracy. Mechanical rigidity is also well captured by the model, with the persistence lengths of ds DNA for various sequences yielding values in the range of 40–60 nm, in good agreement with the corresponding experimental values. The qualitative trend of ionic strength dependence of the persistent length is reproduced despite not including explicit electrostatics of the ions. Quite remarkably, the model with the same parametrization can also reproduce the persistence length of ss DNA, which is shown to be in the range of 2–4 nm, also in agreement with the known experimental measurements.

Besides reproducing the known structural and thermodynamic quantities, the 3SPN model is also capable of addressing problems involving dynamics and kinetic rates of conformational transitions. In general, modeling kinetics of molecular conformational changes is challenging for CG simulations due to the nontrivial ways the dynamics in the reduced description results from the underlying microscopic dynamics. However, 3SPN2 was specifically parametrized by calculating the rates of the association of 25 bp sequences with various degrees of complementarity. In a recent application of the 3SPN2 model [44], the dependence of the effects of sequence, chain length, and solution ionic strength on DNA oligomer hybridization processes was studied. The scaling exponents obtained by fitting simulation data indicate excellent agreement with the corresponding experimental estimates. These results imply that the mechanisms of hybridization are highly dependent on sequence, while the solution ionic strength only modulates the rate of association, having little impact on the hybridization mechanisms.

At last, in an important step toward modeling protein–DNA interactions, the 3SPN2 model has been extended to account for intrinsic shape and local sequence dependent flexibility of DNA [45]. As proteins have no direct access to the identity of bases, they rely on intrinsic sequence dependent curvature and structural deviations of major and minor grooves for recognizing and binding to specific regions [46,47]. Reproducing this aspect of recognition is a crucial step forward for modeling nucleosome positioning and transcription factor–DNA interactions. The extension of the model was accomplished by (1) using sequence-dependent base-step parameters and (2) making the flexibility of the bonded interactions sequence dependent. This latest development is named 3SPN.2C [45]. In 3SPN.2, nonbonded potentials were constructed in such a way that penalized deviations from the B-DNA crystal structure. The sequence dependence in 3SPN.2 is accounted for via sequence dependence of base-stacking and base-pairing energies, while force constants of bond, angle, and dihedral potentials are sequence independent. In contradistinction, 3SPN.2C accounts for sequence dependence more comprehensively and is specifically intended for modeling ds DNA interactions with proteins. The new model was validated by computing sequence dependence of persistence length, melting temperature, and minor/major groove widths, all of which agreed well with the prior experiments. Thus, 3SPN2.C was built on the strengths of the 3SPN2, incorporating local sequence effects that are prerequisite for accurately modeling protein–DNA interactions.

7.3.3 oxDNA: A Nearest Neighbor Anisotropic Top-Down Model by Ouldridge, Louis, and Coworkers

The oxDNA model emerged from the need to have a relatively coarse, physically motivated, and computationally efficient model of DNA for exploring DNA molecules' long timescale mechanical behaviors, which are of primary interest in nanotechnology. The processes that figure prominently in nanotechnological applications mostly deal with strand-breaking and forming events, adsorption on surfaces and diffusion of strands [17]. Hence, it is a reasonable approximation to rely on relatively coarse description of DNA chains if the main goal is to capture the thermodynamic trends of such events. The main strengths of oxDNA appear to be its relative simplicity and computational efficiency, which are achieved by accounting for the basic physics of nucleotide bases while treating electrostatics either implicitly or through short-range mean-field treatment. Due to its focus on reproducing duplex formation, the oxDNA model has found wider applications in nanomaterial studies compared to the other models (Figure 7.3) [17].

In the oxDNA model, each nucleotide is represented as a three-dimensional (3D) rigid body, in contrast to many other models where additional sites for phosphate or sugar are used for maintaining the geometry of the helix. Representing nucleotides as rigid bodies ignores the internal atomic motions of bases, resulting in drastic speedup of sampling of more global large-scale conformational motions. The detailed interaction Hamiltonian, proposed in the original paper [48], consists of hydrogen bonding, cross stacking, coaxial stacking, nearest neighbor stacking, excluded volume, and backbone chain connectivity terms (see Figure 7.4):

$$
\begin{aligned}
V_{\text{total}} = &\sum_{\text{nearest neighbors}} (V_{\text{backbone}} + V_{\text{stack}} + V'_{\text{exc}}) \\
&+ \sum_{\text{other pairs}} (V_{\text{HB}} + V_{\text{cross-stack}} + V_{\text{exc}} + V_{\text{coax-stack}})
\end{aligned}
\tag{7.7}
$$

In the original CG model of oxDNA, electrostatics was not included explicitly and was accounted for effectively via soft excluded volume interactions. The parameterization was done for fixed salt concentration of M = 0.5 M[Na^+], where electrostatic effects are thoroughly screened, hence justifying modeling the interphosphate interactions at the purely excluded volume level. Also, the original model did not capture the differences between the minor and major grooves.

In the latest report, the new and improved version of the original model [18], oxDNA2, was introduced, with one of the major improvements being the introduction of explicit electrostatics interactions modeled via a short-ranged Debye–Huckel term, $V_{\text{DH}}(T, I)$, which depends on temperature T and on ionic strength I via a screening length function, $\lambda_{\text{DH}}(T, I)$:

$$
V_{\text{DH}}(T, I) = \sum_{ij} \frac{q^2_{\text{eff}}}{4\pi\epsilon_0\epsilon_r} \frac{\exp(-r^{b-b}_{ij}/\lambda_{\text{DH}}(T, I))}{r^{b-b}_{ij}}
\tag{7.8}
$$

Here, q_{eff} is the effective charge placed on each backbone site of each of the nucleotides, r^{b-b}_{ij} is the distance between backbone sites i and j, ϵ_0 is the

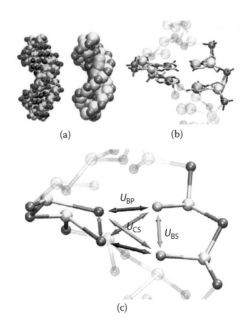

(a)   (b)

(c)

FIGURE 7.3 Comparison of (a) all-atom and (b) CG representation of 3SPN.2 model. (c) The main potentials maintaining DNA base paring are angle-dependent nonbonded interactions acting between the base sites. The green arrows show base stacking potential U_{BS}, the blue arrows the base-pairing potential U_{BP}, and the orange arrows the cross-stacking potential U_{CS}.

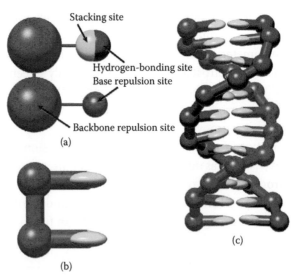

Stacking site

Hydrogen-bonding site
Base repulsion site

Backbone repulsion site
(a)

(b)

(c)

FIGURE 7.4 Interaction sites in the oxDNA model. (a) Key interaction sites of the coarse-grain model are shown. (b) The ellipsoidal representation of the beads, emphasizing that bases in oxDNA are inherently anisotropic with the planarity that is explicitly accounted for in the model. (c) A 12-bp duplex representation.

permittivity of the vacuum, and ϵ_r is the relative permittivity of the water. In contrast to the 3SPN.2 family of models described above, the ϵ_rs dependence on temperature and distances is neglected by setting it to a constant value. Such a choice simplifies calculations and is in the spirit of top-down coarse graining, where the local effects of charge–charge interactions are not rigorously modeled. To parametrize q_{eff} charges, the melting temperatures of duplexes of various lengths were computed as a function of q_{eff} for several salt concentrations. The basic parametrization approach was based on reproducing melting temperatures of short, sequence-averaged duplexes [49] and also stacking transitions of ss DNA. In order to reproduce DNA's mechanical properties, the persistence lengths of ss DNA and ds DNA and the elastic moduli of ds DNA were also used for parameter fitting.

To aid in computational efficiency, the Debye–Huckel term was set to zero at a finite cutoff distance, given by $r_{smooth} = 3\lambda_{DH}$, where $\lambda_{DH}(T, I)$ is the Debye length. The smoothing potential was added to ensure that the electrostatic potential decays smoothly to zero after the cutoff distance without introducing any artifacts in DNA's structural dynamics. The introduction of cutoffs was shown to have a negligible effect on the duplex thermodynamics, while maintaining competitive speed for simulating large nanosize DNA-based structures. Additionally, in the newest oxDNA2, the backbone stacking, hydrogen bonding, and coaxial stacking potentials have been further fine-tuned to reproduce the salt dependence of mechanical and thermodynamic characteristics. In particular, the oxDNA2 model reproduces the experimental trends of salt dependence of duplex formation, melting temperature for duplexes of various lengths as well as the salt dependence of the persistence length and torsional stiffness. The original model was also extended to account for the sequence dependence via distinct interaction strengths for stacking and base-pairing for different sequences [50]. With the added sequence specificity, the model was shown to successfully reproduce the effects of sequence on thermodynamics of stacking transitions, hybridization profiles of duplexes, and various force extension curves.

The original force field parametrization was carried out using small ss and ds DNA molecules. However, parameterization based on small DNA fragments does not guarantee that one can capture the mechanical properties of truly nanoscale objects like DNA origami. To explore this possibility, the original oxDNA was used for computing global helical twists of different DNA origami superstructures composed of approximately 15,000 nucleotides [51]. The original model was found to induce unphysical twists, which was attributed to over-twisting. In the newest version of oxDNA2, the coaxial stacking and backbone potential have been modified to reduce over-twisting without compromising any of the other thermodynamic or mechanical properties. With the modified potential, the agreement with the experimentally obtained global twists is now excellent.

The oxDNA has already found numerous applications in the fields of material science and nanotechnology [7,17]. Among notable applications are the computational studies of DNA nano tweezers, strand displacement reactions [52,53], DNA cruciforms [54], kissing hairpins [55], duplex solutions [56], and nano stars [57,58]. In one of the recent publications, the authors reported a simulation-aided design of self assembling DNA-based vitrimer [59], a gel-like network structure made of DNA supramolecules with self-healing properties that can restructure itself to heal any internal fracture via toehold-mediated strand displacement. Thus, the oxDNA2 models show promise as a useful computational tool for capturing higher level mechanical and thermodynamic properties of DNA objects with reasonable efficiency while neglecting the finer structural details.

7.4 COARSE GRAINING OF NUCLEOSOMES AND THEIR ASSEMBLIES: TOWARD GENOME-SCALE SIMULATIONS OF CHROMATIN

Genomic DNA in eukaryotic cells is stored in a compact yet highly dynamic form inside the micrometer size nucleus, which is a million-fold reduction compared to the contour length of the bare ds genomic DNA [19,60]. The compaction is achieved in large part via generic polymeric coiling of a polyflexible chain. However, that is not sufficient; further compaction occurs by complexation of DNA segments with positively charged histone proteins, resulting in semiregular arrays of nucleosomal particles [61,62]. *In vivo*, these poly-nucleosomal arrays fold into higher order dense fiber-like structures known as the chromatin. The structure of a nucleosomal core particle, which consists of 145–147 bp of DNA wrapped 1.75 turns around a hockey-puck-like octamer of histone proteins, has been resolved by x-ray crystallography [63]. Structural information beyond the level of a single nucleosome is scarce, however, not providing sufficient resolution with regard to the kind of finer structural details customary in protein science. Furthermore, the important intrinsically disordered tails of the histones protruding from the rim of the nucleosomes have not been well characterized structurally due to their highly dynamic nature. Yet, a large body of evidence shows that histone tails are the crucial players in stabilizing the dense high-order folds of chromatin, most likely via attractive intranucleosomal interactions mediated by histone tails [64]. Computer simulations using the CG DNA models, discussed in this chapter, can help shed light on this problem by providing a computational window into some of these elusive structures. The latest advances in the experimental techniques for probing the higher order structures of the chromatin [65–67] have stimulated the computational biophysics community to model the chromatin from the level of individual nucleosomes to the large-scale models of chromatin [4,68–72]. There have been a number of all-atom simulations of nucleosomes (see Refs. [4,72] for review), which were helpful in pointing out the important roles played by the ionic environment [73] and the disordered nature of histone tails [74–76] in stabilizing the nucleosomes. Such all-atom simulations, however, are currently limited to microsecond timescales [77,78] because of the sheer number of particles they have to include. CG simulations, meanwhile, are far ahead in modeling single nucleosomes [79], poyl-nucleosomal systems [79,80], and chromatin fibers on the orders of kilo base pair [81–83] as well as entire chromosomes [84–86]. The number of studies using CG models of nucleosomes has been growing rapidly, where several groups have been spearheading the development of different types of models.

One of the earliest models was developed by the group of Schlick using a discrete charge optimization algorithm [87] (DiSCO). In their model, the nucleosomes are treated as uniformly charged fine discretized surfaces with charge distribution optimized to match the solution of the Poisson–Boltzamn equation of the corresponding all-atom model. The resulting model was applied for studying the effect of tail-mediated nucleosomal aggregation [88–90], chromatin fiber folding [82,83,91,92], the electrostatic environment of the chromatin [93,94], and force induced unfolding of poly-nucleosmal arrays [95]. However, the mean-field level treatment of electrostatics via Debye–Huckel potentials is a serious limitation in this as well as many others models. The implicit treatment of electrostatics ignores some of the important effects of long-ranged electrostatic forces by not accounting for heterogeneous ionic condensation, ion–ion correlations, and entropic effects associated with release of discrete ions [42,73,96]. Also, an ionic environment containing ions of higher valency is known for showing significant departures relative to the mean-field picture. Hence, the electrostatic effects are crucial for highly charged systems like chromatin and one needs to account for them rigorously in computational models in order to have a more accurate picture of chromatin folding and organization.

Another approach for simulating nucleosomes has been pursued by Korolev, Lubartsev, and Nordenskold who developed a poyl-electroyle model of the nucleosome [97–99]. In the first version of their models, the histone core was treated as a uniformly charged sphere with a fully disordered model of histone tails. The ions, however, were modeled explicitly, which is a crucial ingredient of the model, thanks to which the corresponding simulations were early on able to reproduce effects of ion–ion correlations, entropic effects due to the release of monovalent cations, and polyvalent ion-induced condensation patterns of nucleosomes. The latest and more refined version of their model is named the advanced CG nucleosome core particle model [100]. It provides a more detailed description of the histone octamer and the distribution of charges among the amino acids (see Figure 7.5). DNA is represented as a spaced filling model [79] consisting of five-site two base pair elementary units, with four sites corresponding to four phosphate groups and the middle site to the stacked pair of base pairs. The stability of the nucleosome in the model is maintained by imposing harmonic bonds between DNA and histones. Such

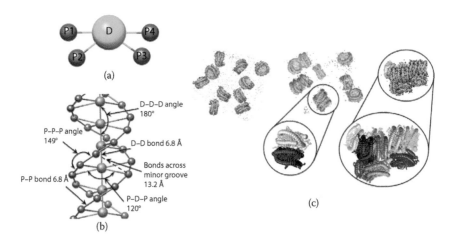

FIGURE 7.5 Representation of the advanced nucleosome core particle model. (a) Elementary unit of DNA consisting of five sites representing two base pairs (green bead) and four phosphate groups (orange-red). (b) The CG model of the ds DNA with bonds and angles set to maintain the canonical B-structure. (c) Snapshots from simulation of the ion-induced aggregation of the nucleosomal particles.

a description is not fully motivated by microscopic physics, and hence, this model does not yet capture finer details of electrostatic, hydrogen bonding, and solvent-mediated interactions that stabilize nucleosomal arrays. However, some other aspects of the nucleosomal behaviors are well reproduced. In particular, the dependence of the intranucleosomal interactions on the nature and concentration of various mono and polyvalent ions is well captured. Also, the poly-nucleosomal aggregates found in simulations match well with the available x-ray diffraction and electron microscopy observations. In another study, Lyubartsev et al. [80] used simulations with the advanced CG model in combination with inverse Monte Carlo techniques (see above) to coarse-grain their model even more, obtaining a super-CG model of nucleosomes consisting of seven beads for each nucleosome. The super-CG model was then used to simulate the aggregation of 5,000 nucleosomal particles induced by the presence of trivalent cations. Comparison of the factors corresponding to small-angle x-ray scattering (SAXS) spectra obtained via simulations were in qualitative agreements with the data obtained from the related experiments.

7.5 CONCLUDING REMARKS

Thanks to the collective efforts of several groups working on the development of CG DNA models, the computational biophysics community now has access to a wide arsenal of tools for biophysical and nanotechnological applications. An important take home message of this chapter is that there is more than one way of modeling different DNA characteristics. Thus, instead of pursuing the goal of designing an all-encompassing universal CG DNA model, a more practical approach may be used in different CG DNA force fields for addressing different types of problems. This chapter focused on three CG DNA models, namely MRG-CG DNA, 3SPN, and oxDNA, all of which have their strengths and weaknesses. We expect all of these models to be used in synergy by many groups, to help us uncover various facets of DNA's complex mechanical, thermal, and dynamic behaviors. In particular, at present, it appears that the 3SPN family of models is particularly well suited for studying protein–DNA interactions, the oxDNA for simulating folding of massive nano-objects, and MRG-CG for systems where the role of long-range electrostatics needs to be addressed with high precision, with highly charged nucleosome particles and chromatin fibers being prime examples.

ACKNOWLEDGMENTS

GAP would like to thank the Beckman Young Investigator Program and Camille Dreyfus Teacher-Scholar Program for the early support of his group's research on DNA coarse graining. He is grateful to the National Science Foundation (Grants: CAREER CHE-0846701, CHE-1363081 and POLS-1206060) for supporting his group's research in theoretical chemistry and computational biophysics.

REFERENCES

1. D Frenkel and B Smit. *Understanding Molecular Simulation: From Algorithms to Applications*. San Diego, CA: Academic Press, vol. 1, 2001.
2. M Tuckerman. *Statistical Mechanics: Theory and Molecular Simulation*. Oxford, UK: Oxford University Press, 2010.
3. PD Dans, J Walther, H Gómez, and M Orozco. Multiscale simulation of DNA. *Curr Opin Struct Biol*, 37:29–45, 2016.
4. C Maffeo, J Yoo, J Comer, DB Wells, B Luan, and A Aksimentiev. Close encounters with DNA. *J Phys Condens Matter*, 26(41):413101, 2014.
5. A Pérez, FJ Luque, and M Orozco. Frontiers in molecular dynamics simulations of DNA. *Acc Chem Res*, 45(2):196–205, 2011.
6. TE Cheatham and DA Case. Twenty-five years of nucleic acid simulations. *Biopolymers*, 99(12):969–977, 2013.
7. JPK Doye, TE Ouldridge, AA Louis, F Romano, P Šulc, C Matek, BEK Snodin, L Rovigatti, JS Schreck, RM Harrison et al. Coarse-graining DNA for simulations of DNA nanotechnology. *Phys Chem Chem Phys*, 15(47):20395–20414, 2013.
8. DA Potoyan, A Savelyev, and GA Papoian. Recent successes in coarse-grained modeling of DNA. *Wiley Interdiscip Rev Comput Mol Sci*, 3(1):69–83, 2013.
9. N Korolev, L Nordenskiöld, and AP Lyubartsev. Multiscale coarse-grained modelling of chromatin components: DNA and the nucleosome. *Adv Colloid Interface Sci*, 232:36–48, 2016.
10. A Savelyev and GA Papoian. Molecular renormalization group coarse-graining of polymer chains: Application to double-stranded DNA. *Biophys J*, 96:4044, Jan 2009.
11. A Savelyev and GA Papoian. Molecular renormalization group coarse-graining of electrolyte solutions: Application to aqueous NaCl and KCl. *J Phys Chem B*, 113(22):7785–7793, June 2009.
12. A Savelyev and GA Papoian. Chemically accurate coarse graining of double-stranded DNA. *Proc Natl Acad Sci U S A*, 107(47):20340–20345, Nov 2010.
13. T Knotts IV, N Rathore, DC Schwartz, and JJ de Pablo. A coarse grain model for DNA. *J Chem Phys*, 126:084901, Jan 2007.

14. EJ Sambriski, DC Schwartz, and JJ de Pablo. A mesoscale model of DNA and its renaturation. *Biophys J*, 96(5):1675–90, Mar 2009.
15. GS Freeman, DM Hinckley, and JJ de Pablo. A coarse-grain three-site-per-nucleotide model for DNA with explicit ions. *J Chem Phys*, 135(16):165104, 2011.
16. DM Hinckley, GS Freeman, JK Whitmer, and JJ de Pablo. An experimentally-informed coarse-grained 3-site-per-nucleotide model of DNA: Structure, thermodynamics, and dynamics of hybridization. *J Chem Phys*, 139(14):144903, 2013.
17. TE Ouldridge. DNA nanotechnology: Understanding and optimisation through simulation. *Mol Phys*, 1:1–15, 2014.
18. BEK Snodin, F Randisi, M Mosayebi, P Sulc, JS Schreck, F Romano, TE Ouldridge, R Tsukanov, E Nir, AA Louis et al. Introducing improved structural properties and salt dependence into a coarse-grained model of DNA. *J Chem Phys*, 142:234901, 2015.
19. B Alberts, A Johnson, J Lewis, M Raff, K Roberts, and P Walter. *Molecular Biology of the Cell*. Garland Science, 5th edition, 2007.
20. A Vologodskii. *Biophysics of DNA*. Cambridge, UK: Cambridge University Press, 2015.
21. AF Jorge, SCC Nunes, TFGG Cova, and AACC Pais. Cooperative action in DNA condensation. *Curr Opin Colloid Interface Sci*, 26:66–74, 2016.
22. R Podgornik, MA Aksoyoglu, S Yasar, D Svenšek, and VA Parsegian. DNA equation of state: In vitro vs in vivo. *J Phys Chem B*, 120(26):6051–6060, 2016.
23. H Schiessel. The physics of chromatin. *J Phys Condens Matter*, 15(19):R699, 2003.
24. JF Marko. Biophysics of protein–DNA interactions and chromosome organization. *Physica A*, 418:126–153, 2015.
25. JJ de Pablo. Coarse-grained simulations of macromolecules: From DNA to nanocomposites. *Annu Rev Phys Chem*, 62:555–574, May 2011.
26. JP Peters and LJ Maher. DNA curvature and flexibility in vitro and in vivo. *Q Rev Biophys*, 43(1):23–63, 2010.
27. AK Mazur and M Maaloum. DNA flexibility on short length scales probed by atomic force microscopy. *Phys Rev Lett*, 112(6):068104, 2014.
28. R Vafabakhsh and T Ha. Extreme bendability of DNA less than 100 base pairs long revealed by single-molecule cyclization. *Science*, 337(6098): 1097–1101, 2012.
29. C Bustamante, Z Bryant, and SB Smith. Ten years of tension: Single-molecule DNA mechanics. *Nature*, 421(6921):423–427, 2003.
30. PA Wiggins, T van der Heijden, F Moreno-Herrero, A Spakowitz, R Phillips, J Widom, C Dekker, and PC Nelson. High flexibility of DNA on short length scales probed by atomic force microscopy. *Nat Nanotechnol*, 1(2):137–141, 2006.
31. B Bonev and G Cavalli. Organization and function of the 3D genome. *Nat Rev Genet*, 17(11):661–678, 2016.

32. MR Hübner, MA Eckersley-Maslin, and DL Spector. Chromatin organization and transcriptional regulation. *Curr Opin Genet Dev*, 23(2):89–95, 2013.

33. Q Bian and AS Belmont. Revisiting higher-order and large-scale chromatin organization. *Curr Opin Cell Biol*, 24(3):359–366, 2012.

34. SV Razin and AA Gavrilov. Chromatin without the 30-nm fiber: Constrained disorder instead of hierarchical folding. *Epigenetics*, 9(5):653–657, 2014.

35. A Naômé, A Laaksonen, and DP Vercauteren. A solvent-mediated coarse-grained model of DNA derived with the systematic newton inversion method. *J Chem Theory Comput*, 10(8):3541–3549, 2014.

36. RH Swendsen. Monte Carlo renormalization group. *Phys Rev Lett*, 42:859–861, 1979.

37. A Lyubartsev and A Laaksonen. Calculation of effective interaction potentials from radial distribution functions: A reverse Monte Carlo approach. *Phys Rev E*, 52(4):3730–3737, Oct 1995.

38. A Savelyev, CK Materese, and GA Papoian. Is DNA's rigidity dominated by electrostatic or nonelectrostatic interactions? *J Am Chem Soc*, 133(48):19290–19293, Dec 2011.

39. A Naômé, A Laakonsen, and DP Vercauteren. A coarse-grained simulation study of the structures, energetics, and dynamics of linear and circular DNA with its ions. *J Chem Theory Comput*, 11(6):2813–2826, 2015.

40. Q Cao, C Zuo, Y Ma, L Li, and Z Zhang. Interaction of double-stranded DNA with a nanosphere: A coarse-grained molecular dynamics simulation study. *Soft Matter*, 7(2):506–514, 2011.

41. P Debye and E Hückel. De la theorie des electrolytes. I. Abaissement du point de congelation et phenomenes associes. *Phys Zeit*, 24(9):185–206, 1923.

42. R Messina. Electrostatics in soft matter. *J Phys Condens Matter*, 21(11): 113102, 2009.

43. AV Dobrynin and M Rubinstein. Theory of polyelectrolytes in solutions and at surfaces. *Prog Polym Sci*, 30(11):1049–1118, 2005.

44. DM Hinckley, JP Lequieu, and JJ de Pablo. Coarse-grained modeling of DNA oligomer hybridization: Length, sequence, and salt effects. *J Chem Phys*, 141(3):035102, 2014.

45. GS Freeman, DM Hinckley, JP Lequieu, JK Whitmer, and JJ de Pablo. Coarse-grained modeling of DNA curvature. *J Chem Phys*, 141, 2014.

46. R Rohs, X Jin, SM West, R Joshi, B Honig, and RS Mann. Origins of specificity in protein-DNA recognition. *Annu Rev Biochem*, 79:233–269, 2010.

47. A Sarai and H Kono. Protein-DNA recognition patterns and predictions. *Annu Rev Biophys Biomol Struct*, 34:379–398, 2005.

48. TE Ouldridge, AA Louis, and JPK Doye. Structural, mechanical, and thermodynamic properties of a coarse-grained DNA model. *J Chem Phys*, 134(8):085101, 2011.

49. J SantaLucia Jr and D Hicks. The thermodynamics of DNA structural motifs. *Annu Rev Biophys Biomol Struct*, 33:415–440, 2004.
50. P Šulc, F Romano, TE Ouldridge, L Rovigatti, JPK Doye, and AA Louis. Sequence-dependent thermodynamics of a coarse-grained DNA model. *J Chem Phys*, 137(13):135101, 2012.
51. H Dietz, SM Douglas, and WM Shih. Folding DNA into twisted and curved nanoscale shapes. *Science*, 325(5941):725–730, 2009.
52. RRF Machinek, TE Ouldridge, NEC Haley, J Bath, and AJ Turberfield. Programmable energy landscapes for kinetic control of DNA strand displacement. *Nat Commun*, 5:5324, 2014.
53. N Srinivas, TE Ouldridge, P Šulc, JM Schaeffer, B Yurke, AA Louis, JPK Doye, and E Winfree. On the biophysics and kinetics of toehold-mediated DNA strand displacement. *Nucleic Acids Res*, 41(22):10641–10658, 2013.
54. C Matek, TE Ouldridge, A Levy, JPK Doye, and AA Louis. DNA cruciform arms nucleate through a correlated but asynchronous cooperative mechanism. *J Phys Chem B*, 116(38):11616–11625, 2012.
55. F Romano, A Hudson, JPK Doye, TE Ouldridge, and AA Louis. The effect of topology on the structure and free energy landscape of DNA kissing complexes. *J Chem Phys*, 136(21):215102, 2012.
56. CD Michele, L Rovigatti, T Bellini, and F Sciortino. Self-assembly of short DNA duplexes: From a coarse-grained model to experiments through a theoretical link. *Soft Matter*, 8(32):8388–8398, 2012.
57. L Rovigatti, F Bomboi, and F Sciortino. Accurate phase diagram of tetravalent DNA nanostars. *J Chem Phys*, 140(15):154903, 2014.
58. L Rovigatti, F Smallenburg, F Romano, and F Sciortino. Gels of DNA nanostars never crystallize. *ACS Nano*, 8(4):3567–3574, 2014.
59. F Romano and F Sciortino. Switching bonds in a DNA gel: An all-DNA vitrimer. *Phys Rev Lett*, 114:078104, 2015.
60. R Phillips, J Kondev, J Theriot, and H Garcia. *Physical Biology of the Cell*. Oxford, UK: Garland Science, 2012.
61. K Luger, ML Dechassa, and DJ Tremethick. New insights into nucleosome and chromatin structure: An ordered state or a disordered affair? *Nat Rev Mol Cell Biol*, 13(7):436–447, 2012.
62. SA Grigoryev and CL Woodcock. Chromatin organization—The 30nm fiber. *Exp Cell Res*, 318(12):1448–1455, 2012.
63. K Luger, AW Mader, RK Richmond, DF Sargent, and TJ Richmond. Crystal structure of the nucleosome core particle at 2.8 a resolution. *Nature*, 389(6648):251–260, Sep 1997.
64. DA Potoyan and GA Papoian. Regulation of the h4 tail binding and folding landscapes via lys-16 acetylation. *Proc Natl Acad Sci*, 109(44):17857–17862, 2012.
65. E Lieberman-Aiden, NL van Berkum, L Williams, M Imakaev, T Ragoczy, A Telling, I Amit, BR Lajoie, PJ Sabo, MO Dorschner et al. Comprehensive mapping of long-range interactions reveals folding principles of the human genome. *Science*, 326(5950):289–293, 2009.

66. SSP Rao, MH Huntley, NC Durand, EK Stamenova, ID Bochkov, JT Robinson, AL Sanborn, I Machol, AD Omer, ES Lander et al. A 3D map of the human genome at kilobase resolution reveals principles of chromatin looping. *Cell*, 159(7):1665–1680, 2014.

67. VI Risca and WJ Greenleaf. Unraveling the 3D genome: Genomics tools for multiscale exploration. *Trends Genet*, 31(7):357–372, 2015.

68. G Bascom and T Schlick. Linking chromatin fibers to gene folding by hierarchical looping. *Biophys J*, 112(3):434–445, 2017.

69. N Korolev, Y Fan, AP Lyubartsev, and L Nordenskiöld. Modelling chromatin structure and dynamics: Status and prospects. *Curr Opin Struct Biol*, 22(2):151–159, 2012.

70. T Schlick, J Hayes, and S Grigoryev. Toward convergence of experimental studies and theoretical modeling of the chromatin fiber. *J Biol Chem*, 287(8):5183–5191, 2012.

71. R Collepardo-Guevara and T Schlick. Insights into chromatin fibre structure by in vitro and in silico single-molecule stretching experiments. *Biochem Soc Trans*, 41(2):494, 2013.

72. M Biswas, J Langowski, and TC Bishop. Atomistic simulations of nucleosomes. *Wiley Interdiscip Rev Comput Mol Sci*, 3(4):378–392, 2013.

73. CK Materese, A Savelyev, and GA Papoian. Counterion atmosphere and hydration patterns near a nucleosome core particle. *J Am Chem Soc*, 131(41): 15005–15013, Oct 2009.

74. M Biswas, K Voltz, JC Smith, J Langowski. Role of histone tails in structural stability of the nucleosome. *PLoS Comput Biol*, 7(12):e1002279, 2011.

75. J Erler, R Zhang, L Petridis, X Cheng, JC Smith, and J Langowski. The role of histone tails in the nucleosome: A computational study. *Biophys J*, 107(12):2902–2913, 2014.

76. D Winogradoff, I Echeverria, DA Potoyan, and GA Papoian. The acetylation landscape of the H4 histone tail: Disentangling the interplay between the specific and cumulative effects. *J Am Chem Soc*, 137(19):6245–6253, May 2015.

77. D Winogradoff, H Zhao, Y Dalal, and GA Papoian. Shearing of the CENP-A dimerization interface mediates plasticity in the octameric centromeric nucleosome. *Sci Rep*, 5:17038, Nov 2015.

78. AK Shaytan, D Landsman, and AR Panchenko. Nucleosome adaptability conferred by sequence and structural variations in histone H2A-H2B dimers. *Curr Opin Struct Biol*, 32:48–57, June 2015.

79. N Korolev, D Luo, AP Lyubartsev, and L Nordenskiöld. A coarse-grained DNA model parameterized from atomistic simulations by inverse Monte Carlo. *Polymers*, 6(6):1655–1675, 2014.

80. AP Lyubartsev, N Korolev, Y Fan, and L Nordenskiöld. Multiscale modelling of nucleosome core particle aggregation. *J Phys Condens Matter*, 27(6): 064111, 2015.

81. O Müller, N Kepper, R Schöpflin, R Ettig, K Rippe, and G Wedemann. Changing chromatin fiber conformation by nucleosome repositioning. *Biophys J*, 107(9):2141–2150, 2014.

82. R Collepardo-Guevara and T Schlick. Chromatin fiber polymorphism triggered by variations of DNA linker lengths. *Proc Natl Acad Sci*, 111(22): 8061–8066, 2014.

83. A Luque, R Collepardo-Guevara, S Grigoryev, and T Schlick. Dynamic condensation of linker histone c-terminal domain regulates chromatin structure. *Nucleic Acids Res*, 42(12):7553–7560, 2014.

84. MD Pierro, B Zhang, EL Aiden, PG Wolynes, and JN Onuchic. Transferable model for chromosome architecture. *Proc Natl Acad Sci*, 113(43): 12168–12173, 2016.

85. CA Brackley, J Johnson, S Kelly, PR Cook, and D Marenduzzo. Simulated binding of transcription factors to active and inactive regions folds human chromosomes into loops, rosettes and topological domains. *Nucleic Acids Res*, 44(8):3503–3512, 2016.

86. A-M Florescu, P Therizols, and A Rosa. Large scale chromosome folding is stable against local changes in chromatin structure. *PLoS Comput Biol*, 12(6):e1004987, 2016.

87. Q Zhang, DA Beard, and T Schlick. Constructing irregular surfaces to enclose macromolecular complexes for mesoscale modeling using the discrete surface charge optimization (disco) algorithm. *J Comput Chem*, 24(16): 2063–2074, Dec 2003.

88. G Arya, Q Zhang, and T Schlick. Flexible histone tails in a new mesoscopic oligonucleosome model. *Biophys J*, 91(1):133–150, 2006.

89. G Arya and T Schlick. Role of histone tails in chromatin folding revealed by a mesoscopic oligonucleosome model. *Proc Natl Acad Sci U S A*, 103(44):16236–16241, Oct 2006.

90. G Arya and T Schlick. A tale of tails: How histone tails mediate chromatin compaction in different salt and linker histone environments. *J Phys Chem A*, 113(16): 4045–4059, Apr 2009

91. O Perišić, R Collepardo-Guevara, and T Schlick. Modeling studies of chromatin fiber structure as a function of DNA linker length. *J Mol Biol*, 403(5): 777–802, Nov 2010.

92. SA Grigoryev, G Arya, S Correll, CL Woodcock, and T Schlick. Evidence for heteromorphic chromatin fibers from analysis of nucleosome interactions. *Proc Natl Acad Sci*, 106(32):13317–13322, Aug 2009.

93. HH Gan and T Schlick. Chromatin ionic atmosphere analyzed by a mesoscale electrostatic approach. *Biophys J*, 99(8):2587–2596, Oct 2010.

94. R Collepardo-Guevara and T Schlick. Crucial role of dynamic linker histone binding and divalent ions for DNA accessibility and gene regulation revealed by mesoscale modeling of oligonucleosomes. *Nucleic Acids Res*, 40(18):8803–8817, 2012.

95. G Ozer, R Collepardo-Guevara, and T Schlick. Forced unraveling of chromatin fibers with nonuniform linker DNA lengths. *J Phys Condens Matter*, 27(6):064113, 2015.

96. AG Cherstvy. Electrostatic interactions in biological DNA-related systems. *Phys Chem Chem Phys*, 13(21):9942–9968, 2011.

97. N Korolev, AP Lyubartsev, and L Nordenskiïld. Computer modeling demonstrates that electrostatic attraction of nucleosomal DNA is mediated by histone tails. *Biophys J*, 90(12):4305–4316, June 2006.

98. N Korolev, A Allahverdi, Y Yang, Y Fan, AP Lyubartsev, and L Nordenskiöld. Electrostatic origin of salt-induced nucleosome array compaction. *Biophys J*, 99(6):1896–1905, Sep 2010.

99. Y Yang, AP Lyubartsev, N Korolev, and L Nordenskiöld. Computer modeling reveals that modifications of the histone tail charges define salt-dependent interaction of the nucleosome core particles. *Biophys J*, 96(6):2082–2094, Mar 2009.

100. Y Fan, N Korolev, AP Lyubartsev, and L Nordenskiöld. An advanced coarse-grained nucleosome core particle model for computer simulations of nucleosome–nucleosome interactions under varying ionic conditions. *PLOS ONE*, 8(2):e54228, 2013.

CHAPTER 8

Coarse-Grained Modeling of Nucleosomes and Chromatin

Lars Nordenskiöld, Alexander P. Lyubartsev, and Nikolay Korolev

CONTENTS

8.1	Introduction	298
8.2	Background to Chromatin Structure and Dynamics	300
	8.2.1 Chromatin Organization	300
	8.2.2 Experimental Background to Physical and Structural Properties of Chromatin	302
8.3	Approaches to Coarse-Graining of Chromatin	308
8.4	CG Modeling of Nucleosomes	309
	8.4.1 Sphere Models with Grafted Histone Tail Chains	313
	8.4.2 Advanced CG NCP Model	316
	8.4.3 Multiscale CG Model of a Single NCP from Boltzmann Inversion	319
8.5	CG Modeling of Chromatin	321
	8.5.1 The Six-Angle Model	322
	8.5.2 The DiSCO Mesoscale Model	323
	8.5.3 The Electrostatic Sphere-Bead Model	326
8.6	Conclusions and Future Perspectives	329
	References	330

8.1 INTRODUCTION

Recently, computer simulation approaches to model chromatin structure and dynamics have become important tools to address fundamental questions related to the biophysics of chromatin. Computer modeling has the advantage that within various levels of approximation, it is possible to separately investigate different physical mechanisms and assess, both qualitatively and quantitatively, their importance to different chromatin structural features and structural transitions. Many of the interesting problems necessitate the modeling of a piece of a chromatin fiber, corresponding to a nucleosome array (Figure 8.1). A purely atomistic approach to modeling the chromatin fiber including solvent water molecules and mobile ions would produce an unmanageable number of particles, so a coarse-grained (CG) approach becomes necessary. Ideally, it would be desirable to perform a systematic coarse-graining based on bridging a fine-grained description of the system (i.e., based on all-atom models) to the CG model where the CG interaction potentials were obtained from a rigorous mapping of the fine-grained description to CG model. This kind of approach is based on constructing effective pair potentials for the interactions among the CG sites from atomistic simulations of a small system representing the large-scale system that is being described by the CG model.

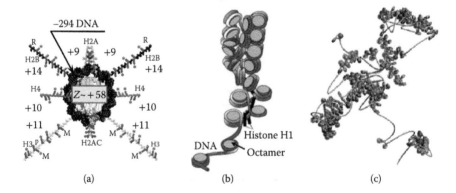

(a)   (b)   (c)

FIGURE 8.1   (a) The basic regular unit of eukaryotic chromatin, the NCP. The negative charge of the DNA (−294e) and the net positive charge of the HO globular domain (+58e) and in the histone tails are indicated to underline the importance of electrostatics. (Adapted from Wolffe, A. P., and J. J. Hayes, *Nucleic Acids Res*, 27, 711–720, 1999. With permission.) (b) NCPs are connected by linker DNA to form a *beads-on-a-string*, 10-nm chromatin fiber, which folds into a 30-nm fiber in a salt-dependent manner. (c) Chromatin fibers can form folded 30-nm structures as well as various self-associated irregular structures. (Adapted from Diesinger, P. M., and D. W. Heermann, *PMC Biophys*, 3, 11, 2010. With permission.)

There are several approaches to such systematic CG modeling, namely structure-based approaches like the inverse Monte Carlo (MC) method (Lyubartsev and Laaksonen 1995; Lyubartsev et al. 2010; Mirzoev and Lyubartsev 2013) (also discussed in Chapter 1) and the iterative Boltzmann inversion (Reith et al. 2003), the force matching method (Ayton et al. 2007; Izvekov and Voth 2005), and the relative entropy minimization methods (Scott Shell 2008). Another group of CG approaches are based on the principles similar to ones used in the construction of the atomistic force fields, by parameterization of interaction potentials using experimental data like thermodynamic properties, e.g., the MARTINI CG force field approach (Marrink et al. 2007). Different systematic CG approaches are described elsewhere in this book.

However, when it comes to chromatin modeling, the CG approaches so far used are still at a rather primitive level and not yet performed in the systematic way outlined above. Most approaches are based on continuum model description of the solvent and in many cases also employing some mean field Poisson–Boltzmann (PB) or Debye–Hückel (DH) description of the electrostatic interactions. This situation is in contrast to the modeling of protein folding (Ayton et al. 2007) and the modeling of lipid self-assembly (Lyubartsev 2005; Lyubartsev et al. 2010; Lyubartsev and Rabinovich 2011), as well as modeling of some protein assemblies like actin fibers and viruses (Saunders and Voth 2012, 2013), which has advanced considerably in recent years and is described elsewhere in this book (for detailed and systematic protein coarse-graining, see Chapters 3 and 4, and for nucleic acids, Chapters 6 and 7).

In this chapter, we review the present status of modeling of chromatin and the relation of these models to experimental studies. We concentrate on CG modeling at the first level of chromatin compaction, namely the nucleosome core particle (NCP) and nucleosome arrays. Higher levels of chromatin folding such as structural organization of chromosomal domains and chromosomes (see, e.g., Naumova et al., 2013 and references cited therein) are not considered.

We discuss some of the approaches and models that have been used in recent years with a main focus on CG computer simulations methods based on producing trajectories that describe the ensemble of states of the system from which various average structural properties can be calculated. We review efforts at modeling the nucleosome (the NCP) and the chromatin fiber that have been conducted in recent years. In the concluding section, the prospect of employing more detailed and rigorous CG modeling of chromatin using systematic multiscale approaches is discussed.

However, we first give the pertinent background on chromatin structure and dynamics with emphasis on the physical–chemical properties of the system.

8.2 BACKGROUND TO CHROMATIN STRUCTURE AND DYNAMICS

8.2.1 Chromatin Organization

The presence of a compact, folded higher order structure of chromatin represses many DNA-dependent activities such as transcription, replication, and repair. Decondensation of chromatin is therefore a prerequisite for normal cellular functions (Horn and Peterson 2002). The studies of higher order chromatin structure, regulation of chromatin compaction, and of the mechanism for unwinding of the DNA from the nucleosome are consequently of great importance in structural and molecular biology and much research has been directed at experimental studies of these questions.

At the first level of its organization, chromatin is highly uniform in most of eukaryotes and can be considered as a linear array of nucleosomes that is formed by DNA and the histone octamer (HO). The HO comprises two copies each of the four core histone proteins: H2A, H2B, H3, and H4. The most regular part of the nucleosome, the NCP, includes 145–147-bp DNA wrapped as a left-handed 1.75 turn super helix around the HO (Davey et al. 2002; Luger et al. 1997; Wolffe 1998). The NCP is essentially a polyanion–polycation complex with a net charge of about −148e, comprising a highly negatively charged central particle (−236e, with −294e from DNA and +58e from the globular part of the HO) to which the eight flexible and positively charged N-terminal tails are attached with net charge +88e (Figure 8.1a).

Linker DNA having a variable length in the range 10–70 bp connects the NCPs to form an array of chromatin in the open, uncondensed "beads-on-a-string" state. Arrays of nucleosomes are folded, yielding a fiber of approximately 30 nm diameter (Figure 8.1b), whose detailed structure is still a matter of debate (Dorigo et al. 2004; Kruithof et al. 2009; Robinson et al. 2006; Robinson and Rhodes 2006; Routh et al. 2008; Schalch et al. 2005; Tremethick 2007). The relevance of the 30-nm fiber to the *in vivo* situation has also recently been a subject of controversy (Daban 2011; Maeshima et al. 2010; Nishino et al. 2012) (and references cited in Nishino et al. 2012). Interarray aggregation leads to further layers of higher order compaction, resulting in chromosomes. Figure 8.1c illustrates the irregular character of chromatin folding and aggregation which is likely to be relevant *in vivo* (Grigoryev et al. 2009; Grigoryev and Woodcock 2012; Maeshima et al. 2010; Scheffer et al. 2011, 2012). In higher organisms, one more family of histone proteins, the linker histones H1, are typically present at an NCP:H1 ratio of 1:1 (Woodcock et al. 2006). The linker histones contribute to the chromatin structure stabilization but it was shown that the nucleosome arrays can reach "native" degree of compaction without histone H1 (McBryant and Hansen 2012; Woodcock et al. 2006).

Eight flexible, highly basic N-terminal histone tails protrude out from the core domain of the NCP (Figure 8.1a). The tails facilitate interactions between neighboring nucleosomes by bridging and electrostatic interactions (Bertin et al. 2007a,c, 2004; Gordon et al. 2005; Kan et al. 2007, 2009; Kan and Hayes 2007; Korolev et al. 2006; Mangenot et al. 2002b; McBryant et al. 2009; Wang and Hayes 2008; Zheng et al. 2005). Similar to DNA, chromatin is a highly negatively charged polyelectrolyte, but the presence of the positively charged histone tails changes the electrostatic properties of the chromatin fiber compared to DNA. Observations *in vitro* show that folding of the array and further interarray aggregation (often called oligomerization or self-association) into tertiary chromatin structures can occur upon increase in monovalent salt or by addition of Mg^{2+} or other multivalent cations (Gordon et al. 2005; Hansen 2002; Kan et al. 2007, 2009; Kan and Hayes 2007; Luger and Hansen 2005; McBryant et al. 2009; Wang and Hayes 2008; Woodcock and Dimitrov 2001; Zheng et al. 2005). These observations indicate an electrostatic mechanism to the compaction of chromatin (Clark and Kimura 1990; van Holde and Zlatanova 1996; van Holde 1988). Solutions of NCPs (where linker DNA is absent) display a similar behavior (Bertin et al. 2004, 2007b,c; de Frutos et al. 2001; Mangenot et al. 2002a,b). Systems of tailless (trypsin treated) NCPs and globular recombinant nucleosome arrays do not show Mg^{2+}-induced folding/aggregation or they exhibit a reduced condensation propensity. These observations imply that tails perform an important role in array folding and aggregations as well as condensation of NCPs (Hansen 2002). Although knowledge on the mechanisms of chromatin unfolding, stretching, and DNA unwrapping from the histone core and the role of the histone tail modifications have advanced in recent years (Allahverdi et al. 2011; Brower-Toland et al. 2005; Zhou et al. 2007), the physical mechanisms that govern these processes are still not entirely understood. Similarly, the detailed 30-nm chromatin fiber structure and its relevance to organization of chromatin *in vivo* are still the matter of much debate (Maeshima et al.

2010; Scheffer et al. 2011; Tremethick 2007). Recent advances in recombinant protein expression and purification techniques have enabled the preparation of large quantities of homogeneous NCPs as well as nucleosome arrays. These *in vitro* prepared systems can be used for systematic biophysical investigations, the result of which can be compared with theoretical modeling.

Before discussing different CG chromatin models and the major results of coarse-graining studies on chromatin fibers and its major element, the NCP, it is useful to make some comments on relevant experimental information on the physical chemical and structural properties of chromatin, which is given below.

8.2.2 Experimental Background to Physical and Structural Properties of Chromatin

Energetics and forces involved in assembly and disassembly of the nucleosome. Since the nucleosome organization is the natural state of chromatin, this implies that under physiological conditions the formation of the nucleosome corresponds to the lowest free energy among all possible arrangements for the DNA and histones (illustrated in Figure 8.2a). However, there exist little experimental and theoretical estimations for the actual depth of the free energy well shown in Figure 8.2a. In our opinion, the major contribution stabilizing the chromatin/nucleosome structure is the nonspecific electrostatic free energy of interaction between the positively charged Lys and Arg of the histones and negative phosphate groups of the DNA.

At first glance, this conclusion about high stability of the nucleosome contradicts many experimental observations (Gottesfeld and Luger 2001; Yager et al. 1989) (and references cited therein) and theoretical models (Marky and Manning 1991, 1995). Data on NCP dissociation at low salt/low NCP concentration give an estimation of the absolute stability of the NCP at about 45–60 kJ/mol (Gottesfeld and Luger 2001). However, these estimates of the free energy of nucleosome formation have been questioned (Thåsröm et al. 2004). Our estimation is that the major contrition to the formation of the NCP originates from the electrostatic contribution and this free energy term is large ($\Delta G^{el} \approx -1500$ kJ/mol) (Figure 8.2b, see also Korolev et al. 2004), is in a good agreement with data by Manning and coworkers (Manning 2003; Manning et al. 1989) and also with early results of proton nuclear magnetic resonance (NMR) studies (Walker 1984). A confirmation of the high stability of the nucleosome due to electrostatic

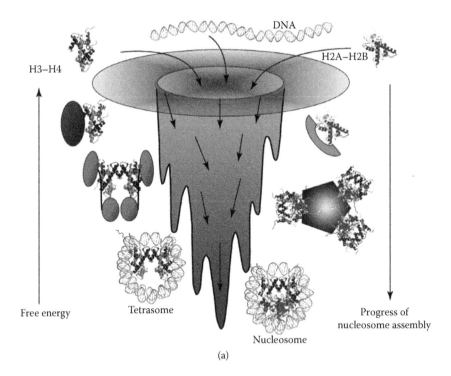

FIGURE 8.2 Assembly and disassembly of the nucleosome and chromatin. (a) Schematic presentation of the energy well for chaperon-assisted *in vivo* assembly of nucleosome from DNA and histones. (Adapted from Das, C. et al., *Trends Biochem. Sci.*, 35, 476–489, 2010. With permission.)

(*Continued*)

forces can be obtained by an estimation of the electrostatic contribution to the process of detachment of the DNA polyion from the positively charged histone core in the absence of proteins in chromatin fiber stretching experiments (Bennink et al. 2001; Brower-Toland et al. 2002). The role of electrostatics in this process has been analyzed (Korolev et al. 2004). Detachment of the DNA from the NCP increases the NCP charge and simultaneously exposes the stretched DNA polyion to the solvent. The model is similar to the simple scheme analyzed by Schiessel (2003) and references therein. The results shown in Figure 8.2 demonstrate that the initial stages of unwinding the first three quarters of the DNA superhelical turn from the NCP core are accompanied by small changes in electrostatic free energy, ΔG^{el}. This allows other factors such as DNA bending or "bendability," (Widom 2001) mutual repulsion of the DNA polyions on the lateral surface of the octamer, formation of ion pairs, specific histone-DNA contacts, etc. to

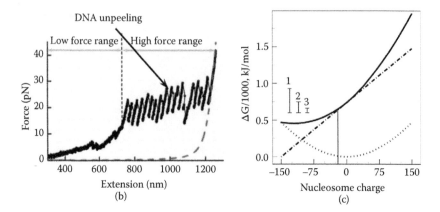

FIGURE 8.2 (*Continued*)   Assembly and disassembly of the nucleosome and chromatin. (b) Force versus extension dependencies obtained for nucleosome array stretching by tweezers. Two stages of the nucleosome disassembly are shown that correspond to detachment of 70-bp DNA at low force (0.75 superhelical turn) and 80 bp at high force (80-bp DNA, roughly one turn of superhelix). (Adapted from Brower-Toland, B. D. et al., *Proc. Natl. Acad. Sci. U.S.A.*, 99, 1960–1965, 2002. With permission.) (c) Estimation of the change in the electrostatic free energy during nucleosome disassembly. The degree of the nucleosome unwinding (abscissa) is measured as the change of the total charge of the NCP (HO + DNA attached to it). The dash-dotted line shows the free energy contribution from DNA released from the octamer; the dotted line is the electrostatic free energy of the NCP approximated as a sphere of 5 nm radius (the total charge of the NCP increases from about $-150e$ to $+150e$); the solid line is the sum of the two terms. The vertical line drawn at the NCP charge $-20e$, shows the position when one full turn of DNA is wrapped around the histone core. Salt concentration and temperature are similar to these applied in the chromatin fiber stretching experiment (Bennink et al. 2001; Brower-Toland et al. 2002) Vertical bars numbered 1–3 are estimations of the energy of DNA bending around the histone core according to (1) (Widom 2001) (320 kJ/mol); (2) (Schiessel 2003) (150 kJ/mol); (3) (Flaus and Owen-Hughes 2003) (62 kJ/mol) and are given to highlight the importance of the electrostatic interactions relative to the DNA bending (see Korolev et al., 2004, 2007 for details) (Adapted from Korolev, N. et al., *Prog. Biophys. Mol. Biol.*, 95, 23–49, 2007.)

contribute and regulate the nucleosome behavior. However, the electrostatic forces become dominant when the electroneutrality point of the NCP is reached. After this point, stripping more DNA from the histones becomes thermodynamically unfavorable since it exposes both the negative charges of the DNA polyion and positive charge of the nucleosome core to solvent. The major contribution to ΔG^{el} is the decrease in the entropy due to the condensation of mobile counterions on the polyelectrolytes. The results in Figure 8.2c explain (Korolev et al. 2004) the appearance of an "all-or-nothing" transition in most of the reported single-nucleosome-fiber experiments (Bennink et al. 2001; Brower-Toland et al. 2002, 2005;

Claudet et al. 2005; Gemmen et al. 2005; Pope et al. 2005) with a minimal size corresponding to the one full turn DNA around the histone core.

Nucleosome–nuclesome stacking interaction as a key element of chromatin compaction. A compulsory feature of condensed nucleosomes and folded compact chromatin structures is the formation of nucleosome stacking, i.e., close NCP–NCP contacts between the relatively flat surfaces of the HO core on both sides of the cylindrical wedge-shaped NCP. This stacking was observed in NCP crystals (Davey et al. 2002; Finch et al. 1977, 1981; Luger et al. 1997; Vasudevan et al. 2010), in NCP liquid crystalline phases (Bertin et al. 2007b; Leforestier et al. 2001; Leforestier and Livolant 1997; Mangenot et al. 2003a,b), in the crystal of the tetranucleosome (Schalch et al. 2005), in folded nucleosome arrays (Dorigo et al. 2004; Robinson et al. 2006; Song et al. 2014), and in cryo-microscopy images of frozen isolated native chromatin (Castro-Hartmann et al. 2010; Scheffer et al. 2012). Figure 8.3a displays NCP–NCP stacking as seen in the crystal structure (left) (Luger et al. 1997), which includes interaction between the histone H4 tail and the acidic patch of the histone H2A of the neighboring NCPs. The middle panel displays NCP–NCP stacking observed by EM (Leforestier et al. 2001). The right-hand panel shows an illustration of NCP–NCP stacking in a recent 11 Å cryo-electron microscopy (EM) structure of a nucleosome array (Song et al. 2014). There are a few estimations of the energy of the attractive nucleosome–nucleosome contact. A chromatin fiber stretching experiment estimated this energy at about 35 kJ/mol at physiological ionic conditions (100 mM K^+, 2 mM Mg^{2+}) (Kruithof et al. 2009); earlier experiments reported a much smaller value of 7 kJ/mol at lower salt (40 mM Na^+) (Cui and Bustamante 2000). Since the nucleosome arrays are polyelectrolytes, one would expect significant dependence of their folding and dynamics on ionic conditions; however, detailed experiential data and models are still lacking. Crystal structures (Luger et al. 1997; Luger and Richmond 1998) and numerous experiments with arrays (Allahverdi et al. 2011; Dorigo et al. 2003; Robinson et al. 2008; Shogren-Knaak et al. 2006) pointed out that the H4 histone tail and the acetylation state of the H4 Lys16 residue play an important role in the stabilization of NCP stacking through internucleosomal contact with the acidic patch in the H2A/H2B dimer. At the same time, the crystal structure of the tetranucleosome (Schalch et al. 2005) and recent cryo-EM data (Song et al. 2014) reported the existence of close NCP–NCP conformations where mediation of the H4 tail is sterically excluded.

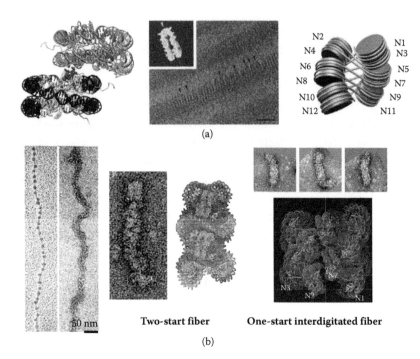

FIGURE 8.3 (a) Nucleosome–nucleosome stacking is a universal primary element of chromatin structure observed *in vivo* and *in vitro*. Left panel: NCP–NCP contact in the crystal (Luger et al. 1997) is mediated by internucleosomal interaction between the H4 histone tail and the acidic domain of the H2A/H2B dimer (purple). Central panel: NCP–NCP stacking observed by EM in concentrated NCP solutions (Adapted from Leforestier, A. et al., *Biophys. J.*, 81, 2114–2421, 2001. With permission.) Right panel: Schematic presentation of the structure of a nucleosome array determined from electron density maps of cryo-EM images. The fiber is formed from an arrangement of a units formed by two tightly stacked nucleosomes. (Adapted from Song, F. et al., *Science*, 344, 376–380, 2014. With permission.) (b) Chromatin extracted from eukaryotic nuclei in the 10-nm beads-on-a-string (left) and 30-nm fiber (right) forms. (Adapted from Alberts, B. et al., *Molecular Biology of the Cell*, Garland Publishing, New York and London, 2007. With permission.) In the center and to the right are EM images and structural models for the 30-nm fiber with different NRLs: two-start cross-linked fibers are observed for short-medium (157–187 bp) NRL and the one-start interdigitated fiber for long NRL (>187 bp). (Adapted from Routh, A. et al., *Proc. Natl. Acad. Sci. U.S.A.*, 105, 8872–8877, 2008; Robinson, P. J. J. et al., *Proc. Natl. Acad. Sci. U.S.A.*, 103, 6506–6511, 2006; and Schalch, T. et al., *Nature*, 436, 138–141, 2005. With permissions.)

(Continued)

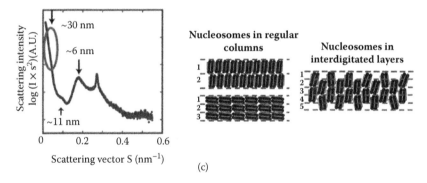

(c)

FIGURE 8.3 (*Continued*)   (c) Left: SAXS spectra of both interphase and mitotic eukaryotic chromatin demonstrates the absence of a 30-nm periodicity expected for the 30-nm fiber (arrow and red oval) whereas side-by-side (11 nm) and face-to-face (stacking, 6 nm) of repetitive nucleosome contacts are clearly seen (arrows). (Adapted from Nozaki, T. et al., *Nucleus*, 4, 349–356, 2013. With permission.) Right: Schematic representation of layers or regularly and irregularly tightly packed nucleosomes as suggested from cryo-EM images of metaphase chromosomes by Daban and coworkers (Adapted from Castro-Hartmann, P. et al., *Biochemistry*, 49, 4043–4050, 2010. With permission.)

Folding and higher order structure of the chromatin fiber. Chromatin extracted from interface nuclei was early shown to exhibit a "beads-on-a-string" 10-nm fiber that condenses into a higher order folded so-called 30-nm fiber structure as illustrated in Figure 8.3b, left-hand panel (Alberts et al. 2007; Rattner and Hamkalo 1979). The detailed structure of the 30-nm fiber remains controversial and different models have been proposed on the basis of *in vitro* studies on reconstituted nucleosome arrays, mainly the two-start and the one-start models, the former being prevalent for long (197 and larger) nucleosome repeat length (NRL), the latter for short (187 and shorter) (Routh et al. 2008). The center of Figure 8.3b illustrates the two-start model (Dorigo et al. 2004; Schalch et al. 2005) and to the right the one-start model is depicted (Kruithof et al. 2009; Routh et al. 2008). The recent cryo-EM study (Figure 8.1a, right hand panel) revealed the structure of a 12-mer nucleosome array with NRL of 187 bp including linker histone H1 (at low salt) with a tetranucleosomal unit with a two-start left-handed zigzag conformation (Song et al. 2014). This structure also confirmed the putative role of the H4 tail of one NCP, located at the interface of the two stacked nucleosomes and interacting with the acidic patch of the surface of the NCP of the other nucleosome. In addition to the role of the H4 tail (Allahverdi et al. 2011), NCP–NCP attraction leading to stacking is induced by the presence of multivalent cations, demonstrating the polyelectrolyte nature of the attractive nucleosome–nucleosome interaction leading to compact chromatin (Fan et al. 2013; Korolev et al. 2010a; Liu et al. 2011).

However, the relevance of the 30-nm fiber to the *in vivo* situation has recently been the subject of controversy (Daban 2011; Maeshima et al. 2010; Nishino et al. 2012) (and references cited in Nishino et al. 2012). Figure 8.3c (right-hand panel) illustrates this issue where SAXS spectra of both interphase and mitotic eukaryotic chromatin did not detect patterns indicative of a 30-nm fiber structure, while side-by-side (11 nm) and face-to-face (stacking, 6 nm) nucleosome–nucleosome contact can be observed. These data support a "melted polymer" model of the *in vivo* chromatin with irregular nucleosome stacking being the main element of the structure. The NCP–NCP stacking was also observed in a different EM study that proposed a platelike chromatin structure (Castro-Hartmann et al. 2010), see Figure 8.3c, right-hand panel.

Even though the existence of the 30-nm fiber *in vivo* remains to be unequivocally established, chromatin in living systems is likely to be dynamic and heterogeneous in its structural features. Chromatin structure, even for well-defined *in vitro* systems, depends critically not only on the ionic conditions modulating the electrostatic interactions and on the NRL, but also on other factors such as presence of linker histone, histone variants, and nature and degree of histone tail posttranslational modifications. To investigate all these factors in CG, computer modeling is particularly challenging as it requires rigorous force fields that captures electrostatic, hydrophobic, specific (hydrogen bonding), as well as solvent interactions.

8.3 APPROACHES TO COARSE-GRAINING OF CHROMATIN

As noted above, coarse-graining of chromatin and nucleosomes has, to the best of our knowledge, at this point of time not yet been achieved on the basis of a systematic multiscale approach, starting from a fine-grained atomistic modeling. However, some attempts in this direction have been initiated, e.g., in the recent work by Voltz et al. (2008, 2012). This approach will be discussed below. Another attempt (also discussed below) is represented by recent CG modeling of the NCP and DNA (Fan et al. 2013; Korolev et al. 2014), using HO and DNA CG models, which as shown for the DNA model, can form the basis for systematic multiscale coarse-graining by the inverse MC method (Korolev et al. 2014).

The so far published CG studies of chromatin and NCPs may be classified as mesoscale CG models whereby large biopolymer units or assemblies are represented by one bead (Rippe et al. 2012). Within these models, the HO including the tails may be represented by one bead, often merged with the DNA wrapping it to form one bead for the NCP and using a chain of beads to represent linker DNA. Refined models with explicit representation of the DNA wrapped around it and with explicit tails as a chain of flexible beads have also been introduced. The shape of the nucleosome (i.e., effect on excluded volume) may be described as a sphere, spheroid, cylinder, or more irregular shape (Langowski and Heermann 2007; Rippe et al. 2012). The interactions between the nucleosome beads are often represented by some effective *ad hoc* potentials reflecting a given salt content of the system, with the linker DNA interaction described with a DH approach (Kepper et al. 2008, 2011; Rippe et al. 2012; Stehr et al. 2008). More refined descriptions with explicit (salt-dependent) fractional charges on the various units and the use of salt-dependent DH interactions were also introduced (Schlick et al. 2011; Sun et al. 2005). We have used CG mesoscale approaches focusing on electrostatics and the effects of multivalent cations on NCP aggregation and chromatin folding, using explicit mobile ions at varying salt (Fan et al. 2013; Korolev et al. 2006, 2010a, 2012b; Yang et al. 2009). All of the above approaches use the continuum description of water or more or less ad hoc effective potentials meant to include all effects of electrostatics, water, and other short-ranged interactions. A summary of the features captured and omitted by various recent coarse-graining approaches modeling NCP and chromatin is presented in Table 8.1. From this table, one can clearly see how far the current state of chromatin modeling is compared to real situation.

8.4 CG MODELING OF NUCLEOSOMES

Here, we focus on CG simulation studies of mononucleosomes that investigated the attractive interaction between NCPs leading to phase separation into ordered structures induced by multivalent cations. Much of this work as well as different analytical approaches to the problem of NCP–NCP attraction were discussed in recent reviews (Korolev et al. 2010b, 2012a). We will consider three models based on a spherical-bead representation of the nucleosome core as well as an advanced CG NCP model with detailed description of the shape and charge distribution of the NCP. The multiscale CG simulation of a single NCP based on an atomistic simulation (Voltz et al. 2008, 2012) is also discussed.

TABLE 8.1 Modeling Various Features of Chromatin Structure and Dynamics in Recent Literature

Feature	Nucleosome		Chromatin		
	"One-Bead Residue" NCP[a]	"Explicit Ions" NCP[b]	"Six-Angle" Model[c]	"DiSCO-Based" Model[d]	"Explicit Ions" Array[e]
1 Structure/excluded volume factors					
1.1 NCP structure	++	++	+	++	+/−
1.2 Linker DNA	−[f]	−[f]	+	+	+
1.3 Flexible histone tails	+	+	−	+/−	+
1.4 Linker H1 histone	−[f]	−	+/−	+	−
2 Internal degrees of freedom					
2.1 Linker DNA bending, twisting, and stretching	−[f]	−	+	+	−
2.2[g] Core DNA mobility and bending, twisting, and stretching	+	+/−	+/−	−	−
2.3[h] Histone core domain mobility	+	−/+	−	−	−
3 Electrostatic forces					
3.1[i] Ion-ion interactions	+/−	+	−	−	+
3.2[j] Screening/charge neutralization	+/−	+	+/−	+/−	+
3.2[k] Mobile ion correlations	+/−	+	−	−	+

(Continued)

TABLE 8.1 (*Continued*) Modeling Various Features of Chromatin Structure and Dynamics in Recent Literature

Feature	Nucleosome			Chromatin	
	"One-Bead Residue" NCP[a]	"Explicit Ions" NCP[b]	"Six-Angle" Model[c]	"DiSCO-Based" Model[d]	"Explicit Ions" Array[e]
3.3[l] Mobile ion valency and structure	–	+	–	–	+
3.4[m] Tail–tail bridging and correlations	++[f]	+	–	+/–	+
4 Nucleosome–nucleosome interaction					
4.1[n] NCP–NCP effective potential	–[f]	–[o]	+	–[o]	–[o]
5 Other specific features					
5.1 Atomistic details: molecular structure of solvent, dipole–dipole, hydrogen bond, and LJ terms	+/–	–	–	–	–
5.2 Dependence on DNA sequence	+/–	–	–	–	–
5.3 Modeling histone modifications	–	+/–	–	–	+/–

Note: Factors contributing to chromatin structure and dynamics, described according to the nature of the forces and interactions and to what these are considered in various chromatin modeling approaches. In this table, only the latest simulation-based approaches (MC, Brownian, or MD) are listed; chromatin models based on structural or analytical analysis as well as earlier versions of simulation-based models are not presented. The ability of different modeling approaches to describe each of the contributions is indicated. An ideal (not yet achieved) chromatin model would have "+" or "++" for a majority of entries and should include "++" ratings on the description of electrostatic interactions. At the same time, any subdivision of features like presented below is arbitrary and the necessity for details in the modeling of various aspects of chromatin forces and interactions depends on the problem that is the focus of attention in a given approach.

(Continued)

TABLE 8.1 (*Continued*)   Modeling Various Features of Chromatin Structure and Dynamics in Recent Literature

[a] Models a single NCP with a one bead per aa and nucleotide residue resolution (Voltz et al. 2008, 2012).

[b] Based on explicit modeling of mobile ions and electrostatics with a one bead per amino acid and five-bead per two DNA base pairs resolution (Fan et al. 2013)

[c] Developed by Wedemann, Langowski, Rippe, and coworkers over the years from a "two-angle" to a "six-angle" model (Ehrlich et al. 1997; Ettig et al. 2011; Kepper et al. 2008, 2011; Rippe et al. 2012; Stehr et al. 2008, 2010).

[d] Developed by Schlick and coworkers over the years (Arya and Schlick 2006, 2007; Arya 2009; Arya et al. 2006; Beard and Schlick 2001a,b; Collepardo-Guevara and Schlick 2011, 2012, 2013, 2014; Gan and Schlick 2010; Grigoryev et al. 2009; Luque et al. 2014; Perisic et al. 2010; Schlick and Perisic 2009; Sun et al. 2005; Zhang et al. 2003).

[e] Based on explicit modeling of mobile ions and electrostatics (Korolev et al. 2006, 2010a; Liu et al. 2011; Yang et al. 2009).

[f] The model describes a single NCP.

[g] Possibility of movement and detachment of the nucleosomal DNA from the histone core (e.g., by a mechanical force to model a single-molecule force spectroscopy experiment) and modeling of the DNA mechanical properties similar to the linker DNA.

[h] Possibility for movement and dissociation of the building blocks of the HO (H2A/H2B dimers and (H3/H4)$_2$ tetramer).

[i] Coulomb interactions between pairs of charged particles or domains.

[j] If mobile ions are not modeled explicitly, this entry indicates whether any estimation of the influence from the salt environment by screening of electrostatic interactions is included (typically using DH model).

[k] Dynamic correlations of the fluctuating ion cloud associated with neighboring polyions (like NCP or linker DNA).

[l] Ability of the model to explicitly take into account charge and chemical structure of low-molecular-weight species.

[m] If flexible histone tails are present (listed as 1.3) and if they are charged and there is a charge on NCP/DNA, then the tails can contribute to chromatin folding through formation of bridges and by tail–tail correlations (similar to the mobile ion correlations in 3.2).

[n] The use of an effective potential to describe attractive NCP–NCP interaction; in a more complete model, this potential is redundant.

[o] Nucleosome–nucleosome interaction is partly redundant as electrostatic interactions are included.

8.4.1 Sphere Models with Grafted Histone Tail Chains

Sphere with grafted tails in DH approximation. In the work by Muhlbacher et al. (2006a,b), the NCP is modeled as a sphere with eight attached polymer chains. The sphere is a CG representation of the NCP without the tails, namely the HO with DNA wrapped around it. The sphere has a central negative charge that represents the net charge of the DNA–octamer complex. The sphere radius is set to 52.5 Å and the eight histone tails are modeled by flexible grafted chains where each chain consists of 28 monomers of size 3.5 Å and where each third monomer has a positive unit charge, the others being neutral. All hard cores were modeled with a repulsive Lennard-Jones (LJ) potential and the tail chains are connected with harmonic bonds. Electrostatic interactions are considered by the standard DH approach with an inverse screening length that depends on the monovalent salt concentration. They furthermore use an effective value of the net negative charge for the central bead to account for charge renormalization. In other words, explicit mobile ions are not present and hence ion-correlation effects are not taken into account.

Results from the above described NCP model, performed with molecular dynamics (MD) simulations in a continuum representation of the solvent water and using a Langevin thermostat, are illustrated in Figure 8.4a (Muhlbacher et al. 2006b). The figure shows the interaction potential as a function of separation between two NCP colloids for different values of the fractional charge of the histone tail, which illustrates the tuning of the strength of the interaction for tail charge reduction that mimics histone tail acetylations. The snapshot also illustrates the tail bridging that contributes to the attraction between the like-charged NCPs. The results of this model illustrate that in the presence of high concentration of monovalent salt, this attraction between mononucleosomes may explain the major mechanism of salt-driven phase separation of NCPs to ordered phases as observed in experiments by Livolant and coworkers (Bertin et al. 2007b; Livolant et al. 2007; Mangenot et al. 2003a).

Sphere with grafted tails and explicit ions. Although the above model of the NCP as a sphere with grafted tail chains illustrated and explained monovalent salt-induced NCP aggregation, the description based on the DH treatment of electrostatics cannot capture effects of multivalent ions. The NCP aggregation induced by such ions as Mg^{2+}, Co(hexammine)$^{3+}$ (CoHex^{3+}), and spermine^{4+} is initiated at very low concentration of added cation (in the range of μM to mM, depending on cation charge). The change

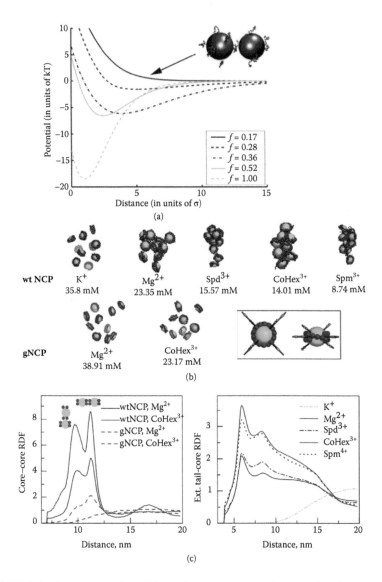

FIGURE 8.4 Coarse-grained modeling of interactions in solutions of nucleosome core particles (NCPs). (a) The NCP–NCP interaction potential becomes attractive and includes mediation of the histone tails. (Adapted from Muhlbacher, F. et al., *Phys. Rev. E*, 74, 031919, 2006b. With permission.) (b) Snapshots illustrating the dependence of NCP aggregation on the ionic environment and presence of the histone tails. The NCP model shown in the insert was according to Liu et al. (2011). (c). The left panel shows the influence of cation charge and histone tails on the NCP–NCP interaction. Two types of NCP–NCP close contacts were observed in solutions of wt NCP with Mg^{2+} and $CoHex^{3+}$ cations. The right panel shows RDFs of external tails in systems of aggregated NCPs illustrating tails bridging between NCPs contributing to attraction. (Adapted from Liu, Y. et al., *J. Mol. Biol.*, 414, 749–764, 2011. With permission.)

in screening due to the small increase in ionic strength caused by this will have minimal effect on the interaction potential, which will be repulsive at the relevant experimental conditions. Korolev et al. introduced a spherical-bead model with charged flexible grafted tails within a continuum description, but with explicit representation of all mobile ions due to added salt (Korolev et al. 2006; Yang et al. 2009). Essentially, the coarse-graining is the same as the above described DH model, but with somewhat different parameter choices. The major difference lies in the treatment of the electrostatic interaction by consideration of explicit mobile ions in the simulations. The size, charge, and distribution of the tails relative to the core were built to mimic real NCP. MD simulations employing a dielectric continuum and a Langevin thermostat illustrated that attractive ion–ion correlation effects due to fluctuations in the ion cloud and the attractive entropic and energetic tail-bridging effects contribute to cation-induced aggregation. In agreement with experimental data, the increase of monovalent salt content from salt-free to physiological concentration led to the formation of NCP aggregates and similarly, in the presence of Mg^{2+}, the NCPs formed condensed systems.

The above described approaches, even considering explicit mobile ions, arguably suffer from considerable simplification, in particular the simple spherical description of the shape of the nucleosome core, as well as its lack of discrete charge distribution. However, the results with a more refined model in this respect (see below) show that the general features and conclusions from the sphere model with respect to the electrostatic mechanism of cation-induced aggregation of NCPs are not largely affected by the simplifications. These models, which can be considered the simplest possible representation of the NCP that still incorporates the major aspects of the electrostatics of the system (when using explicit mobile ions), highlight the underlying physical mechanism responsible for the nucleosome aggregation.

Sphere-bead NCP model. Figure 8.4b illustrates a refined spherical-bead NCP model that was introduced as a follow up-to the earlier simple sphere model with grafted tail chains (Liu et al. 2011; Yang et al. 2009). The NCP was approximated as a combination of three different beads. A central sphere represents the globular part of the HO. Twenty-five particles model the DNA wrapped around the HO and eight strings of connected beads attached to the central HO bead describe the histone tails. The number of charged particles in each tail was 9, 14, 11, and 10 to match the charge of the H2A, H2B, H3, and H4 tails. Each DNA bead (effective radius 10 Å) represents six bp of DNA wrapped on the surface of the central bead, which has a radius 35 Å. The DNA thus forms a 1.75-superhelix that mimics the DNA wrapping around the HO. Harmonic bonds between particles maintained the stability of the NCP structure. The −12e charge of each DNA bead was reduced to −9.44e (which is equal to that of the sum of the charges on 147-bp DNA (−294e) plus the globular part of the HO (+58e). This approximation and its justifications are further discussed below. The described combination of all these bead sizes and positions closely represents the crystallographic structure of the NCP.

Figure 8.4b and c illustrates some results from this model with snapshots and relevant radial distribution functions (RDFs), obtained from MD simulations of systems of 10 NCPs in dielectric continuum with the presence of explicit mobile ions employing a Langevin thermostat, investigating the effect of truncation of the tails. The results nicely reproduce the experimental behavior of normal and tailless NCPs under the influence of added salt with varying cation charge (Liu et al. 2011). Typical polyelectrolyte behavior was observed as manifested by the inability of monovalent cations to induce aggregation and the considerably stronger (as compared to divalent ions like Mg^{2+}) attractions between NCPs in the presence of the spermidine^{3+}, CoHex^{3+}, and spermine^{4+}.

8.4.2 Advanced CG NCP Model

We recently developed a CG model of DNA that is used to build a more detailed CG description of the NCP, see Figure 8.5a (Fan et al. 2013). Benchmarking and validation of the parameter values by comparison with several physical properties from experimental data were performed (Fan et al. 2013). Furthermore, the parameters of the DNA model that determines the flexibility of free DNA were subsequently optimized using the inverse MC method and benchmarked against experimental data (Korolev et al. 2014).

The model is based on the crystal structure of the NCP with appropriate modeling of the geometry and charge distribution in the NCP as illustrated in Figure 8.5. The integrity of the HO core and the 370 DNA beads wrapped around the core is maintained by assigning harmonic bonds corresponding to equilibrium distances between each DNA bead and its nearest HO core amino acid particle. The beads belonging to the DNA and the HO are thus held together in a single NCP by a network of harmonic bond and angle interactions as described in detail in Fan et al. (2013). The histone tails are modeled as strings of linearly connected beads. To balance the

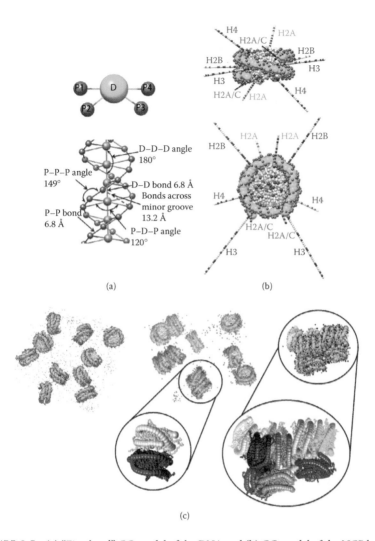

(a)

(b)

(c)

FIGURE 8.5 (a) "Five-bead" CG model of the DNA and (b) CG model of the NCP based on the five-bead CG DNA and the HO modeled with one bead per amino acid resolution according to Fan et al. (2013). The central green particle (bead D) approximates the pentose and base atoms of a two base pair DNA fragment. The orange beads represent the phosphate groups at the positions of an idealized B-form double helix (bead P). The geometry with bonds and angles were set to maintain the B-form conformation. Figure (b) shows side and top views of the CG NCP model. Gray spheres represent space-filling approximation of the HO core with the negatively and positively charged amino acids highlighted in magenta and cyan, respectively; DNA is modeled as in (a). Histone tails are shown as light blue (neutral), blue (positive), and red (negative) particles each representing one amino acid. (c) Increase of the cation charge from monovalent (K^+) to trivalent ($CoHex^{3+}$) results in a change in the NCP–NCP interaction from repulsive in the presence of K^+ to slightly attractive in the presence of Mg^{2+} to attractive with aggregation in the presence of $CoHex^{3+}$. (Adapted from Fan, Y. et al., *PLOS ONE*, 8, e54228, 2013. With permission.)

net negative charge of the NCP and achieve electroneutrality in the system, mobile ions are included in the simulation box. Figure 8.5a illustrates the design of DNA and the NCP in the model. DNA is modeled in sufficient detail to take into account the double helical nature and the detailed phosphate charge distribution. All the amino acids of the HO as well as in the histone tails are represented by a single spherical bead and explicitly including the net charge for each relevant amino acid. As a first step, we used this approach in continuum model simulations of NCP self-assembly (Fan et al. 2013) (see below). We used a five-site two base pair elementary DNA unit for the representation of the double helical B-form DNA oligonucleotide. The solvent was modeled by a continuous dielectric constant and ions at appropriate variable salt conditions are explicitly present in the simulations (represented by gray dots in the snapshots in Figure 8.5c). Thus, the total potential describing all interactions is a sum of electrostatic Coulombic, repulsive excluded volume LJ type of interactions and intramolecular bond and angle potentials as described in detail (Fan et al. 2013).

The MD method with a Langevin thermostat as implemented in the ESPResSo package (Limbach et al. 2006) was used for all simulations. Figure 8.5c shows results (Fan et al. 2013) illustrating snapshots for simulations in the presence of monovalent counterions (left), divalent Mg^{2+} (middle), and trivalent $CoHex^{3+}$ counterions (right). In agreement with experiments, in the presence of monovalent counterions (like K^+ or Na^+) the particles repel each other and form a stable solution. However, the presence of $CoHex^{3+}$ leads to distinct phase separation. This validation shows that the approach and model reproduce the expected salt-dependent self-assembly of NCPs, induced by multivalent cations (Liu et al. 2011). Interestingly, even for such a small test system consisting of 10 particles, nucleosome–nucleosome stacking is observed. The structural details of the stacked NCPs were analyzed and remarkably good agreement with NCP–NCP stacking parameters observed in EM and x-ray scattering studies of NCP self-assembly to columnar phases (Bertin et al. 2007b, Leforestier and Livolant 1997; Mangenot et al. 2003a,b) were obtained.

The present model maintains the integrity of the NCP by a number of harmonic bonds between the DNA and the histone core components whereas in the real NCP no such bonds exist between these components and DNA is held to the core by electrostatic forces and by the formation of a large number hydrogen bonds. The continuum description is also an approximation to the real solvent mediated potentials between the charged units in the system.

8.4.3 Multiscale CG Model of a Single NCP from Boltzmann Inversion

This model was applied to a single NCP and is illustrated in Figure 8.6a. The coarse-graining of the atomistic crystal structure of the NCP used a bead model based on a united atom representation of DNA and the HO, following approaches developed and used for CG modeling of protein folding.

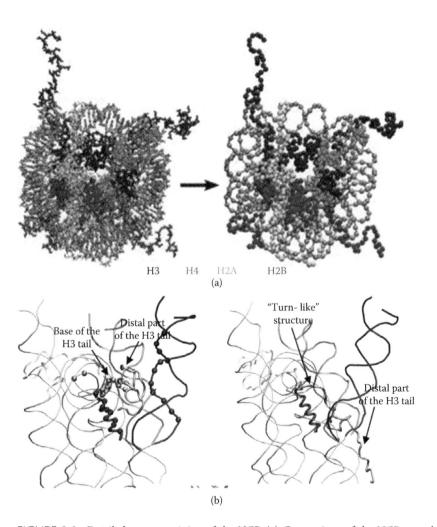

H3   H4   H2A   H2B

(a)

(b)

FIGURE 8.6 Detailed coarse-graining of the NCP. (a) Comparison of the NCP crystal structure (left) and with the CG model developed by Voltz et al. (2008). (b) Dynamic behavior of the H3 histone tail in simulations of the NCP. (Adapted from Voltz, K. et al., *J. Comp. Chem.*, 29, 1429–1439, 2008; Voltz, K. et al., *Biophys. J.*, 102, 849–858, 2012. With permission.)

The authors used an approach based on the so-called Go model (Bahar and Jernigan 1997; Go and Taketomi 1978), which was previously employed in a study of the ribosome with the one bead concept extended to nucleotides (Trylska et al. 2005), using the Boltzmann inversion procedure to obtain effective potentials (Reith et al. 2003). The force field includes bonded and nonbonded interactions, where the number of nonnearest neighbor interactions for the nonbonded terms depends on the structural features of the given amino acids. As shown in Figure 8.6a, one bead centered on the C_α atom represents each amino acid and another type of bead centered on the phosphate group models the nucleotide (P). The force field for these C_α–C_α and P–P interactions were described by harmonic potentials and the parameters of these distance-dependent potentials were obtained by running short (5 ns) all-atom simulations of the full NCP and then applying the Boltzmann inversion to extract the effective CG potentials from the corresponding RDFs in the all-atom MD simulations. The parameters for the C_α–P interactions on the other hand cannot be accurately obtained by Boltzmann inversion. This means that the electrostatic interactions between DNA phosphates and charged HO amino acid units are not well represented, which limits the accuracy of results pertaining to the charged histone tail DNA interactions. In spite of this limitation, the structural features of the resulting CG simulation did rather well reproduce the properties compared to an all-atom simulation of the NCP.

In Figure 8.6b, results of this approach are illustrated (Voltz et al. 2012). The dynamics of the DNA flexibility and detachment from the HO core and the role of the H3 histone tail were investigated. The results suggested an active role of the H3 tail by abrogating the contact of the nucleosomal DNA end to the nucleosome core after detachment. The simulations indicated "a long-lived open state in which the rearrangement of the H3 tail induces the formation of a turn-like structure at the base of the tail, thus preventing the DNA from reforming native contacts with the histones" (Voltz et al. 2012). These observations were supported by explicit solvent all-atom MD simulations.

Although the approach in the work by Voltz et al. is promising and represents the only real systematic coarse-graining of the NCP based on all-atom simulations that has been made so far, the problems with electrostatic interactions inherent in the Boltzmann inversion approach (Hess et al. 2006; Lyubartsev et al. 2010) means that results on the role of the charged histone tails must be interpreted with some care.

8.5 CG MODELING OF CHROMATIN

Figure 8.7 illustrates the three major types of CG mesoscale chromatin models that have been applied to study different aspects of chromatin structure and dynamics. These approaches emphasize different aspects of the physical properties of the chromatin fiber depending on what problem is the main focus in the modeling (see Table 8.1). The two-angle model (Ehrlich et al. 1997; Wedemann and Langowski 2002) later developed to the six-angle model (Ettig et al. 2011; Kepper et al. 2008, 2011; Rippe et al. 2012; Stehr et al. 2008, 2010) illustrated in Figure 8.7a was the first CG computer simulation model of chromatin. The model includes a description of the mechanical properties of the linker DNA and a DH approximation for the electrostatic interactions between linker DNA segments. Explicit

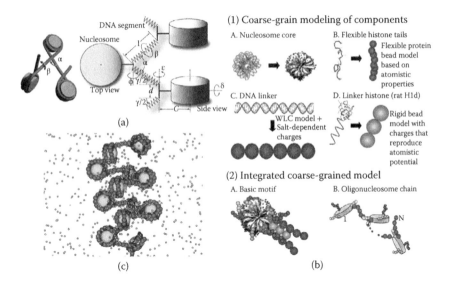

FIGURE 8.7   CG models of nucleosome arrays. (a) The two-angle model (left) was initially developed to account for mutual positions of the nucleosome in the fiber (Ehrlich et al. 1997) and later developed to the detailed two-angle model (right) (Ettig et al. 2011; Kepper et al. 2011; Rippe et al. 2012; Stehr et al. 2010). (Adapted from Iyer, B. V. S. et al., *BMC Biophys.*, 4, 8, 2011; Stehr, R. et al., *Biophys. J.*, 98, 1028–1037, 2010. With permission.) (b) Coarse-grained model of a chromatin fiber developed by Schlick and coworkers (Beard and Schlick 2001a,b; Collepardo-Guevara and Schlick 2014; Schlick et al. 2011). (Adapted from Arya, G., and T. Schlick, *Proc. Natl. Acad. Sci. U.S.A.*, 103, 16236–16241, 2006. With permission.) (c) CG model of the fiber developed by Nordenskiöld and coworkers which includes explicit mobile ions. (From Korolev, N. et al., *Biophys. J.*, 99, 1896–1905, 2010a; Korolev, N. et al., *Soft Matter*, 8, 9322–9333, 2012a.)

contributions of the histone tails are not included and the presence of the linker histone is estimated by changing the entry–exit angle parameters of the model. The discrete surface charge optimization (DiSCO) approach (Arya and Schlick 2006, 2007; Arya 2009; Arya et al. 2006; Beard and Schlick 2001a,b; Collepardo-Guevara and Schlick 2011, 2012, 2013, 2014; Gan and Schlick 2010; Grigoryev et al. 2009; Luque et al. 2014; Perisic et al. 2010; Rippe et al. 2012; Schlick and Perisic 2009; Sun et al. 2005; Zhang et al. 2003) depicted in Figure 8.7b models the NCP shape and charge distribution by a set of discrete surface charges and uses charged spherical (excluded volume) particles to model linker DNA, the flexible histone tails, and the linker histone. Electrostatic interactions are approximated by a DH approach using salt-dependent scalable charges calculated separately for the NCP, tails, and linker histone to match the electrostatic potential obtained from solution of the PB equation for given salt conditions.

8.5.1 The Six-Angle Model

The "six-angle" model of chromatin fiber was developed over the years in the Langowski, Rippe, and Wedemann laboratories from the "two-angle" model (Figure 8.7a, left) (Ehrlich et al. 1997). At its present state, it makes a realistic account for various structural and mechanical properties of the chromatin fiber, see Table 8.1 (Figure 8.7a, right) (Kepper et al. 2008, 2011; Rippe et al. 2012; Stehr et al. 2008, 2010; Wedemann and Langowski 2002). The NCP is described by a complex spherocylindrical force potential. The NCPs are connected by variable segments of linker DNA corresponding to different NRLs. Two angles defining, respectively, the entry–exit directions of the two linker DNAs and the orientation of the two cylindrical NCPs (Figure 8.7a, left) were combined with four more angles describing the path of the linker DNAs and the mutual orientation of the cylindrical NCPs in the three-NCP domain (Figure 8.7a, right) (Stehr et al. 2010). Additionally, to define the geometry of the structure of the DNA stem formed in the presence of linker histone, two distances were defined (respectively from the NCP center to the start of the linker DNA and the gap between the linker DNA leaving and entering the chromatosome). The linker DNA contributes to the total energy of the chromatin fiber through its stretching, bending, and torsional rigidity and through linker DNA—linker DNA electrostatic repulsion that was approximated by a linear charge density and calculated within the DH approximation at 100 mM monovalent salt (Kepper et al. 2011). The excluded volume of the linker DNA is defined by a soft repulsive potential. The NCP–NCP infraction is described by multimodal S-functions, which take into account the NCP–NCP distance and orientation and that can be attractive or repulsive reflecting the energetically favorable NCP–NCP stacking (Stehr et al. 2010). In the latest development, unwinding of the DNA from the core was included to model single-molecule chromatin stretching experiments (Kepper et al. 2011).

MC simulations of nucleosome arrays of considerable length (100 nucleosomes) were carried out to investigate the structure of nucleosome arrays as a function of various geometric constrains for linker DNA (modeling absence and presence of the linker histone) (Kepper et al. 2008; Wedemann and Langowski 2002), variation of the nucleosome–nucleosome interaction potential (Kepper et al. 2008; Stehr et al. 2008; Wedemann and Langowski 2002), and change in NRL (Stehr et al. 2010). Figure 8.8a shows several snapshots obtained for the arrays with different NRLs and for two variations of the entry/exit structure of the linker DNA representing presence and absence of the linker histone (Kepper et al. 2011; Stehr et al. 2010). Particularly important is the recent analysis of the phase diagram for the stability of various modes of nucleosome packaging (Stehr et al. 2010) applying a systematic approach developed by Depken and Schiessel (2009). The results of the modeling are in good agreement with experimental data. Snapshots from the simulations and EM images frequently match each other with packing densities of the simulated fibers coinciding with the measured values. In is interesting to note that the simulations predict substantial disordering and irregularities in fiber packing and this variability increases with increase of the NRL.

The same researchers also studied the mechanical behavior of the fibers under stretching by modeling the conditions of single-molecule experiments (Aumann et al. 2006; Kepper et al. 2011; Langowski 2006; Wocjan et al. 2009). The aim of these simulations was to determine the contributions of the fiber geometry, DNA bending and twisting, and nucleosome–nucleosome interactions to the mechanical behavior of the fiber under tension. However, at present more efforts may have to be devoted to improve both the modeling as well as analysis in order to reach an understanding of the mechanical behavior of chromatin fibers.

8.5.2 The DiSCO Mesoscale Model

The DiSCO mesoscale approach was developed over several years and is summarized in the supporting information of the recent articles by

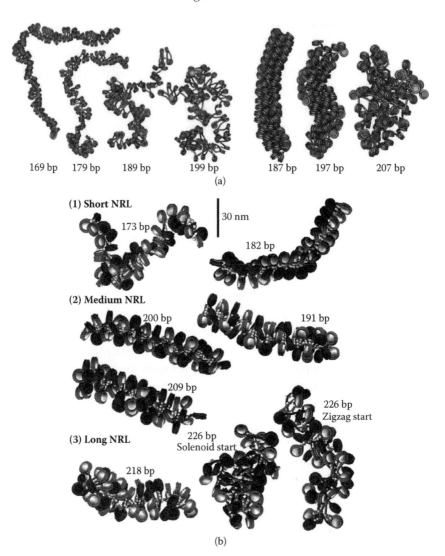

169 bp   179 bp   189 bp   199 bp   187 bp   197 bp   207 bp

(a)

(1) Short NRL

173 bp   30 nm

182 bp

(2) Medium NRL

200 bp   191 bp

209 bp

226 bp
Zigzag start

(3) Long NRL   226 bp
Solenoid start

218 bp

(b)

FIGURE 8.8 Conformations of chromatin fibers as a function of NRL obtained in Monte Carlo simulations using (a) the "two-angle" model (Kepper et al. 2011) and (b) the approach developed by Schlick and coworkers (Perisic et al. 2010). (Adapted from Kepper, N. et al., *Biopolymers*, 95, 435–447, 2011; Perisic, O. et al., *J. Mol. Biol.*, 403, 777–802, 2010. With permission.)

Collepardo-Guevera and Schlick (Collepardo-Guevara and Schlick 2011, 2012, 2013, 2014; Luque et al. 2014) and in the work by Arya et al. (Arya 2009; Arya and Schlick 2006, 2007, 2009; Arya et al. 2006). In brief, it uses a dielectric continuum model of the solvent, treating electrostatics by assigning effective fractional charges to all components, and accounting for salt by Coulombic interactions between all fractional charges using DH potentials. It treats the various fiber components in somewhat different ways, assigning fractional charges and excluded volume LJ potentials separately to the NCP, the histone tails, linker DNA, and the linker histone. The chromatin fiber is then built from these components.

The NCP atomic structure, excluding histone tails, is used together with a nonlinear PB approach and fractional charges according to the Assisted Model Building with Energy Refinement AMBER force field, to calculate the electric field at a given salt concentration. Using the DiSCO approach, the NCP (without tails) is then described by an irregular surface having 300 charged beads that generates an electric field screened by the appropriate DH potential (and having excluded volume properties described by a suitable LJ potential). The values of these fractional charges are optimized so that the field generated from this DiSCO model of the NCP matches the PB field obtained from solution of the PB equation around NCP with AMBER force field atomistic charges in continuum dielectric media (see Figure 8.7b). Following this, flexible histone tails are introduced with a two-step procedure starting from the atomistic sequence that is mapped to the subunit model of the tail, which is then further course-grained to a protein bead model. Each protein bead represents five amino acids and each have an excluded volume diameter 18 Å (see Figure 8.7b). The force constants describing the torsion, bending, and stretching of these beads were optimized individually for each tail. The electrostatic interactions of the tails are described by effective charges that are assigned by comparison with the field generated from the PB calculation of fully atomistic tails to obtain reasonable matching at different monovalent salt concentrations (Arya et al. 2006). Then the linker DNA is described using a discrete elastic chain model with spherical beads of diameter 30 Å using an LJ potential with elastic force constants for stretching, bending, and torsion assigned to reproduce the experimental flexibility of free DNA. The effective charges on each DNA bead that determine their DH mediated interactions are assigned salt-dependent variable values using a charge renormalization procedure (Stigter 1977). This linker DNA is attached to the NCP, at the appropriate entry–exit points, taking phase and orientation of the nucleosomal DNA into account. In later work, this model also incorporated linker histone through a rigid three-bead model with the chain of these beads placed on the dyad axis of each nucleosome and having suitable fractional charges of the beads obtained from the DiSCO approach

(Arya 2009). Finally, in some applications this DiSCO chromatin model was used to study effects of presence of divalent cations like Mg^{2+} (Arya 2009). It is well known that the PB/DH approaches are not suitable for the description of electrostatic interactions in the presence of multivalent ions where ion correlations are important. To account for this, the developers of this model resorted to a phenomenological approach whereby the effects of the presence of Mg^{2+} were introduced by scaling the DH screening parameter for DNA–DNA linker interactions and by reducing the linker DNA persistence length.

A main advantage of the DiSCO chromatin model is the rather detailed description of the NCP (excluding tails) that in principle allows for the well-known close stacking between nucleosomes. On the other hand, the large histone tail bead size seems to limit this stacking and the use of the PB/DH approach neglects ion-correlation effects that become highly important in the presence of multivalent ions.

The DiSCO mesoscale model was used in a range of applications studying how various aspects of chromatin fiber structure depends on different factors such as concentration of monovalent salt, and presence of linker histone and NRL (Schlick et al. 2012). Figure 8.8b (Perisic et al. 2010) shows a recent example of the dependence of fiber compaction on NRLs and this work together with a later study (Collepardo-Guevara and Schlick 2014) suggested a highly polymorphic chromatin structure as a result of variations in NRL.

8.5.3 The Electrostatic Sphere-Bead Model

This model (Korolev et al. 2010a) is illustrated in Figure 8.7c and is essentially an extension of the NCP sphere-bead model with grafted histone tail chains shown in Figure 8.4b (lower right-hand side). Twelve NCPs are connected with linker DNA that is described by a connection of charged beads at the entry–exit points of the core to build a chromatin fiber corresponding to a NRL of 177 bp in a dielectric continuum with explicit presence of mobile ions representing counterions and added salt. The NCP model description is given in Section 8.4.1 and a full account of all details can be found in the supplementary information material to the paper of Korolev et al. (2010a), see also Table 8.1.

In brief, to construct the array from the above NCP sphere-bead model (Section 8.4.1), a string of five beads of 1.0 nm radius each having a charge $-12e$ to model a 30-bp linker DNA fragment is used to connect 12 NCPs. The total charge of the DNA attached to the histone core plus linker DNA is −3492e, which combined with the +1056e charge of the histone tails results in a net charge of the array of −2436e. Excluded volume is taken care of by a short-range repulsive LJ potential for each type of particle pair. The rigidity of the DNA is modeled with a bending potential for each instance of three consecutive DNA beads. Linker DNA torsion was not included in this model. The mobile ion effective hydration radii were obtained using extensively validated data from earlier MC simulations (Korolev et al. 1999, 2001a,b). The charged amino groups in spermidine^{3+} and spermine^{4+} had radii of 0.25 nm and were connected by harmonic bonds with an equilibrium distance of 0.47 nm. MD simulations with a Langevin thermostat were carried out with the ESPResSo package (Limbach et al. 2006).

In the model of DNA wrapped around the HO core, the original −12e charge of each DNA bead (due to the 6 bp it represents) was reduced to −9.44e (which is equal to that of the sum of the charges on 147-bp DNA (−294e) plus the globular part of the HO (+58e). To justify this assumption, the work by Korolev et al. (2010a) carried out a detailed analysis of the NCP atomic structure and made a number of test simulations with variations of the NCP models where a negative or positive charge of the central particles was applied, which was shown to have minor effects of the behavior of the model in the actual simulations. The spatial distribution of the HO core Lys and Arg amino acid positive side chain charges relative to the phosphate groups of the DNA wrapped around the histone core was investigated and it was found that these contacts can be counted as residues neutralizing the charge on the DNA since the charges are concentrated on the lateral surface of the histone core and direct the DNA wrapping. Hence, the net charge of the histone core appears to be neutral or even somewhat negative. There is an "acidic islet" (Chodaparambil et al. 2007; Luger et al. 1997; Luger and Richmond 1998) on each of the surfaces of the NCP "cylinder," which is formed by a cluster of 5 Glu and Asp amino acids of the H2A histones. Within the present description of the HO core as a spherical particle, the effects of the acidic patch can be approximately tested by assigning a negative charge of −10e to the central core. Estimates of the sensitivity of the simulations to the choice of the charge of the central particle was made by testing models with a neutral, positive (using a DH approximation for the positive core), or negatively charged (−10e) central particle (the charge of each of the DNA beads was changed accordingly). It was found that the results obtained using these different models are similar. However, it should be noted that to capture the putative effects of the interaction between the histone H4 tail and the H2A acidic patch (Luger et al. 1997; Luger and Richmond 1998), more advanced coarse-graining must be developed.

Arguably, one of the most serious approximations in this nucleosome array model is the coarse description of the shape and charge distribution of the HO core particle, which due to its spherical shape excludes the face-to-face "stacking" between the almost flat surfaces of the real wedge-shaped cylinder-like NCPs. Another serious approximation in this, as in all other present chromatin models, is the lack of hydration effects, i.e., the continuum description of the aqueous solvent with a uniform dielectric constant. There are other simplifications such as the absence of hydrogen bonding and hydrophobic effects, which likewise are absent in other approaches. These deficiencies can, in principle, be rectified by a systematic multiscale coarse-graining of the system based on the NCP model described in Section 8.4.2 and such work is presently under way. However, in spite of these serious limitations, this chromatin model that includes the explicit presence of small mobile ions and accounts for the effect of ion correlations and ion release reproduces the qualitative features of multivalent-induced chromatin compaction (see below examples). To the best of our knowledge, this is the only published chromatin model incorporating explicit mobile ions, which therefore can be used as a basis for a discussion of the general polyelectrolyte-driven chromatin compaction.

Figure 8.9 shows results obtained from this chromatin model. The data demonstrate that the folding transition of the 12–177 nucleosome array observed in the simulations takes place at concentrations of Mg^{2+} and $CoHex^{3+}$ similar to those observed experimentally. From the simulation trajectory, the hydrodynamic sedimentation coefficient can be calculated (Figure 8.9c and d). Compact folded structures result in larger sedimentation coefficients compared to extended ones. The intensity of the maximum in the external tail-core RDFs gives information on the probability that a tail on a given NCP is located on a neighboring NCP (instead of on its own core), giving a measure of tail to NCP bridging. These results (Korolev et al. 2010a) together with the data in Figure 8.9c and d, show that folding of the chromatin fiber is associated with bridging of tails extending to nearby nucleosomes. This effect is mediated by the increased presence of multivalent ions. In the presence of a monovalent salt, the bridging is minor at millimolar concentrations, while in the presence of multivalent ions, strong attraction between nucleosomes leading to compaction is observed.

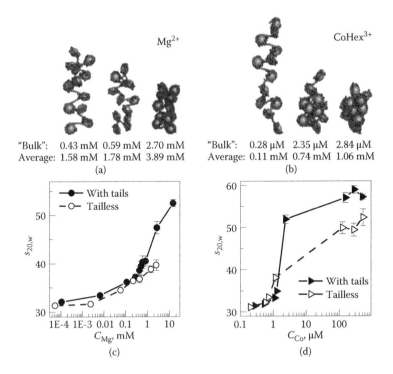

FIGURE 8.9 Folding of the chromatin fiber upon addition of multivalent cations. Snapshots of the fiber at different concentrations of (a) Mg^{2+} and (b) $CoHex^{3+}$. Change of sedimentation coefficient, $s_{20,w}$, calculated from the simulation trajectories as a function of (c) Mg^{2+} and (d) $CoHex^{3+}$ concentrations. Results for both wild type and tailless arrays are shown. (Adapted from Korolev, N. et al., *Biophys. J.*, 99, 1896–1905, 2010a. With permission.)

8.6 CONCLUSIONS AND FUTURE PERSPECTIVES

Table 8.1 summarizes the various factors that are important in determining chromatin structure and dynamics and how these factors are taken into account in different modeling approaches. In spite of the advances in chromatin modeling summarized in this table, a major challenge modeling such a biomacromolecular assembly as chromatin is that the adequate description of the system in principle requires a rigorous treatment of electrostatics, a molecular description of water, as well as taking into account specific interactions such as hydrogen bonding and hydrophobic effects. Such effects are largely absent in many of the approaches presently at hand and described in Table 8.1 (or only some of the effects included). Therefore, it is clear that there is a need for further development of the existing CG approaches and for the introduction of alternative methods.

In our opinion, multiscale CG simulations based on systematic coarse-graining starting from atomistic simulations hold considerable promise for improving present-day chromatin models. Another approach, which directly employs atomistic simulations including water, such as steered MD (SMD) simulations (Ettig et al. 2011), can also be used for rigorous chromatin modeling, but is limited by the temporal and spatial scales that can be simulated.

CG computer simulations of chromatin are highly challenging. However, many studies as described above have been presented that significantly advanced the understanding of the forces and interactions important for chromatin structure and dynamics. On the basis of these advances and with the possibilities and developments of alternative methods such as systematic multiscale CG simulations, considerable progress is expected within the near future.

REFERENCES

Alberts, B., A. Johnson, J. Lewis, M. Raff, K. Roberts, and P. Walter. 2007. *Molecular Biology of the Cell*. New York, NY and London: Garland Publishing.

Allahverdi, A., R. Yang, N. Korolev, Y. Fan, C. A. Davey, C. F. Liu, and L. Nordenskiöld. 2011. The effects of histone H4 tail acetylations on cation-induced chromatin folding and self-association. *Nucleic Acids Res.* 39:1680–1691.

Arya, G. 2009. Energetic and entropic forces governing the attraction between polyelectrolyte-grafted colloids. *J. Phys. Chem. B* 113:15760–15770.

Arya, G., and T. Schlick. 2006. Role of histone tails in chromatin folding revealed by a mesoscopic oligonucleosome model. *Proc. Natl. Acad. Sci. U.S.A.* 103:16236–16241.

Arya, G., and T. Schlick. 2007. Efficient global biopolymer sampling with end-transfer configurational bias Monte Carlo. *J. Chem. Phys.* 126:044107.

Arya, G., and T. Schlick. 2009. A tale of tails: How histone tails mediate chromatin compaction in different salt and linker histone environments. *J. Phys. Chem. A* 113:4045–4059.

Arya, G., Q. Zhang, and T. Schlick. 2006. Flexible histone tails in a new mesoscopic oligonucleosome model. *Biophys. J.* 91:133–150.

Aumann, F., F. Lankas, M. Caudron, and J. Langowski. 2006. Monte Carlo simulation of chromatin stretching. *Phys. Rev. B* 73:041927.

Ayton, G. S., W. G. Noid, and G. A. Voth. 2007. Multiscale modeling of biomolecular systems: In serial and in parallel. *Curr. Opin. Struct. Biol.* 17:192–198.

Bahar, I., and R. L. Jernigan. 1997. Inter-residue potentials in globular proteins and the dominance of highly specific hydrophilic interactions at close separation. *J. Mol. Biol.* 266:195–214.

Beard, D. A., and T. Schlick. 2001a. Computational modeling predicts the structure and dynamics of chromatin fiber. *Structure* 9:105–114.

Beard, D. A., and T. Schlick. 2001b. Modeling salt-mediated electrostatics of macro-molecules: The discrete surface charge optimization algorithm and its application to the nucleosome. *Biopolymers* 58:106–115.

Bennink, M. L., S. H. Leuba, G. H. Leno, J. Zlatanova, B. G. de Grooth, and J. Greve. 2001. Unfolding individual nucleosomes by streyching single chromatin fibers with optical tweezers. *Nature Struct. Biol.* 8:606–610.

Bertin, A., D. Durand, M. Renouard, F. Livolant, and S. Mangenot. 2007a. H2A and H2B tails are essential to properly reconstitute nucleosome core particles. *Eur. Biophys. J.* 36:1083–1094.

Bertin, A., A. Leforestier, D. Durand, and F. Livolant. 2004. Role of histone tails in the conformation and interaction of nucleosome core particles. *Biochemistry* 43:4773–4780.

Bertin, A., S. Mangenot, M. Renouard, D. Durand, and F. Livolant. 2007b. Structure and phase diagram of nucleosome core particles aggregated by multivalent cations. *Biophys. J.* 93:3652–3663.

Bertin, A., M. Renouard, J. S. Pedersen, F. Livolant, and D. Durand. 2007c. H3 and H4 histone tails play a central role in the interactions of recombinant NCPs. *Biophys. J.* 92:2633–2645.

Brower-Toland, B. D., C. L. Smith, R. C. Yeh, J. T. Lis, C. L. Peterson, and M. D. Wang. 2002. Mechanical disruption of individual nucleosomes reveals a reversible multistage release of DNA. *Proc. Natl. Acad. Sci. U.S.A.* 99:1960–1965.

Brower-Toland, B. D., D. A. Wacker, R. M. Fulbright, J. T. Lis, W. L. Kraus, and M. D. Wang. 2005. Specific contributions of histone tails and their acetylation to the mechanical stability of nucleosomes. *J. Mol. Biol.* 346:135–146.

Castro-Hartmann, P., M. Milla, and J.-R. Daban. 2010. Irregular orientation of nucleosomes in the well-defined chromatin plates of metaphase chromosomes. *Biochemistry* 49:4043–4050.

Chodaparambil, J. V., A. J. Barbera, X. Lu, K. M. Kaye, J. C. Hansen, and K. Luger. 2007. A charged and contoured surface on the nucleosome regulates chromatin compaction. *Nat. Struct. Mol. Biol.* 14:1105–1107.

Clark, D. J., and T. Kimura. 1990. Electrostatic mechanism of chromatin folding. *J. Mol. Biol.* 211:883–896.

Claudet, C., D. Angelov, P. Bouvet, S. Dimitrov, and J. Bednar. 2005. Histone octamer instability under single molecule experiment conditions. *J. Biol. Chem.* 280:19958–19965.

Collepardo-Guevara, R., and T. Schlick. 2011. The effect of linker histone's nucleosome binding affinity on chromatin unfolding mechanisms. *Biophys. J.* 101:1670–1680.

Collepardo-Guevara, R., and T. Schlick. 2012. Crucial role of dynamic linker histone binding and divalent ions for DNA accessibility and gene regulation revealed by mesoscale modeling of oligonucleosomes. *Nucleic Acids Res.* 40:8803–8817.

Collepardo-Guevara, R., and T. Schlick. 2013. Insights into chromatin fibre structure by in vitro and in silico single-molecule stretching experiments. *Biochem. Soc. Trans.* 41:494–500.

Collepardo-Guevara, R., and T. Schlick. 2014. Chromatin fiber polymorphism triggered by variations of DNA linker lengths. *Proc. Natl. Acad. Sci. U.S.A.* 111:8061–8066.

Cui, Y., and C. Bustamante. 2000. Pulling a single chromatin fiber reveals the forces that maintain its higher order structure. *Proc. Natl. Acad. Sci. U.S.A.* 97:127–132.

Daban, J.-R. 2011. Electron microscopy and atomic force microscopy studies of chromatin and metaphase chromosome structure. *Micron* 42:733–750.

Das, C., J. K. Tyler, and M. E. A. Churchill. 2010. The histone shuffle: Histone chaperones in an energetic dance. *Trends Biochem. Sci.* 35:476–489.

Davey, C. A., D. F. Sargent, K. Luger, A. W. Maeder, and T. J. Richmond. 2002. Solvent mediated interactions in the structure of nucleosome core particle at 1.9 Å resolution. *J. Mol. Biol.* 319:1097–1113.

de Frutos, M., E. Raspaud, A. Leforestier, and F. Livolant. 2001. Aggregation of nucleosomes by divalent cations. *Biophys. J.* 81:1127–1132.

Depken, M., and H. Schiessel. 2009. Nucleosome shape dictates chromatin fiber structure. *Biophys. J.* 96:777–784.

Diesinger, P. M., and D. W. Heermann. 2010. Monte Carlo simulations indicate that chromatin nanostructure is accessible by light microscopy. *PMC Biophys.* 3:11.

Dorigo, B., T. Schalch, K. Bystricky, and T. J. Richmond. 2003. Chromatin fiber folding: Requirement for the histone H4 N-terminal tail. *J. Mol. Biol.* 327:85–96.

Dorigo, B., T. Schalch, A. Kulangara, S. Duda, R. R. Schroeder, and T. J. Richmond. 2004. Nucleosome arrays reveal the two-start organization of the chromatin fiber. *Science* 306:1571–1573.

Ehrlich, L., C. Münkel, G. Chirico, and J. Langowski. 1997. A Brownian dynamics model for the chromatin fiber. *Comput. Appl. Biosci.* 13:271–279.

Ettig, R., N. Kepper, R. Stehr, G. Wedemann, and K. Rippe. 2011. Dissecting DNA-histone interactions in the nucleosome by molecular dynamics simulations of DNA unwrapping. *Biophys. J.* 101:1999–2008.

Fan, Y., N. Korolev, A. P. Lyubartsev, and L. Nordenskiöld. 2013. An advanced coarse-grained nucleosome core particle model for computer simulations of nucleosome-nucleosome interactions under varying ionic conditions. *PLOS ONE* 8:e54228.

Finch, J. T., R. S. Brown, T. J. Richmond, B. Rushton, L. C. Lutter, and A. Klug. 1981. X-ray diffraction study of a new crystal form of the nucleosome core showing higher resolution. *J. Mol. Biol.* 145:757–769.

Finch, J. T., L. C. Lutter, D. Rhodes, R. S. Brown, B. Rushton, M. Levitt, and A. Klug. 1977. Structure of nucleosome core particles of chromatin. *Nature* 269:29–36.

Flaus, A., and T. Owen-Hughes. 2003. Mechanisms for nucleosome mobilization. *Biopolymers* 68:563–578.

Gan, H. H., and T. Schlick. 2010. Chromatin ionic atmosphere analyzed by a mesoscale electrostatic approach. *Biophys. J.* 99:2587–2596.

Gemmen, G. J., R. Sim, K. A. Haushalter, P. C. Ke, J. T. Kadonaga, and D. E. Smith. 2005. Forced unraveling of nucleosomes assembled on heterogeneous DNA using core histones, NAP-1, and ACF. *J. Mol. Biol.* 351:89–99.

Go, N., and H. Taketomi. 1978. Respective roles of short- and long-range interactions in protein folding. *Proc. Natl. Acad. Sci. U.S.A.* 75:559–563.

Gordon, F., K. Luger, and J. C. Hansen. 2005. The core histone N-terminal tail domains function independently and additively during salt-dependent oligomerization of nucleosomal arrays. *J. Biol. Chem.* 280:33701–33706.

Gottesfeld, J. M., and K. Luger. 2001. Energetics and affinity of the histone octamer for defined DNA sequences. *Biochemistry* 40:10927–10933.

Grigoryev, S. A., and C. L. Woodcock. 2012. Chromatin organization—The 30 nm fiber. *Exp. Cell Res.* 318:1448–1455.

Grigoryev, S. A., G. Arya, S. Correll, C. L. Woodcock, and T. Schlick. 2009. Evidence for heteromorphic chromatin fibers from analysis of nucleosome interactions. *Proc. Natl. Acad. Sci. U.S.A.* 106:13317–13322.

Hansen, J. C. 2002. Conformational dynamics of the chromatin fiber in solution: Determinants, mechanisms, and functions. *Annu. Rev. Biophys. Biomol. Struct.* 31:361–392.

Hess, B., C. Holm, and N. van der Vegt. 2006. Modeling multibody effects in ionic solutions with a concentration dependent dielectric permittivity. *Phys. Rev. Lett.* 96:147801.

Horn, P. J., and C. L. Peterson. 2002. Chromatin higher order folding: Wrapping up transcription. *Science* 297:1824–1827.

Iyer, B. V. S., M. Kenward, and G. Arya. 2011. Hierarchies in eukaryotic genome organization: Insights from polymer theory and simulations. *BMC Biophys.* 4:8.

Izvekov, S., and G. A. Voth. 2005. Multiscale coarse-graining method for biomolecular systems. *J. Phys. Chem. B* 109:2469–2473.

Kan, P.-Y., T. L. Caterino, and J. J. Hayes. 2009. The H4 tail domain participates in intra- and internucleosome interactions with protein and DNA during folding and oligomerization of nucleosome arrays. *Mol. Cell. Biol.* 29:538–546.

Kan, P.-Y., and J. J. Hayes. 2007. Detection of interactions between nucleosome arrays mediated by specific core histone tail domains. *Methods* 41:278–285.

Kan, P.-Y., X. Lu, J. C. Hansen, and J. J. Hayes. 2007. The H3 tail domain participates in multiple interactions during folding and self-association of nucleosome arrays. *Mol. Cell. Biol.* 27:2084–2091.

Kepper, N., R. Ettig, R. Stehr, S. Marnach, G. Wedemann, and K. Rippe. 2011. Force spectroscopy of chromatin fibers: Extracting energetics and structural information from Monte Carlo simulations. *Biopolymers* 95:435–447.

Kepper, N., D. Foethke, R. Stehr, G. Wedemann, and K. Rippe. 2008. Nucleosome geometry and internucleosomal interactions control the chromatin fiber conformation. *Biophys. J.* 95:3692–3705.

Korolev, N., A. Allahverdi, A. P. Lyubartsev, and L. Nordenskiöld. 2012a. The polyelectrolyte properties of chromatin. *Soft Matter* 8:9322–9333.

Korolev, N., A. Allahverdi, Y. Yang, Y. Fan, A. P. Lyubartsev, and L. Nordenskiöld. 2010a. Electrostatic origin of salt-induced nucleosome array compaction. *Biophys. J.* 99:1896–1905.

Korolev, N., L. Di, A. P. Lyubartsev, and L. Nordenskiöld. 2014. A coarse-grained DNA model parameterized from atomistic simulations by inverse Monte Carlo. *Polymers* 6:1655–1675.

Korolev, N., Y. Fan, A. P. Lyubartsev, and L. Nordenskiöld. 2012b. Modelling chromatin structure and dynamics: Status and prospects. *Curr. Opin. Struct. Biol.* 22:151–159.

Korolev, N., A. P. Lyubartsev, and A. Laaksonen. 2004. Electrostatic background of chromatin fiber stretching. *J. Biomol. Struct. Dyn.* 22:215–226.

Korolev, N., A. P. Lyubartsev, A. Rupprecht, and L. Nordenskiöld. 1999. Competitive binding of Mg^{2+}, Ca^{2+}, Na^+, and K^+ to DNA in oriented DNA fibers: Experimental and Monte Carlo simulation results. *Biophys. J.* 77:2736–2749.

Korolev, N., A. P. Lyubartsev, A. Rupprecht, and L. Nordenskiöld. 2001b. Competitive substitution of hexammine cobalt(III) for Na^+ and K^+ ions in oriented DNA fibers. *Biopolymers* 58:268–278.

Korolev, N., A. P. Lyubartsev, and L. Nordenskiöld. 2006. Computer modeling demonstrates that electrostatic attraction of nucleosomal DNA is mediated by histone tails. *Biophys. J.* 90:4305–4316.

Korolev, N., A. P. Lyubartsev, and L. Nordenskiöld. 2010b. Cation-induced polyelectrolyte-polyelectrolyte attraction in solutions of DNA and nucleosome core particles. *Adv. Colloid Interface Sci.* 158:32–47.

Korolev, N., A. P. Lyubartsev, L. Nordenskiöld, and A. Laaksonen. 2001a. Spermine: An "invisible" component in the crystals of *B*-DNA. A grand canonical Monte Carlo and molecular dynamics simulation study. *J. Mol. Biol.* 308:907–917.

Korolev, N., O. V. Vorontsova, and L. Nordenskiöld. 2007. Physicochemical analysis of electrostatic foundation for DNA-protein interactions in chromatin transformations. *Prog. Biophys. Mol. Biol.* 95:23–49.

Kruithof, M., F. T. Chien, A. Routh, C. Logie, D. Rhodes, and J. van Noort. 2009. Single-molecule force spectroscopy reveals a highly compliant helical folding for the 30-nm chromatin fiber. *Nat. Struct. Mol. Biol.* 16:534–540.

Langowski, J. 2006. Polymer chain models of DNA and chromatin. *Eur. Phys. J. E* 19:241–249.

Langowski, J., and D. W. Heermann. 2007. Computational modeling of the chromatin fiber. *Semin. Cell Dev. Biol.* 18:659–667.

Leforestier, A., and F. Livolant. 1997. Liquid crystalline ordering of nucleosome core particles under macromolecular crowding conditions: Evidence for a discotic columnar hexagonal phase. *Biophys. J.* 73:1771–1776.

Leforestier, A., J. Dubochet, and F. Livolant. 2001. Bilayers of nucleosome core particles. *Biophys. J.* 81:2114–2421.

Limbach, H. J., A. Arnold, B. A. Mann, and C. Holm. 2006. ESPResSo—An extensible simulation package for research on soft matter systems. *Comp. Phys. Comm.* 174:704–727.

Liu, Y., C. Lu, Y. Yang, Y. Fan, R. Yang, C.-F. Liu, N. Korolev, and L. Nordenskiöld. 2011. Influence of histone tails and H4 tail acetylations on nucleosome–nucleosome interactions. *J. Mol. Biol.* 414:749–764.

Livolant, F., S. Mangenot, A. Leforestier, A. Bertin, M. de Frutos, E. Raspaud, and D. Durand. 2007. Are liquid crystalline properties of nucleosomes involved in chromosome structure and dynamics? *Phil. Trans. R. Soc. Lond. A* 364:2615–2633.

Luger, K., and J. C. Hansen. 2005. Nucleosome and chromatin fiber dynamics. *Curr. Opin. Struct. Biol.* 15:188–196.

Luger, K., A. W. Mader, R. K. Richmond, D. F. Sargent, and T. J. Richmond. 1997. Crystal structure of the nucleosome core particle at 2.8 Å resolution. *Nature* 389:251–260.

Luger, K., and T. J. Richmond. 1998. The histone tails of the nucleosome. *Curr. Opin. Genet. Dev.* 8:140–146.

Luque, A., R. Collepardo-Guevara, S. Grigoryev, and T. Schlick. 2014. Dynamic condensation of linker histone C-terminal domain regulates chromatin structure. *Nucleic Acids Res.* 42(12):7553–7560. doi:10.1093/nar/gku1491.

Lyubartsev, A. P. 2005. Multiscale modeling of lipids and lipid bilayers. *Eur. Biophys. J.* 35:53–61.

Lyubartsev, A. P., and A. Laaksonen. 1995. Calculation of effective interaction potentials from radial distribution functions: A reverse Monte Carlo approach. *Phys. Rev. E* 52:3730–3737.

Lyubartsev, A. P., A. Mirzoev, L.-J. Chen, and A. Laaksonen. 2010. Systematic coarse-graining of molecular models by the Newton inversion method. *Faraday Discuss.* 144:43–56.

Lyubartsev, A. P., and A. L. Rabinovich. 2011. Recent development in computer simulations of lipid bilayers. *Soft Matter* 7:25–39.

Maeshima, K., S. Hihara, and M. Eltsov. 2010. Chromatin structure: Does the 30-nm fibre exist in vivo? *Curr. Opin. Cell Biol.* 22:291–297.

Mangenot, S., A. Leforestier, D. Durand, and F. Livolant. 2003a. Phase diagram of nucleosome core particles. *J. Mol. Biol.* 333:907–916.

Mangenot, S., A. Leforestier, D. Durand, and F. Livolant. 2003b. X-ray diffraction characterization of the dense phases formed by nucleosome core particles. *Biophys. J.* 84:2570–2584.

Mangenot, S., A. Leforestier, P. Vachette, D. Durand, and F. Livolant. 2002a. Salt-induced conformation and interaction changes of nucleosome core particles. *Biophys. J.* 82:345–356.

Mangenot, S., E. Raspaud, C. Tribet, L. Belloni, and F. Livolant. 2002b. Interactions between isolated nucleosome core particles. A tail bridging effect? *Eur. Phys. J. E* 7:221–231.

Manning, G. S. 2003. Is a small number of charge neutralizations sufficient to bend nucleosome core DNA onto its superhelical ramp? *J. Amer. Chem. Soc.* 125:15087–15092.

Manning, G. S., K. K. Ebralidze, A. D. Mirzabekov, and A. Rich. 1989. An estimate of the extent of folding of nucleosomal DNA by laterally asymmetric neutralization of phosphate groups. *J. Biomol. Struct. Dyn.* 6:877–889.

Marky, N. L., and G. S. Manning. 1991. The elastic residence of DNA can induce all-or-none structural transitions in the nucleosome core particle. *Biopolymers* 31:1543–1557.

Marky, N. L., and G. S. Manning. 1995. A theory of DNA dissociation from the nucleosome. *J. Mol. Biol.* 254:50–61.

Marrink, S. J., H. J. Risselada, S. Yefimov, D. P. Tieleman, and A. H. de Vries. 2007. The MARTINI force field: Coarse grained model for biomolecular simulations. *J. Am. Chem. Soc.* 111:7812–7824.

McBryant, S. J., and J. C. Hansen. 2012. Dynamic fuzziness during linker histone action. In *Fuzziness: Structural Disorder in Protein Complexes.* M. Fuxreiter and P. Tompa, editors. New York, NY and Austin, TX: Landes Bioscience/Springer ScienceScience+Business Media, 15–26.

McBryant, S. J., J. Klonoski, T. C. Sorensen, S. S. Norskog, S. Williams, M. G. Resch, J. A. Toombs, 3rd, S. E. Hobdey, and J. C. Hansen. 2009. Determinants of histone H4 N-terminal domain function during nucleosomal array oligomerization: Roles of amino acid sequence, domain length, and charge density. *J. Biol. Chem.* 284:16716–16722.

Mirzoev, A., and A. P. Lyubartsev. 2013. MagiC: Software package for multiscale modeling. *J. Chem. Theory Comput.* 9:1512–1520.

Muhlbacher, F., C. Holm, and H. Schiessel. 2006a. Controlled DNA compaction within chromatin: The tail-bridging effect. *Europhys. Lett.* 73:135–141.

Muhlbacher, F., H. Schiessel, and C. Holm. 2006b. Tail-induced attraction between nucleosome core particles. *Phys. Rev. E* 74:031919.

Naumova, N., M. Imakaev, G. Fudenberg, Y. Zhan, B. R. Lajoie, L. A. Mirny, and J. Dekker. 2013. Organization of the mitotic chromosome. *Science* 342: 948–953.

Nishino, Y., M. Eltsov, Y. Joti, K. Ito, H. Takata, Y. Takahashi, S. Hihara, A. S. Frangakis, N. Imamoto, T. Ishikawa, and K. Maeshima. 2012. Human mitotic chromosomes consist predominantly of irregularly folded nucleosome fibres without a 30-nm chromatin structure. *EMBO J.* 31:1644–1653.

Nozaki, T., K. Kaizu, C. G. Pack, T. Tamura, T. Tani, S. Hihara, T. Nagai, K. Takahashi, and K. Maeshima. 2013. Flexible and dynamic nucleosome fiber in living mammalian cells. *Nucleus* 4:349–356.

Perisic, O., R. Collepardo-Guevara, and T. Schlick. 2010. Modeling studies of chromatin fiber structure as a function of DNA linker length. *J. Mol. Biol.* 403:777–802.

Pope, L. H., M. L. Bennink, K. A. van Leijenhorst-Groener, D. Nikova, J. Greve, and J. F. Marko. 2005. Single chromatin fiber stretching reveals physically distinct populations of disassembly events. *Biophys. J.* 88:3572–3583.

Rattner, J. B., and B. A. Hamkalo. 1979. Nucleosome packing in interphase chromatin. *J. Cell. Biol.* 81:453–457.

Reith, D., M. Putz, and F. Muller-Plathe. 2003. Derived effective mesoscale potentials from atomistic simulations. *J. Comp. Chem.* 24:1624–1636.

Rippe, K., R. Stehr, and G. Wedemann. 2012. Monte Carlo simulations of nucleosome chains to identify factors that control DNA compaction and access. In *Innovations in Biomolecular Modeling and Simulations.* T. Schlick, editor. London: Royal Society of Chemistry, 198–235.

Robinson, P. J. J., W. An, A. Routh, F. Martino, L. Chapman, R. G. Roeder, and D. Rhodes. 2008. 30 nm chromatin fibre decompaction requires both H4-K16 acetylation and linker histone eviction. *J. Mol. Biol.* 381:816–825.

Robinson, P. J. J., L. Fairall, V. A. T. Huynh, and D. Rhodes. 2006. EM measurements define the dimensions of the "30-nm" chromatin fiber: Evidence for a compact, interdigitated structure. *Proc. Natl. Acad. Sci. U.S.A.* 103:6506–6511.

Robinson, P. J. J., and D. Rhodes. 2006. Structure of the '30 nm' chromatin fibre: A key role for the linker histone. *Curr. Opin. Struct. Biol.* 16:336–343.

Routh, A., S. Sandin, and D. Rhodes. 2008. Nucleosome repeat length and linker histone stoichiometry determine chromatin fiber structure. *Proc. Natl. Acad. Sci. U.S.A.* 105:8872–8877.

Saunders, M. G., and G. A. Voth. 2012. Coarse-graining of multiprotein assemblies. *Curr. Opin. Struct. Biol.* 22:144–150.

Saunders, M. G., and G. A. Voth. 2013. Coarse-graining methods for computational biology. *Annu. Rev. Biophys.* 42:73–93.

Schalch, T., S. Duda, D. F. Sargent, and T. J. Richmond. 2005. X-ray structure of a tetranucleosome and its implications for the chromatin fibre. *Nature* 436:138–141.

Scheffer, M. P., M. Eltsov, J. Bednar, and A. S. Frangakis. 2012. Nucleosomes stacked with aligned dyad axes are found in native compact chromatin in vitro. *J. Struct. Biol.* 178:207–214.

Scheffer, M. P., M. Eltsov, and A. S. Frangakis. 2011. Evidence for short-range helical order in the 30-nm chromatin fibers of erythrocyte nuclei. *Proc. Natl. Acad. Sci. U.S.A.* 108:16992–16997.

Schiessel, H. 2003. The physics of chromatin. *J. Phys. Condens. Mat.* 15:R699–R774.

Schlick, T., R. Collepardo-Guevara, L. A. Halvorsen, S. Jung, and X. Xiao. 2011. Biomolecular modeling and simulation: A field coming of age. *Q. Rev. Biophys.* 44:191–228.

Schlick, T., J. Hayes, and S. Grigoryev. 2012. Toward convergence of experimental studies and theoretical modeling of the chromatin fiber. *J. Biol. Chem.* 287:5183–5191.

Schlick, T., and O. Perisic. 2009. Mesoscale simulations of two nucleosome-repeat length oligonucleosomes. *Phys. Chem. Chem. Phys.* 11:10729–10737.

Scott Shell, M. 2008. The relative entropy is fundamental to multiscale and inverse thermodynamic problems. *J. Chem. Phys.* 129:144108.

Shogren-Knaak, M. A., H. Ishii, J.-M. Sun, M. Pazin, J. R. Davie, and C. L. Peterson. 2006. Histone H4-K16 acetylation controls chromatin structure and protein interactions. *Science* 311:844–847.

Song, F., P. Chen, D. Sun, M. Wang, L. Dong, D. Liang, K.-M. Xu, P. Zhu, and C. Li. 2014. Cryo-EM study of the chromatin fiber reveals a double helix twisted by tetranucleosomal units. *Science* 344:376–380.

Stehr, R., N. Kepper, K. Rippe, and G. Wedemann. 2008. The effect of internucleosomal interaction on folding of the chromatin fiber. *Biophys. J.* 95: 3677–3691.

Stehr, R., R. Schöpflin, R. Ettig, N. Kepper, K. Rippe, and G. Wedemann. 2010. Exploring the conformational space of chromatin fibers and their stability by numerical dynamic phase diagrams. *Biophys. J.* 98:1028–1037.

Stigter, D. 1977. Interactions of highly charged colloidal cylinders with applications to double-stranded DNA. *Biopolymers* 16:1435–1448.

Sun, J., Q. Zhang, and T. Schlick. 2005. Electrostatic mechanism of nucleosomal array folding revealed by computer simulation. *Proc. Natl. Acad. Sci. U.S.A.* 102:8180–8185.

Thåsröm, A., J. M. Gottesfeld, K. Luger, and J. Widom. 2004. Histone-DNA binding free energy cannot be measured in dilution-driven dissociation experiments. *Biochemistry* 43:736–741.

Tremethick, D. J. 2007. Higher-order structures of chromatin: The elusive 30 nm fiber. *Cell* 128:651–654.

Trylska, J., V. Tozzini, and J. A. McCammon. 2005. Exploring global motions and correlations in the ribosome. *Biophys. J.* 89:1455–1463.

van Holde, K., and J. Zlatanova. 1996. What determines the folding of the chromatin fiber? *Proc. Natl. Acad. Sci. U.S.A.* 93:10548–10555.

van Holde, K. E. 1988. *Chromatin*. New York, NY: Springer-Verlag.

Vasudevan, D., E. Y. Chua, and C. A. Davey. 2010. Crystal structures of nucleosome core particles containing the '601' strong positioning sequence. *J. Mol. Biol.* 403:1–10.

Voltz, K., J. Trylska, N. Calimet, J. C. Smith, and J. Langowski. 2012. Unwrapping of nucleosomal DNA ends: A multiscale molecular dynamics study. *Biophys. J.* 102:849–858.

Voltz, K., J. Trylska, V. Tozzini, V. Kurkal-Siebert, J. Langowski, and J. Smith. 2008. Coarse-grained force field for the nucleosome from self-consistent multiscaling. *J. Comp. Chem.* 29:1429–1439.

Walker, I. O. 1984. Differential dissociation of histone tails from core chromatin. *Biochemistry* 23:5622–5628.

Wang, X., and J. J. Hayes. 2008. Acetylation mimics within individual core histone tail domains indicate distinct roles in regulating stability of higher-order chromatin structure. *Mol. Cell. Biol.* 28:227–236.

Wedemann, G., and J. Langowski. 2002. Computer simulation of the 30-nanometer chromatin fiber. *Biophys. J.* 82:2847–2859.

Widom, J. 2001. Role of DNA sequence in nucleosome stability and dynamics. *Q. Rev. Biophys.* 34:269–324.

Wocjan, T., K. Klenin, and J. Langowski. 2009. Brownian dynamics simulation of DNA unrolling from the nucleosome. *J. Phys. Chem. B* 113:2639–2646.

Wolffe, A. P. 1998. *Chromatin: Structure and Function*. San Diego, CA: Academic Press.

Wolffe, A. P., and J. J. Hayes. 1999. Chromatin disruption and modification. *Nucleic Acids Res.* 27:711–720.

Woodcock, C. L., and S. Dimitrov. 2001. Higher-order structure of chromatin and chromosomes. *Curr. Opin. Genet. Dev.* 11:130–135.

Woodcock, C. L., A. I. Skoultchi, and Y. Fan. 2006. Role of linker histone in chromatin structure and function: H1 stoichiometry and nucleosome repeat length. *Chromosome Res.* 14:17–25.

Yager, T. D., C. T. McMurray, and K. E. van Holde. 1989. Salt-induced release of DNA from nucleosome core particles. *Biochemistry* 28:2271–2281.

Yang, Y., A. P. Lyubartsev, N. Korolev, and L. Nordenskiöld. 2009. Computer modeling reveals that modifications of the histone tail charges define salt-dependent interaction of the nucleosome core particles. *Biophys. J.* 96:2082–2094.

Zhang, Q., D. A. Beard, and T. Schlick. 2003. Constructing irregular surfaces to enclose macromolecular complexes for mesoscale modeling using the discrete surface charge optimization (DiSCO) algorithm. *J. Comp. Chem.* 24:2063–2074.

Zheng, C., X. Lu, J. C. Hansen, and J. J. Hayes. 2005. Salt-dependent intra- and internucleosomal interactions of the H3 tail domain in a model oligonucleosomal array. *J. Biol. Chem.* 280:33552–33557.

Zhou, J., J. Y. Fan, D. Rangasamy, and D. J. Tremethick. 2007. The nucleosome surface regulates chromatin compaction and couples it with transcriptional repression. *Nat. Struct. Mol. Biol.* 14:1070–1076.

CHAPTER 9

Modeling of Genomes

Naoko Tokuda and Masaki Sasai

CONTENTS

9.1 Two Basic Issues in Genome Modeling 341
9.2 A Genome Model in Interphase Budding Yeast 345
    9.2.1 Equation of Motion 347
    9.2.2 Potential Representing the Nuclear Constraints 348
    9.2.3 Worm-Like Chain Potential 350
    9.2.4 Knowledge-Based Potential 353
    9.2.5 Simulation 354
9.3 Fluctuating Genome Structure in Interphase Budding Yeast 354
9.4 Diploid Genomes 359
9.5 Summary 362
    Acknowledgments 363
    References 363

9.1 TWO BASIC ISSUES IN GENOME MODELING

A genome is not an abstract one-dimensional sequence of alphabet but a physical entity embedded in three-dimensional (3D) space. Observations based on the chromatin-conformation-capture (3C) methods have revealed that chromatins in fly (Sexton et al. 2012), mouse (Dixon et al. 2012), and human (Dixon et al. 2012) comprise arrays of looped regions termed "topologically associating domains (TADs)." A TAD is about several hundred kilobase (kb) to 1 megabase (Mb) in size, in which genes and regulatory regions tend to associate with each other due to the topological constraints of the looped chromatin structure. Therefore, the existence of TADs suggests a close relationship between the genome 3D structure and gene regulation. It has also been revealed that TAD is not a rigid static structure but a dynamical state; observation by fluorescence in situ hybridization (FISH) has shown large conformational fluctuation of TADs in mouse embryonic stem cells, which correlates or anticorrelates with the fluctuations of gene expression levels (Giorgetti et al. 2014). Moreover, in the scale of 100 Mb, each chromosome occupies a localized area or a territory in nucleus, and the transcription activity seems large at the interface of these areas (Cremer and Cremer 2010). The structure of the chromosome territory is also not rigid but dynamically fluctuates (Zidovska et al. 2013). Based on these observations, we can imagine a scenario where the multiscale dynamical movement of genome DNA in the nucleus regulates the accessibility of various factors to DNA, which, in turn, regulates transcription and cell behaviors. In order to check the validity of the hypothesis on the relationship between genome movement and transcription, which we call *dynamical genome hypothesis*, computational modeling of the 3D structure and movement of the genome are cruicial. In this chapter, we explain recent techniques to simulate the 3D structure and dynamics of the genome of a model organism, *Saccharomyces cerevisiae* (budding yeast). Although the yeast genome does not contain distinct TADs as observed in mammalian genomes, the yeast genome offers a platform to test various simulation techniques and is a starting point for modeling the larger and more complex genomes of higher organisms.

In recent years, intense efforts have been focused on constructing the 3D genome structures of various organisms using the data obtained by the

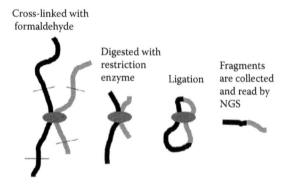

FIGURE 9.1 A schematic illustration of the procedure in the chromatin-conformation-capture (3C)-based method. From left to right: cells are first fixed with formaldehyde. Spatially proximal chromatins are linked with fixed proteins. Then, chromatins are digested into fragments with a restriction enzyme such as HindIII. Fragments are ligated under a diluted condition to create a circular dimer. The dimer is sheared into fragments, whose sequence is read by next-generation sequencing (NGS). The read sequence is mapped to the genome data to identify two loci which were adjacent in nucleus before fixation.

3C-based methods. Figure 9.1 is a schematic illustration of the 3C-based observation. With such 3C-based methods, one can estimate the number of pairs of DNA fragments, $F(\mu i; \nu j)$, in which the ith site in the μth chromosome and the jth site in the νth chromosome reside in spatial proximity in the observed ensemble of N_{cell} cells. Although the recent improvement of the signal/noise ratio opened a way to single-cell observation with $N_{cell} = 1$ (Nagano et al. 2013), N_{cell} has been typically larger than 10^4 in the hitherto reported data. With Hi-C methods, which are extensions of the 3C method to the genome-wide scale (Lieberman-Aiden et al. 2009), the genome-wide matrix of encounter frequency, $\mathbf{F} = \{F(\mu i; \nu j)\}$, is obtained with suffices μi and νj running through the genome. Because $F(\mu i; \nu j)$ should be correlated with $1/r(\mu i; \nu j)$, where $r(\mu i; \nu j)$ is the average spatial distance between two sites, one can obtain the distance-geometry matrix $\mathbf{R} = \{r(\mu i; \nu j)\}$, from which the 3D model of chromatins can be constructed.

A fundamental difficulty in constructing the 3D model structure by the 3C-based methods is the problem of diploid cells. A diploid nucleus of higher organism has pairs of homologous chromosomes, each pair comprising the maternal copy α of the chromosome and the paternal copy β of the same chromosome. Because the 3C-based methods do not distinguish α from β, the observed probability $F(\mu i; \nu j)$ is a sum of four contributions:

$$F(\mu i; \nu j) = F(\mu \alpha i; \nu \alpha j) + F(\mu \alpha i; \nu \beta j) + F(\mu \beta i; \nu \alpha j) + F(\mu \beta i; \nu \beta j) \quad (9.1)$$

For constructing the structure model of a diploid genome, the decomposition of $F(\mu i; \nu j)$ into four contributions is necessary (Kalhor et al. 2012). In order to avoid the difficulty of this decomposition, most modeling studies have focused on either haploid cells in which each nucleus has only single copies of individual chromosomes (Duan et al. 2010; Tanizawa et al. 2010), or on single chromosome modeling based on the intrachromosome (i.e., *cis*) encounter frequency $F(\mu i; \mu j)$ (Baù et al. 2011; Nagano et al. 2013; Giorgetti et al. 2014); in the latter case, one can assume $F(\mu \alpha i; \mu \alpha j)$, $F(\mu \beta i; \mu \beta j) >> F(\mu \alpha i; \mu \beta j)$, $F(\mu \beta i; \mu \alpha j)$, so that the calculated structure from $F(\mu i; \mu j)$ is the average of the structures α and β. Note that we can not assume $F(\mu \alpha i; \nu \alpha j)$, $F(\mu \beta i; \nu \beta j) >> F(\mu \alpha i; \nu \beta j)$, $F(\mu \beta i; \nu \alpha j)$ for the general case of *trans* interactions with $\mu \neq \nu$. In this chapter, we mostly explain the method for modeling a haploid cell but, in the last part of the chapter, we briefly discuss the decomposition in diploid cells for general cases including both *trans* and *cis* interactions to construct a whole genome model of a higher organism.

However, to examine the dynamical genome hypothesis, not only the static average structure but also the dynamical movement of the genome should be analyzed. The second important issue in modeling genomes is how to simulate the dynamical movement with computational models. Although there has not yet been a systematic method to derive the equation of motion of the genome at the coarse-grained level from the microscopic foundation, we can learn how to resolve this problem from previous studies on coarse-grained modeling of proteins. The average structure of a protein can be constructed from the distance-geometry matrix **R** obtained from nuclear magnetic resonance (NMR) measurement, which is similar to the case of the 3C-based measurement. **R** for a protein is a quantity averaged over an ensemble of molecules in solution; therefore, the protein structure constructed from **R** is the average structure of many molecules. Based on this **R**, an important assumption has been made in the definition of interactions among amino-acid residues to simulate the dynamics of protein structure; stable interactions lasting for more than a microsecond in a dynamically fluctuating protein are those found in the average structure of the protein. These interactions are called "native interactions" and the assumption of dominance of native interactions over the nonnative interactions should be validated if the protein has been evolutionarily designed to have microscopic interactions that are consistent with the whole protein structure. The validity of this assumption has been checked by comparisons between simulations and experiments (Onuchic and Wolynes 2004).

The assumption of the ideal consistency was called the "consistency principle" (Gō 1983) and the more quantitative remark on microscopic interactions that involves inconsistency to some extent was called the "minimum frustration principle" (Bryngelson and Wolynes 1987; Ferreiro et al. 2014). Guided by this minimum frustration principle, dynamical fluctuation of proteins around their average structures can be regarded as fluctuating motions under the influence of interaction potentials, each of which has its minimum at the pairwise distance between amino-acid residues in the average structure; these interaction potentials have been called Gō-like potentials (Onuchic and Wolynes 2004). As explained in other chapters in this book, with this picture, even large structural deformations including folding/unfolding processes and the functioning process of protein complexes have been successfully simulated with coarse-grained models. Similarly, in problems of genomes, we may be able to assume that factors bound to DNA in specific ways mediate stable interactions between sites in the genome and these interactions are consistent with the whole average structure of the genome. Then, we can expect that the dynamical structural fluctuation around the average structure derived from the 3C-based method is simulated using the interaction potentials, each of which has a minimum at the distance between two sites found in the average genome structure.

As shown in Section 9.2, the simulated structural fluctuations of the genome in yeast cell is large; therefore, the genome state is not a "folded state" but is more similar to a "molten globule" or liquid-like state. However, the movement of chromosomes is not completely random but shows heterogeneous distributions, which generates the biased movement of each part of the genome in the nucleus. We can expect that interaction potentials derived from the 3C-based method appropriately describe these biases as the Gō-like potentials describe structural fluctuations in molten globules in a site-specific manner. In this chapter, the suitability of this assumption is discussed by comparing the simulated data with experiments.

9.2 A GENOME MODEL IN INTERPHASE BUDDING YEAST

A straightforward application of the idea described in the previous section is to construct computational models of genomes in haploid cells. Among the haploid cells, budding yeast is a model organism, for which data of intensive analyses have been accumulated. The genome-wide Hi-C data, $\mathbf{F} = F(\mu i; \nu j)$, with a few kb resolution for the interphase budding yeast was reported by Duan et al. (2010), who estimated $\mathbf{R}$ from $\mathbf{F}$ in the following way. They first showed that the *cis*-encounter frequency $F(\mu i; \mu j)$ is anticorrelated with the distance along the sequence $|i - j|$. The yeast nucleus is small in size (1 μm radius) and the configuration of each chromosome is rather strongly constrained by the connection between the centromere of the chromosome and the spindle pole body (SPB) located near the nuclear envelope (see Figure 9.2), and also by the interaction between the telomeres of the chromosome and the nuclear envelope. Thus, each chromosome tends to show an extended conformation, so that the spatial distance between the two sites on the same chromatin is approximately proportional to the distance along the sequence. The estimated distance is 110–150 bp/nm. Using the intermediate value 130 bp/nm, the average relation between $r(\mu i; \mu j)$ and $F(\mu i; \mu j)$ was obtained in Supplementary Figure 17 of Duan et al. (2010). Applying this average relation to *trans* interactions, the observed $\mathbf{F}$ was converted to $\mathbf{R}$ ranging over the whole genome. In this chapter, the thus obtained $\mathbf{R}$ is used as the basis for deriving the interaction potential U_{HiC}.

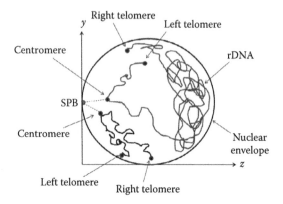

FIGURE 9.2 Interactions between chromosomes and envelope of budding yeast nucleus. Structures of Chr 4 and Chr 12 are schematically illustrated. Two ends of each chromosome are right and left telomeres which interact attractively with nuclear envelope. Centromeres of chromosomes are linked through microtubles to spindle pole body (SPB), which is located near the nuclear envelope. rDNA in the right arm of Chr 12 is confined in the nucleolus, which resides on the opposite side of the nucleus to SPB.

We should note that the present use of **F** or **R** is based on approximations and should be improved. We need to take care of the biases in the Hi-C data, which may prevent the straightforward use of **F**. The possible biases come from the GC content, length, and mappability of DNA fragments used in the Hi-C observation (Yaffe et al. 2011). Although various normalization methods to diminish these biases in the data were proposed and demonstrated to be helpful to enhance the contrasts in **F** (Cournac et al. 2012; Hu et al. 2012), it is not yet clarified how these normalizations affect the accuracy of the 3D modeling. Estimation of the effects of the biases in the 3D modeling was investigated for the case of budding yeast (Tokuda et al. 2017). In Tokuda et al. 2017, the model was further improved from the version explained in this chapter. Another point to be remarked is that, in the method proposed by Duan et al. (2010), the same relationship between **F** and **R** was assumed everywhere in the genome. Because the nuclear envelope should give different constraints on the conformation of chromatins depending on the spatial distance between the genome locus and the nuclear envelope, the relation between **F** and **R** could be modulated depending on where the locus resides in the nucleus. Errors arising from the spatial inhomogeneity in the **F–R** relation could be reduced by a self-consistent modification of each interaction potential through an iterative comparison of the calculated conformation with the 3C-based data (Giorgetti et al. 2014; Zhang et al. 2015). Although these sophistications are important, we here explain the vanilla version of the method to construct U_{HiC} for budding yeast without using normalization or iterative improvement procedures. The method explained in this chapter follows the one introduced by Tokuda et al. (2012), but the potential U_{nucleus} representing the constraints due to the interactions between chromatins and the nuclear envelope was improved and changed from the corresponding potential in Tokuda et al. (2012).

9.2.1 Equation of Motion

We represent a chromosome by a chain of beads and springs. Therefore, 16 chromosomes in interphase haploid budding yeast are represented by 16 chains. In order to use the Hi-C data of Duan et al., whose resolution is 1–3 kb, we assume that each bead represents a bin of 3-kb DNA segments. The number of beads in each chromosome ranges from 78 for Chromosome 1 (Chr 1) to 806 for Chr 12, and the total number of beads in the genome is 4460. The movement of these chains is simulated by using the Langevin equation of motion,

$$m\frac{d^2\mathbf{r}_i^{\mu}}{dt^2} = -\frac{\partial U}{\partial \mathbf{r}_i^{\mu}} - \zeta\frac{d\mathbf{r}_i^{\mu}}{dt} + \mathbf{w}_i^{\mu} \qquad (9.2)$$

where $\mathbf{r}_i^{\mu}$ is a vector representing the position of the ith bead in the μth chromosome with $\mu = 1 - 16$, and $\mathbf{w}_i^{\mu}$ is a vector where each component is Gaussian white noise, satisfying

$$< \mathbf{w}_{i\alpha}^{\mu}(t)\mathbf{w}_{j\beta}^{\nu}(t') > = 2\zeta T\delta(t - t')\delta_{ij}\delta_{\mu\nu}\delta_{\alpha\beta} \qquad (9.3)$$

where α and $\beta = x, y$, and z represent the components of vectors. Here, we used the unit $k_B = 1$. It has been observed that the amplitude of movement of chromosomes depends on the adenosine triphosphate (ATP) synthesis (Heun et al. 2001). Therefore, the genome movement should be a nonequilibrium process driven by the consumption of ATP. However, there is no large complex of motor proteins in yeast nucleus, and the genome movement is dominated by random fluctuations. In Equations 9.2 and 9.3, we conveniently represent these random fluctuations by thermal noise at temperature T, which is not necessarily the physical temperature of the cell but includes the nonequilibrium fluctuations implicitly. The effects of the nonequilibrium feature would be more important for the process of DNA repair in which the ATP consumption plays important roles (Neumann et al. 2012). However, in this chapter, we do not explicitly consider these nonequilibrium effects for simplicity but we leave the treatment of position- and time-dependent fluctuations of $T(\mathbf{r}, t)$ as a future problem. U in Equation 9.2 comprises three parts:

$$U = U_{\text{nucleus}} + U_{\text{chain}} + U_{\text{HiC}} \tag{9.4}$$

where U_{nucleus} represents interactions between chromatins and nuclear envelope, U_{chain} represents the connectivity, rigidness, and exclusion interactions of chromatin chains, and U_{HiC} is a Gō-like potential that represents the average structure inferred from the Hi-C data. In the following three subsections, we explain them one by one.

9.2.2 Potential Representing the Nuclear Constraints

The nucleus of budding yeast can be modeled by a sphere of 1 µm radius (Berger et al. 2008). Here, we represent the nucleus with a sphere whose center is at $\mathbf{r}_{\text{center}} = (0, 1000, 1000)$ in units of nanometers. As shown in Figure 9.2, the movement of yeast chromosomes is constrained by the structure of the nucleus. The ribosomal DNA (rDNA) of Chr 12 is localized at one side of the nucleus and, together with the synthesized rRNA and other factors, rDNA forms a nucleolus. At the opposite side to this nucleolus, the SPB is located near the nuclear envelope. We represent the SPB by a point at $(0, 1000, 10)$. The centromeres of chromosomes are connected to SPB by microtubules, and the telomeres of the chromosomes tend to be near the nuclear envelope. These constraints are represented by the potential

$$U_{\text{nucleus}} = U_{\text{cen}} + U_{\text{envelope}} \tag{9.5}$$

The linkage between the centromeres and SPB is represented by the following potential:

$$U_{\text{cen}} = \sum_{\mu} \frac{h}{2} \left(\frac{l^{\mu} - l_0^{\mu}}{2} \right)^2 \tag{9.6}$$

where l^{μ} is the distance between the SPB and the centromere of the µth chromosome in the simulated genome structure. l_0^{μ} is the distance in a model structure constructed by Duan et al. (2010). The spring constant was set to be small so as to allow length variation of the linker microtubules: $h/T = 0.3$.

U_{envelope} represents interactions between chromatins and the nuclear envelope:

$$U_{\text{envelope}} = U_{\text{chr-env}} + U_{\text{tel-env}} + U_{\text{rDNA-env}} \tag{9.7}$$

the $U_{\text{chr-env}} = \sum_{\mu} \sum_{i} U_{\text{chr-env}}(r_i^{\mu})$ corresponds to the repulsive interactions between chromatins and nuclear envelope representing the effects of confinement of the chromosomes within the nucleus:

$$U_{\text{chr-env}}(r_i^{\mu}) = \begin{cases} 2\varepsilon \left(\frac{R_i^{\mu}-R_0}{u-R_0} \right)^{12}, & (R_i^{\mu} > R_0) \\ 0, & (R_i^{\mu} \le R_0) \end{cases} \tag{9.8}$$

where i does not belong to telomeres, subtelomere regions, or Chr 12. $R_i^{\mu} = |r_i^{\mu} - r_{\text{center}}|$ is the radial distance of the site i of the chromosome μ from the nucleus center, and we used $\varepsilon/T = 1$, $R_0 = 800$ nm, and $u = 1000$ nm.

In the yeast nucleus, the telomeres and subtelomere regions of the chromosomes are attracted to the nuclear envelope. This attractive interaction is represented by the potential, $U_{\text{tel-env}} = \sum_{\mu} \sum_{i \in (\text{tel. \& subtel.})} U_{\text{tel-env}}(r_i^{\mu})$, which contains a long-range attractive part and a short-range repulsive part as

$$U_{\text{tel-env}}(r_i^{\mu}) = \begin{cases} 2\varepsilon \left[\left(\frac{R_i^{\mu}-R_0}{u-R_0} \right)^{12} - c_1 \right], & (R_i^{\mu} > R_3) \\ -c_0 \varepsilon, & (R_2 < R_i^{\mu} \le R_3) \\ -c\varepsilon \left(\frac{R_i^{\mu}-R_1}{R_2} \right), & (R_1 < R_i^{\mu} \le R_2) \\ 0, & (R_i^{\mu} \le R_1) \end{cases} \tag{9.9}$$

where i is in the telomere or subtelomere regions. Telomeres are the first and last ends of each chromosome chain, and subtelomeres are defined as seven consecutive beads lying before or after the telomere bead along the chain. Here, we use $R_1 = 400$ nm, $R_2 = 800$ nm, $R_3 = 950$ nm, and $c = 3$. c_0 and c_1 are constants to connect parts of $U_{\text{tel-env}}(r_i^{\mu})$ continuously. The attractive force is represented by a piecewise linear slope between R_1 and R_2. The statistical tendency of the telomere behavior is represented here in a mean-field way for the large value of $R_2 - R_1 = 400$ nm.

The right arm region of Chr 12 spanning from 450 kb to 1815 kb is regarded as rDNA. We model the localization of rDNA on one side of the nucleus through the attractive interactions between sites in rDNA and the nuclear envelope, $U_{\text{rDNA-env}} = \sum_\mu \sum_{i \in \text{rDNA}} U_{\text{rDNA-env}}(\mathbf{r}_i^\mu)$, as

$$
U_{\text{rDNA-env}}(\mathbf{r}_i^\mu) =
\begin{cases}
2\varepsilon\left[\left(\dfrac{R_i^\mu - R_0}{u - R_0}\right)^{12} - d_1\right], & (R_i^\mu > R_3') \\
-d_0\varepsilon, & (R_2' < R_i^\mu \leq R_3') \\
-d\varepsilon\left(\dfrac{R_i^\mu - R_1'}{R_2'}\right), & (R_1' < R_i^\mu \leq R_2') \\
0, & (R_i^\mu \leq R_1')
\end{cases}
\tag{9.10}
$$

Here, d_0 and d_1 are constants to connect parts of $U_{\text{rDNA-env}}(\mathbf{r}_i^\mu)$ continuously. We use $R_1' = 800$ nm, $R_2' = 900$ nm, $R_3' = 950$ nm, and $d = 9$. In the calculated results presented in this chapter, we also assumed an attractive potential similar to the one used in Equation 9.9 for the site i in the regions of Chr 12 other than rDNA, telomeres, or subtelomere regions, but this additional attractive potential did not play significant roles and hence can be dropped.

9.2.3 Worm-Like Chain Potential

In Equation 9.4, U_{chain} is the potential representing the physical properties of chromatin chains, which comprises three terms:

$$
U_{\text{chain}} = U_{\text{ex}} + U_{\text{spring}} + U_{\text{bend}}
\tag{9.11}
$$

U_{ex} comprises the Lennard–Jones-type potentials representing the exclusion-volume interactions between chromatins,

$$
U_{\text{ex}} = \sum_{\mu > \nu} \sum_{ij} U_{\text{ex}}(\mu i; \nu j) + \sum_\mu \sum_{j \geq i+2} U_{\text{ex}}(\mu i; \mu j)
\tag{9.12}
$$

with

$$
U_{\text{ex}}(\mu i; \nu j) =
\begin{cases}
4\varepsilon\left(\left(\dfrac{a}{r_{ij}^{\mu\nu}}\right)^{12} - \left(\dfrac{a}{r_{ij}^{\mu\nu}}\right)^{6} + \dfrac{1}{4}\right), & (r_{ij}^{\mu\nu} \leq 2^{\frac{1}{6}}a) \\
0, & (r_{ij}^{\mu\nu} > 2^{\frac{1}{6}}a)
\end{cases}
\tag{9.13}
$$

where a is the effective thickness of the chromatin and $r_{ij}^{\mu\nu} = |\mathbf{r}_i^\mu - \mathbf{r}_j^\nu|$. Because the repulsive force between chromatins works when the ion atmospheres surrounding two chromatins overlap, a should be the sum of the two lengths: $a = l_{\text{ions}} + l_{\text{vdW}}$, where $l_{\text{ions}} \leq 10$ nm represents the thickness of a layer of counter ions surrounding the chromatin, and l_{vdW} is the length defined by the van der Waals radii of the atoms constituting the chromatins. Recent cryo-electron microscopy (Eltsov et al. 2008; Maeshima et al. 2010) and x-ray (Nishino et al. 2012; Maeshima et al. 2014) measurements showed the absence of regular 30-nm chromatin fibers in the nuclei and suggested that the irregular 10-nm fibers are basic constituents. With such irregular 10-nm fibers, the chain at the present level of coarse-graining should look like a bundle of irregular fibers with l_{vdW} approximately 30 nm. In the present model, we assume $a = a_0 = 30$ nm for the interaction between sites that do not belong to the rDNA region, but this assumption does not imply that the model supports either the picture of dominance of 30-nm fibers or the picture of absence of regular 30-nm fibers. We assume that the repulsion due to the overlap of disordered bundles of fibers or ion atmospheres is soft, so that the rather small value of $\varepsilon/T = 1$ is used to represent the strength of U_{LJ}. In the nucleolus, rDNA is packed together with condensed rRNA and other factors. The effects of those RNAs and factors should be modeled assuming that the thickness of chromatin fibers of rDNA is much larger than a_0 (Wong et al. 2012). For the interaction between the sites of rDNA, we therefore use

$$
U_{\text{ex}}(\mu i; \nu j) =
\begin{cases}
4\varepsilon \left(\left(\dfrac{a}{r_{ij}^{\mu\nu}} \right)^2 - \left(\dfrac{a}{r_{ij}^{\mu\nu}} \right) + \dfrac{1}{4} \right), & (r_{ij}^{\mu\nu} \leq 3a) \\[4mm]
0, & (r_{ij}^{\mu\nu} > 3a)
\end{cases}
\tag{9.14}
$$

with $a = a_r = 150$ nm. For the interaction between a site of rDNA and a site not in rDNA, we use

$$
U_{\text{ex}}(\mu i; \nu j) =
\begin{cases}
4\varepsilon \left(\left(\dfrac{a}{r_{ij}^{\mu\nu}} \right)^2 - \left(\dfrac{a}{r_{ij}^{\mu\nu}} \right) + \dfrac{1}{4} \right), & (r_{ij}^{\mu\nu} \leq 2a) \\[4mm]
0, & (r_{ij}^{\mu\nu} > 2a)
\end{cases}
\tag{9.15}
$$

with $a = (a_0 + a_r)/2 = 90$ nm.

U_{spring} represents the connectivity between neighbor beads along the chain,

$$
U_{\text{spring}} = \sum_{\mu} \sum_{i} \left(\theta(r_i^{\mu}) U_{\text{FENE}}(r_i^{\mu}) + (1 - \theta(r_i^{\mu})) U_{\text{long}}(r_i^{\mu}) + U_{\text{short}}(r_i^{\mu}) \right)
$$

$$
\tag{9.16}
$$

where $r_i^\mu = r_{i\,i+1}^{\mu\mu}$. U_{FENE} is the finitely extensible nonlinear elastic (FENE) potential (Rosa and Everaers 2008; Rosa et al. 2010; Kremer and Grest 1990):

$$U_{\text{FENE}} = -\frac{1}{2}kR_0^2 \log\left[1 - \left(\frac{r_i^\mu}{R_0}\right)^2\right] \tag{9.17}$$

Following Rosa et al. (2010), we used $R_0 = 1.5\sigma_0$, where σ_0 is the typical distance between two neighbor beads. When we assume that the chromatin is packed at 130 bp/nm (Bystricky et al. 2004), then $\sigma_0 = 3\,\text{kb}/(130\,\text{bp/nm}) = 23\,\text{nm}$, so that $R_0 = 34.5\,\text{nm}$. k was chosen to be $k/T = 3.5\sigma_0^{-2}$ in order to create the calculated average distance between neighbor beads σ_0. Because U_{FENE} diverges when r_i^μ approaches R_0, U_{FENE} is switched to a milder repulsive potential $U_{\text{long}}(r_i^\mu) = (bk/\sigma_0^2)(r_i^\mu)^5$ to avoid numerical instability using a switching function,

$$\theta(r_i^\mu) = \begin{cases} 1, & \text{for } r_i^\mu < R_0', \\ 0, & \text{for } r_i^\mu \geq R_0' \end{cases} \tag{9.18}$$

We used $R_0 > R_0' = 30\,\text{nm}$ and $b = 0.42$ to enable smooth switching. Moreover, to avoid numerical instability due to collision of the two beads, a mild short-range repulsive potential is introduced as

$$U_{\text{short}}(r_i^\mu) = \begin{cases} 4\varepsilon\left(\left(\frac{R'}{r_i^\mu}\right)^2 - \left(\frac{R'}{r_i^\mu}\right) + \frac{1}{4}\right), & (r_i^\mu \leq 2R') \\ 0, & (r_i^\mu > 2R') \end{cases} \tag{9.19}$$

where $R' = a_0$ when i or j is not the site in rDNA, and $R' = a_r$, otherwise. U_{bend} in Equation 9.11 represents the bending potential,

$$U_{\text{bend}} = \sum_\mu \sum_i U_b(\phi_i^\mu) \tag{9.20}$$

where ϕ_i^μ is the angle between $\mathbf{r}_i^\mu - \mathbf{r}_{i-1}^\mu$ and $\mathbf{r}_{i+1}^\mu - \mathbf{r}_i^\mu$. Because the nucleosome remodeling may allow a chromatin to bend, which should give rise to the kink of the chromatin chain, we use a kinkable potential that saturates for $\phi > \pi/2$ (Rosa et al. 2010; Wiggins et al. 2005) as

$$U_b(\phi_i^\mu) = \begin{cases} k_\phi(1 - \cos\phi_i^\mu), & (0.1 \leq \cos\phi_i^\mu \leq 1), \\ 0.9k_\phi, & (-1 \leq \cos\phi_i^\mu < 0.1) \end{cases} \tag{9.21}$$

The boundary value of 0.1 in Equation 9.21 was chosen to fit U_b to the potential in Rosa et al. (2010). The dispersion $< \phi^2 >$ should be related to the persistent length l_p and the distance σ_0 as $< \phi^2 > \approx 2\sigma_0/l_p$ (Hagerman 1988). Using estimations of $l_p \approx$ 170–220 nm for chromatins (Bystricky et al. 2004) and $\sigma_0 \approx$ 23 nm, we obtain $\sqrt{< \phi^2 >} \approx$ 26–30°. k_ϕ in Equation 9.21 was selected to be $k_\phi/T = 2.0$, to enable the simulated value of $\sqrt{< \phi^2 >}$ to fall in this range.

9.2.4 Knowledge-Based Potential

In order to represent the tendency that chromatin chains take the average configurations inferred from the Hi-C measurement, we introduced a knowledge-based Gō-like potential,

$$U_{\text{HiC}} = \sum_{\mu > \nu} \sum_{ij} U_{\text{Go}}(\mu i; \nu j) + \sum_{\mu} \sum_{j > i} U_{\text{Go}}(\mu i; \mu j) \qquad (9.22)$$

where

$$U_{\text{Go}}(\mu i; \nu j) = -\frac{\xi r_0}{\sqrt{2\pi \left(r(\mu i; \nu j)\right)^2}} \exp\left(-\frac{\left(r_{ij}^{\mu\nu} - r(\mu i; \nu j)\right)^2}{2 \left(r(\mu i; \nu j)\right)^2}\right) \qquad (9.23)$$

Here, $r(\mu i; \nu j)$ is the average distance between μi and νj inferred from the Hi-C data. For a variable to represent the length scale, we used $r_0 = 10$ nm. Then, the parameter to represent the energy scale is ξ/T. In Section 9.3, we will discuss the results of two cases of weak ($\xi/T = 10$) and strong ($\xi/T = 100$) Gō-like interactions.

The genome structure is stabilized by the integration of multiple tendencies, each of which is represented by the individual potentials in Equation 9.4: physical properties of chromatin chains, which are represented by U_{chain} in the model, interactions between chromatins and nuclear landmarks, which are represented by U_{nucleus}, and specific interactions among loci in the genome. The potential energy to stabilize the observed average structure, U_{HiC}, should show the lowest value at a given structure $\{\mathbf{r}_i^\mu\}$ when all these effects are consistent with each other at $\{\mathbf{r}_i^\mu\}$. Among these various effects, U_{HiC} highlights specific chromatin–chromatin interactions that are not represented by the other two terms, U_{chain} and U_{nucleus}; therefore, the larger ξ/T should emphasize the larger role of specific interactions among loci in the genome.

9.2.5 Simulation

We first started from the structure of Duan et al. (2010) and lowered the energy U by relaxing the structure. The obtained relaxed structure was used as the initial conformation for the Langevin dynamics calculation. When the distance between the two sites was smaller than a_0, we use the obtained distance a_0' in this initial conformation as a in Equation 9.13. Starting from this initial conformation, the Langevin dynamics calculation was performed using $m = T = 1$. The time t has the dimension of length L, and the friction constant ζ has the dimension L^{-1}. We used the time step $\Delta t = 0.01$ and $\zeta = 10^{-5}$ in order to sample the different structures efficiently. The first 10^4 steps were used for equilibration, and the subsequent 5×10^4 steps were used for sampling. A total of 39 or 40 trajectories were calculated with different random number realizations and the statistical analyses were performed using the conformations obtained from these runs.

9.3 FLUCTUATING GENOME STRUCTURE IN INTERPHASE BUDDING YEAST

The simulated genome in the budding yeast nucleus shows large structural fluctuations. However, because of the smallness in size of yeast nucleus, the chromatin movement is considerably constrained by the interactions with the nuclear envelope. Figure 9.3 demonstrates a snapshot of the simulated genome movement. The centromeres of the chromosomes gather around the SPB which is located near the nuclear envelope (left side of the figure), and the nucleolus is on the opposite side, where the rDNA region of Chr 12 is spread. The telomeres are distributed in between the centromeres

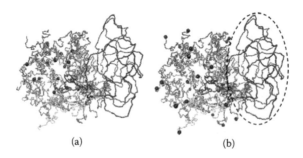

(a)   (b)

FIGURE 9.3   A snapshot of the simulated genome movement. (a) Centromeres are shown with black spheres. (b) Telomeres are shown with black spheres. The approximate region of rDNA of Chr 12 is surrounded by a dashed line.

and nucleolus. This arrangement of centromeres, telomeres, and nucleolus has been called "Rabl-like configuration," which has been confirmed by fluorescence measurements (Berger et al. 2008) and computational modeling based on the Hi-C data (Duan et al. 2010). The present simulation shows that the chromosomes largely move while they maintain the Rabl-like arrangement.

The distribution of each part of the genome can be shown by projecting the 3D structure onto the two-dimensional (2D) plane. In Figure 9.4, the probability to find the locus in a circular tube that runs through the point $(x = 0, y, z)$, i.e., a tube running along the circle of radius $|y|$ lying on the plane of constant z and being centered at $(0, 0, z)$, is plotted on the 2D plane of (y, z). Note that these 2D figures are symmetric about the z axis by definition. Figure 9.4 shows the projected distributions of centromeres, rDNA, and the left telomere of Chr 7 (Tel 7L) for $\xi/T = 100$ and 10. These distributions spread widely in the nucleus showing large structural fluctuations but they occupy their own characteristic regions. In particular, the simulated distributions with $\xi/T = 100$ reasonably reproduce the distributions observed with fluorescent probes (Berger et al. 2008).

The results on the distribution of telomere–telomere distance in two cases, $\xi/T = 100$ and 10, are compared in Figure 9.5 with the experimental data (Bystricky et al. 2005). We find that the results of $\xi/T = 100$

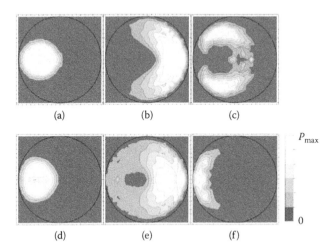

FIGURE 9.4 Two-dimensional representation of probability distributions of parts of budding yeast genome. Distributions of (a,d) centromeres, (b,e) rDNA, and (c,f) the left telomere of Chr 7 (Tel 7L). Results obtained with different strength of the Gō-like potentials, (a–c) $\xi/T = 100$ and (d–f) $\xi/T = 10$.

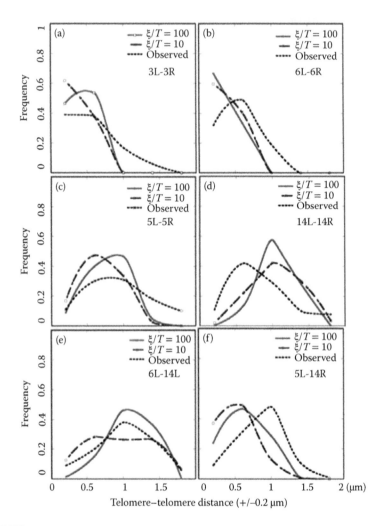

FIGURE 9.5 Telomere–telomere distance distributions. Data simulated with $\xi/T = 100$ (solid line) and $\xi/T = 10$ (dashed and dotted line) are compared with the data observed with the fluorescently labeled proteins (dotted line) (Bystricky et al. 2005). Distributions of distances between (a) Tel 3L and Tel 3R, (b) Tel 6L and Tel 6R, (c) Tel 5L and Tel 5R, (d) Tel 14L and Tel 14R, (e) Tel 6L and Tel 14L, and (f) Tel 5L and Tel 14R. Points obtained by binning data over $\pm0.2\,\mu m$ are connected by smooth lines.

reasonably explain the experimental data. Tjong et al. (2012) and Wong et al. (2012) constructed the 3D models of the budding yeast genome structure without using the Hi-C data and showed that these models reproduce the Rabl-like structure of the yeast genome. These models consider only the exclusion interactions among chromatin chains and interactions between chromatins and the nuclear envelope; therefore, these models are similar to the $\xi/T = 0$ case in our model. However, because the interactions between chromatins and the nuclear envelope are symmetrical around the z axis in Figure 9.2, the structures produced using these models do not show biases in the xy plane; therefore, these structures are isotropic around the axis connecting SPB and the center of the nucleus. Anisotropy around the axis is important when considering the correlation between two loci in the genome and hence should determine the motion of the genome. As shown in Figure 9.5, $\xi/T = 100$ better explains the experimental data on the two-point correlation between the telomeres than $\xi/T = 10$. The specific interactions among loci in the genome, which are anisotropic around the axis, are important for explaining structural correlation and dynamics.

Figure 9.6 shows examples of trajectories of simulated distances between telomeres and the nuclear envelope. The simulated distances fluctuate within 0–500 nm; the telomeres tend to be located near the nuclear envelope but do not stick on it. These on/off fluctuations to/from the envelope are similar to the behaviors experimentally observed using fluorescence probes (Hediger et al. 2002).

Application of this genome model to various biological problems should provide useful insights. For example, if we regard a mutation as a

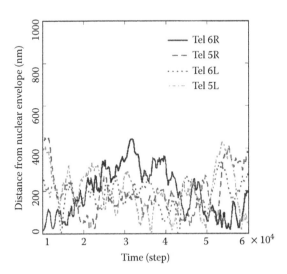

FIGURE 9.6 Examples of trajectories showing the temporal variation of distances between telomeres and the nuclear envelope. Distances between envelope and Tel 6R (solid line), Tel 5R (dashed line), Tel 6L (dotted line), and Tel 5L (dashed and dotted line). $\xi/T = 100$.

perturbation to the system, we can analyze its effects of mutation on the genome dynamics by simulation. Shown in Figure 9.7 is an example in which the simulated distributions of Tel 14L and Tel 6R in the wild-type and *yku70 esc1* double mutant strains are shown, and in Figure 9.8, they are compared with the observed data (Taddei et al. 2004). Because Yku70 and Esc1 are factors that cause the telomeres to bind to the nuclear envelope, the interactions between the telomeres and the nuclear envelope, represented by $U_{\text{tel-env}}$ in Equation 9.9, turn to 0 in the simulated *yku70 esc1* mutant strain. As shown in Figures 9.7 and 9.8, the telomeres are distributed with larger probability at the region separated from the envelope in the mutant than in the wild-type strain. This simulation provides useful information on how such structural and dynamical change on mutation affects the gene

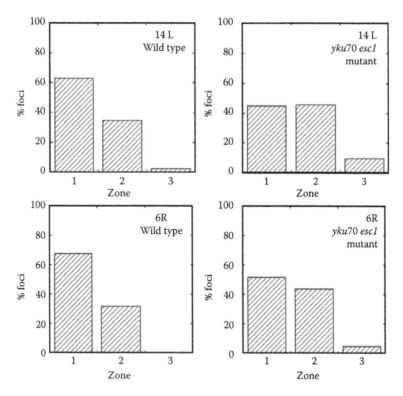

FIGURE 9.7 The simulated probability to find telomeres in zones 1, 2, and 3. These zones are defined by using spheres which share the same center as the nucleus (Taddei et al. 2004). Zone 3 is the region within the sphere of radius 0.578 μm, zone 2 is the region between the sphere of radius 0.816 μm and the sphere of radius 0.578 μm, and zone 1 is the region outside of the sphere of radius 0.816 μm. Results of the simulated distributions of Tel 14L (top) and Tel 6R (bottom) are shown for the wild-type (left) and *yku70 esk1* double mutant (right) strains.

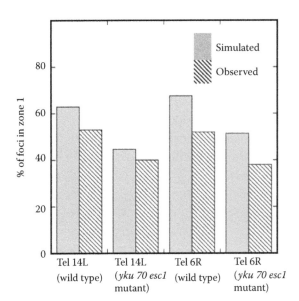

FIGURE 9.8 Simulated and experimentally observed (Taddei et al. 2004) probabilities to find Tel 14L and Tel 6R in zone 1 for the wild-type and *yku70 esk1* double mutant (right) strains are compared.

expression activity (Taddei et al. 2009; Tokuda et al. 2017). It is important to further analyze the relationship between the dynamical structural fluctuation and gene expression activity to examine the dynamical genome hypothesis.

9.4 DIPLOID GENOMES

Simulation of the diploid genome is based on the observation that chromosomes are localized in their own territories in the nucleus. In the small nucleus of budding yeast, such territories are not formed and different chromosomes are mixed in space. However, in the larger mammalian nuclei, the importance of the chromosome territories has been experimentally confirmed (Cremer and Cremer 2010). When the chromosomes are confined in such territories, chromosomes intermingle with each other only at the boundaries of these territories; therefore, one can represent the chromosomes in a coarse-grained way using large beads or spheres not overlapping with each other. The diploid problem is regarded as a difficulty in arranging these spheres in the nucleus. Kalhor et al. (2012) developed a model of human lymphoblastoid nucleus based on this assumption. They summarized each region of the chromatins into a coarse-grained unit by clustering data **F** into groups of regions and represented those individual regions by spheres. The radius R_i of the ith sphere was set to be proportional to $l_i^{1/3}$, where l_i is the length of the sequence of the ith clustered region in the genome. In the Kalhor et al. model, l_i was distributed around 10 Mb and the whole genome was described by $2N$ spheres with $N = 428$ being the number of spheres in the haploid genome.

Kalhor et al. simulated an ensemble of $N_{\text{cell}} = 10{,}000$ nuclei having the same genome by modifying the position of the center of the spheres $\{\mathbf{x}_i\}$ to optimize the cost function $S = \sum_{m=1}^{N_{\text{cell}}} S_m$, where S_m is defined by

$$
S_m(\{\mathbf{x}_i\}) = \sum_{i=1}^{2N} u^{\text{nuc}}(\mathbf{x}_i) + \sum_{i=1}^{2N-1} \sum_{j>i}^{2N} u^{\text{exc}}(\mathbf{x}_i,\ \mathbf{x}_j)
$$

$$
+ \sum_{i=1}^{2N-1} \sum_{j>i}^{2N} \alpha_{ij} w(\mathbf{x}_i, \mathbf{x}_j) u^{\text{con}}(\mathbf{x}_i, \mathbf{x}_j) \tag{9.24}
$$

Here, the spheres $i = 1 - N$ constitute a haploid copy of chromosomes in a nucleus, and the spheres $i = N + 1 - 2N$ constitute another haploid copy of the same chromosomes. Note that the notations in Equation 9.24 are slightly different from those used in the original paper of Kalhor et al. (2012).

The diploid problem was considered in the Kalhor et al. model with α_{ij} in Equation 9.24. For *cis* interactions with $j = i + 1$, α_{ij} was set to be $\alpha_{ij} = 1$, but for *cis* interactions with $j \geq i + 2$ or for *trans* interactions, distances $d_{i,j} = |\mathbf{x}_i - \mathbf{x}_j|$, $d_{i,j+N} = |\mathbf{x}_i - \mathbf{x}_{j+N}|$, $d_{i+N,j} = |\mathbf{x}_{i+N} - \mathbf{x}_j|$, and $d_{i+N,j+N} = |\mathbf{x}_{i+N} - \mathbf{x}_{j+N}|$ were compared, and the pair giving the minimum of these four distances was selected. Then, a and b were assigned as the indices designating the chosen pair. α_{ij} was defined as

$$
\alpha_{ij} = \begin{cases} 1, & \text{for } i = a \text{ and } j = b \\ 0, & \text{otherwise} \end{cases} \tag{9.25}
$$

The 3C-based data **F** was reflected by the factor $w(\mathbf{x}_i, \mathbf{x}_j)$ in Equation 9.24. By sorting **F** into clusters, the coarse-grained matrix $\mathbf{a} = \{a_{ij}\}$ representing the contact probability between two spheres was obtained. The threshold distance d_{ij}^{act} was defined as an increasing function of a_{ij}, whose precise functional form was obtained by Kalhor et al. by fitting the simulation results to the observed **F**. $w(\mathbf{x}_i, \mathbf{x}_j)$ was set to be

$$
w(\mathbf{x}_i, \mathbf{x}_j) = \begin{cases} 1, & \text{for } d_{i,j} < d_{ij}^{\text{act}} \\ 0, & \text{otherwise} \end{cases} \tag{9.26}
$$

Because d_{ij}^{act} is an increasing function of a_{ij}, $w(\mathbf{x}_i, \mathbf{x}_j)$ in Equation 9.26 represents the tendency that spheres with a larger contact frequency in the observed 3C-based data have a higher probability to come close to each other in the simulation. By using

$$u^{con}(\mathbf{x}_i, \mathbf{x}_j) = \begin{cases} \frac{1}{2}k_{con}\left(d_{i,j} - d_{ij}^0\right)^2, & \text{for } d_{i,j} > d_{ij}^0 \\ 0, & \text{otherwise} \end{cases} \tag{9.27}$$

the third term in Equation 9.24 decreases when the selected pair of spheres come close to each other. Here, $d_{ij}^0 = 1.1 \times (R_i + R_j)$.

u^{nuc} in Equation 9.24 represents the confinement of the spheres within the nucleus:

$$u^{nuc}(\mathbf{x}_i) = \begin{cases} \frac{1}{2}k_{nuc}\left(d_{i0} - (R_{nuc} - R_i)\right)^2, & \text{for } d_{i0} > R_{nuc} - R_i \\ 0, & \text{otherwise} \end{cases} \tag{9.28}$$

where k_{nuc} is a constant, R_{nuc} is the radius of the nucleus, and d_{i0} is the distance between the center of the sphere i and the nuclear center. u^{exc} represents the exclusion of the spheres from each other:

$$u^{exc}(\mathbf{x}_i, \mathbf{x}_j) = \begin{cases} \frac{1}{2}k_{exc}\left(d_{i,j} - (R_i + R_j)\right)^2, & \text{for } d_{i,j} < R_i + R_j \\ 0, & \text{otherwise} \end{cases} \tag{9.29}$$

where k_{exc} is a constant.

In this way, Kalhor et al. selected one of the four contributions, $F(\mu\alpha i; v\alpha j)$, $F(\mu\alpha i; v\beta j)$, $F(\mu\beta i; v\alpha j)$, and $F(\mu\beta i; v\beta j)$, in Equation 9.1 to be identical to $F(\mu i; v j)$ and set the other three contributions to be 0 for each simulated nucleus. This choice strongly depends on the initial configuration of the simulation. Kalhor et al. used random N_{cell} configurations of $\{\mathbf{x}_i\}$ as the initial configurations and examined the distribution of the resulting $\{\mathbf{x}_i\}$ after minimizing the cost S. This distribution was expected to represent the distribution of the structural fluctuation among the ensemble of the observed nuclei.

Because the typical size of TAD distributes around several hundred kb, a simulation with a finer resolution needs to be developed to examine the relationship between transcriptional regulation and genome structural fluctuation. Starting from the present coarse-grained representation at a 10-Mb scale, one may be able to construct models with a finer resolution by using the same selection result among the four contributions of maternal and paternal copies of chromosomes as determined in the present coarse-grained level simulation for individual nuclei. However, in order to simulate not only the static distribution but also the dynamical fluctuation of the genome structure, further development of a suitable model potential to describe the chromatin movement is needed. A more careful treatment of the partition into four contributions is important for constructing reasonable model potentials to simulate the dynamical fluctuation of the genome structure. Comparisons of the simulated distribution of the genome structure with the single-cell data from Hi-C measurments or fluorescent probes measurments by super-resolution microscopy should be helpful to establish such reasonable model potentials for dynamical simulations.

9.5 SUMMARY

In this chapter, we explained a method for coarse-grained modeling of the budding yeast genome. Using the observed Hi-C data, one can construct a model capturing the features of fluctuating structures and dynamical movement of the genome; furthermore, the effects of a mutation on the genome structure can be analyzed. It is important and intriguing to use this model to analyze the correlation between the gene regulation dynamics and the genome structural dynamics to examine the validity of the dynamical genome hypothesis. Applications of the model to DNA repair and DNA duplication are also important research avenues. In this way, the application of the model to quantitative analyses of biological processes of haploid genomes is within our view.

In contrast, modeling of diploid genomes is still at its infancy level, and a breakthrough in treating the diploid problem is necessary to develop a method to examine the dynamical genome hypothesis in higher organisms. However, it should be emphasized that a rich field of computational cell biology has just initiated; multiscale techniques of simulation have to be developed to investigate the problems of the phenomena with hierarchically distributed scales, ranging from the local dynamical fluctuations of chromatins, formation, resolution, and interactions among TADs, and dynamics of chromosome territories to the whole-nuclear-scale movement of chromosomes. Combined with experimental techniques such as super-resolution microscopy and 3C-based measurements and informatic methods based on the genome-wide databases on epigenomic patterns and expression levels, computer simulations based on coarse-grained modeling could play a central role in this new field of quantitative biology.

ACKNOWLEDGMENTS

The authors thank Shin Fujishiro for his fruitful discussions. The computations explained in this chapter were performed by the super-computer system of the Research Organization of Information and Systems at the National Institute of Genetics. The research presented in this chapter was supported by Japan Society for the Promotion of Science (JSPS) Grants-in-Aid for scientific research 23654147, 24244068, and 26610130 to M.S. and by The Hori Sciences and Arts Foundation to N.T.

REFERENCES

Baú, D., Sanyal, A., Lajoie, B. R et al., 2011. The three-dimensional folding of the α-globin gene domain reveals formation of chromatin globules. *Nat Struct Mol Biol* 18: 107–114.

Berger, A. B., Cabal, G. G., Fabre, E et al., 2008. High-resolution statistical mapping reveals gene territories in live yeast. *Nat Methods* 5: 1031–1037.

Bryngelson, J. D., and Wolynes, P. G., 1987. Spin glasses and the statistical mechanics of protein folding. *Proc Natl Acad Sci U S A* 84: 7524–7528.

Bystricky, K., Heun, P., Gehlen, L., Langowski, J., and Gasser, S. M., 2004. Long-range compaction and flexibility of interphase chromatin in budding yeast analyzed by high-resolution imaging techniques. *Proc Natl Acad Sci U S A* 101: 16495–16500.

Bystricky, K., Laroche, T., van Houwe, G., Blaszczyk, M., and Gasser, S. M., 2005. Chromosome looping in yeast: Telomere pairing and coordinated movement reflect anchoring efficiency and territorial organization. *J Cell Biol* 168: 375–387.

Cournac, A., Marie-Nelly, H., Marbouty, M., Koszul, R., and Mozziconacci, J., 2012. Normalization of a chromosomal contact map. *BMC Genomics* 13: 436.

Cremer, T., and Cremer, M., 2010 Chromosome territories. *Cold Spring Harb Perspect Biol* 2: a003889.

Dixon, J. R., Selvaraj, S., Yue, F et al., 2012. Topological domains in mammalian genomes identified by analysis of chromatin interactions. *Nature* 485: 376–380.

Duan, Z., Andronescu, M., Schutz, K et al., 2010, A three-dimensional model of the yeast genome. *Nature* 465: 363–367.

Eltsov, M., MacLellan, K. M., Maeshima, K., Frangakis, A. S., and Dubochet, J., 2008. Analysis of cryo-electron microscopy images does not support the existence of 30-nm chromatin fibers in mitotic chromosomes in situ. *Proc Natl Acad Sci U S A* 105: 19732–19737.

Ferreiro, D. U., Komives, E. A., and Wolynes, P. G., 2014. Frustration in biomolecules. http://arxiv.org/abs/1312.0867.

Giorgetti, L., Galupa, R., Nora, E. P et al., 2014. Predictive polymer modeling reveals coupled fluctuations in chromosome conformation and transcription. *Cell* 157: 950–963.

Gō, N., 1983. Theoretical studies of protein folding. *Ann Rev Biophys Bioeng* 12: 183–210.

Hagerman, P. J., 1988. Flexibility of DNA. *Ann Rev Biophys Biophys Chem* 17: 265–286.

Hediger, F., Neumann, F. R., Van Houwe, G., Dubrana, K., and Gasser, S. M., 2002. Live imaging of telomeres: yKu and Sir proteins define redundant telomere-anchoring pathways in yeast. *Curr Biol* 12: 2076–2089.

Heun, P., Laroche, T., Shimada, K., Furrer, P., and Gasser, S. M., 2001. Chromosome dynamics in the yeast interphase nucleus. *Science* 294: 2181–2186.

Hu, M., Deng, K., Selvaraj, S., Qin, Z., Ren, B., and Liu, J. S., 2012. HiCNorm: Removing biases in Hi-C data via Poisson regression. *Bioinformatics* 28: 3131–3133.

Kalhor, R., Tjong, H., Jayathilaka, N., Alber, F., and Chen, L., 2012. Genome architectures revealed by tethered chromosome conformation capture and population-based modeling. *Nat Biotechnol* 30: 90–98.

Kremer, K., and Grest, G. S., 1990. Dynamics of entangled linear polymer melts: A moleculardynamics simulation. *J Chem Phys* 92: 5057–5086.

Lieberman-Aiden, E., van Berkum, N. L., Williams L et al., 2009. Comprehensive mapping of long-range interactions reveals folding principles of the human genome. *Science* 326: 289–293.

Maeshima, K., Hihara, S., and Eltsov, M., 2010. Chromatin structure: Does the 30-nm fiber exist in vivo? *Curr Opin Cell Biol* 22: 291–297.

Maeshima, K., Imai, R., Tamura, S., and Nozaki, T., 2014. Chromatin as dynamic 10-nm fibers. *Chromosoma* 123: 225–237.

Nagano, T., Lubling, Y., Stevens, T. J et al., 2013. Single-cell Hi-C reveals cell-to-cell variability in chromosome structure. *Nature* 502: 59–64.

Neumann, F. R., Dion, V., Gehlen, L. R et al., 2012. Targeted INO80 enhances sub-nuclear chromatin movement and ectopic homologous recombination. *Genes Dev* 26: 369–383.

Nishino, Y., Eltsov, M., Joti, Y et al., 2012. Human mitotic chromosomes consist predominantly of irregularly folded nucleosome fibers without a 30-nm chromatin structure. *EMBO J* 31: 1644–1653.

Onuchic, J. N., and Wolynes, P. G., 2004. Theory of protein folding. *Curr Opin Struct Biol* 14: 70–75.

Rosa, A., Becker, N. B., and Everaers, R., 2010. Looping probabilities in model interphase chromosomes. *Biophys J* 98: 2410–2419.

Rosa, A., and Everaers, R., 2008. Structure and dynamics of interphase chromosomes. *PLoS Comput Biol* 4: e1000153.

Sexton, T., Yaffe, E., Kenigsberg, E et al., 2012. Three-dimensional folding and functional organization principles of the *Drosophila* genome. *Cell* 148: 458–472.

Taddei, A., Hediger, F., Neumann, F. R., Bauer, C., and Gasser, S. M., 2004. Separation of silencing from perinuclear anchoring functions in yeast Ku80, Sir4 and Esc1 proteins. *EMBO J* 23: 1301–1312.

Taddei, A., van Houwe, G., Nagai, S., Erb, I., van Nimwegen, E., and Gasser, S. M., 2009. The functional importance of telomere clustering: Global changes in gene expression result from SIR factor dispersion. *Genome Res* 19: 611–625.

Tanizawa, H., Iwasaki, O., Tanaka, A et al., 2010. Mapping of long-range associations throughout the fission yeast genome reveals global genome organization linked to transcriptional regulation. *Nucleic Acids Res* 38: 8164–8177.

Tjong, H., Gong, K., Chen, L., and Alber, F., 2012. Physical tethering and volume exclusion determine higher-order genome organization in budding yeast. *Genome Res* 22: 1295–1305.

Tokuda, N., and Sasai, M., 2017. Heterogeneous spatial distribution of transcriptional activity in budding yeast nuclei. *Biohys J* 112: 491–504.

Tokuda, N., Terada, T. P., and Sasai, M., 2012. Dynamical modeling of three-dimensional genome organization in interphase budding yeast. *Biophys J* 102: 296–304.

Wiggins, P. A., Phillips, R., and Nelson, P. C., 2005. Exact theory of kinkable elastic polymers. *Phys Rev E* 71: 021909.

Wong, H., Marie-Nelly, H., Herbert, S et al., 2012. A predictive computational model of the dynamic 3D interphase yeast nucleus. *Curr Biol* 22: 1881–1890.

Yaffe, E., and Tanay, A., 2011. Probabilistic modeling of Hi-C contact maps eliminates systematic biases to characterize global chromosomal architecture. *Nat Genet* 43: 1059–1065.

Zhang, B., and Wolynes, P. G., 2015. Topology, structures, and energy landscapes of human chromosomes. *Proc Natl Acad Sci USA* 112: 6062–6067.

Zidovska, A., Weitz, D. A., and Mitchison J. T., 2013. Micron-scale coherence in interphase chromatin dynamics. *Proc Natl Acad Sci U S A* 110: 15555–15560.

CHAPTER 10

Mechanics of Viruses

Olga Kononova, Artem Zhmurov, Kenneth A. Marx, and Valeri Barsegov

CONTENTS

10.1 Introduction	368
10.2 Viruses and Virus-Like Particles	373
10.2.1 Cowpea Chlorotic Mottle Virus	374
10.2.2 Encapsulin	375
10.3 Computational Methods Used to Describe Virus Mechanics	377
10.3.1 Finite Element Analysis, Normal Mode Analysis, and Atomistic MD Simulations	377
10.3.2 Simplified (Coarse-Grained) Models	378
10.4 Native Topology-Based Coarse-Grained Modeling of Virus Mechanics	379
10.4.1 Nanoindentation *In Silico*: Rationale for Coarse-Grained Modeling of Virus Shells	379
10.4.2 SOP Model of Virus Particle	381
10.4.3 Example: SOP Model Parameterization for CCMV Shell	382
10.4.4 SOP-GPU Software	383
10.5 Nanomanipulation *In Vitro* and *In Silico*	384
10.5.1 Single-Particle Dynamic Force Spectroscopy	384
10.5.2 Nanoindentation *In Silico* Method	384
10.6 Biomechanics of CCMV and Encapsulin Shells	387
10.6.1 Equilibrium Properties of Virus Shells Are a Sum of Properties of Its Structure Elements	387
10.6.2 Nanomanipulations *In Silico* Reveal Global Collapse Transitions in Virus Shells	387
10.6.3 Force-Deformation Assays *In Silico* Allows One to Probe Elastic Properties of Virus Shells	392
10.6.4 Structure Collapse Transition in Virus Particles Is a Shape-Changing Process	393

10.6.5 Nanoindentations *In Silico* Allow One to Probe
Thermodynamics of Mechanical Transitions in Viruses 394

10.6.6 Nanoindentations *In Silico* Illuminate Dynamic
Interplay between the Out-of-Plane and In-Plane
Displacements 396

10.6.7 (Ir)reversibility of Deformation and (In)Elastic
Properties of Virus Shells Is Directly Correlated 398

10.7 Models of Mechanical Deformation of Virus Particles 400

10.7.1 Identifying Degrees of Freedom for Virus Particles'
Deformation 400

10.7.2 Local Curvature Change of the Indenter-Particle
Contact Area 404

10.7.3 Global Mechanical Deformation of Virus Shell Structure 406

10.8 Concluding Remarks 408

References 410

10.1 INTRODUCTION

Ubiquitous in nature, viruses are biological infectious nanostructures that, in most instances, consist of just a nucleic acid packaged within a protein shell. Depending upon the specific virus, the nucleic acid consists of DNA or RNA in either single- or double-stranded form. The protein shell enclosing the nucleic acid may simply consist of a small number of proteins called capsid proteins, or in the case of certain classes of viruses, the host membrane may also form part of the outer structure. Viruses themselves contain no metabolic machinery. Instead, they rely upon infection of their host species and subsequent subversion of the host metabolic pathways by genes encoded in their small nucleic acid genome to carry out all viral functions, including viral replication, in order to propagate.

Viruses appear to be ubiquitous, existing as specific infectious agents for all species of bacteria, archaebacteria, animals, and plants. Perhaps surprisingly, for all the variety of viruses found in nature, there appear to be just two types of general three-dimensional (3D) structure classes. The first accomplishes nucleic acid enclosure via their interaction with capsid proteins organized in a helical array. A classic example of this is the first virus studied ultrastructurally, the tobacco mosaic virus (TMV), which also represents the first biological structure studied with the newly developed technique of electron microscopy in 1939 [1]. The second general type of virus structure is based upon a small number of identical capsid proteins self-assembling into icosahedral or quasi-spherical geometry nanostructures, which then package the nucleic acid(s). As well-studied examples, the classic icosahedral bacteriophages λ, T4, and T7 were some of the first bacterial viruses investigated, since they represented convenient model systems for the emerging field of molecular biology. Given their regular 3D structures, viruses possess an advantage to investigators over other large nucleoprotein assemblies; they self-assemble with ease and can be crystallized. Therefore, a number of filled and empty viruses have been crystallized and studied with the high-resolution techniques of x-ray crystallography and cryoelectron microscopy [2].

Viruses can be classified by their genome type (single-stranded or double-stranded RNA or DNA) and the pathway they employ for mRNA synthesis. For both animals and plants as hosts, there exists a full spectrum of viruses that cover these different class types. However, one significant difference between animal and plant viruses involves their mechanism of infectivity. Animal cell viruses typically infect a specific host species via a route involving molecular recognition of the host. This is carried out by some specific recognition event involving selective interaction of the host cell surface with a viral capsid protein, glycoprotein, or lipid component. By contrast, plant viruses appear to lack this host infectivity mechanism based upon molecular recognition (Figure 10.1). Lacking a specific molecular recognition-based mechanism for host cell entry is undoubtedly a result of the fact that plant cells, unlike animal cells, are enclosed within a rigid cell wall and cuticle, features that represent significant physical barriers to breach. It is thought that plant cell viruses rely upon physical injury to the plant and/or damage by invertebrates, like insects, to facilitate infection. These differences in viral infectivity are very likely correlated with differences in the capsid structure, dynamics, and energetics-based mechanical properties. Therefore, an understanding of viral infectivity becomes a motivation for investigating different capsids' energetics-based mechanical and dynamic properties.

In some ways, viral shells closely resemble, and may be the evolutionary origin, of a number of native cellular structures that serve a range of biological functions, including increased efficiency of biochemical catalysis. Examples of these systems include the carboxysome, a polyhedral 80–140 nm bacterial compartment comprised of thousands of self-assembling protein subunits; it is involved in carbon fixation which occurs in most cyanobacteria and chemotrophic bacteria [3]. The carboxysome packages a large number of the relatively inefficient enzyme, ribulose

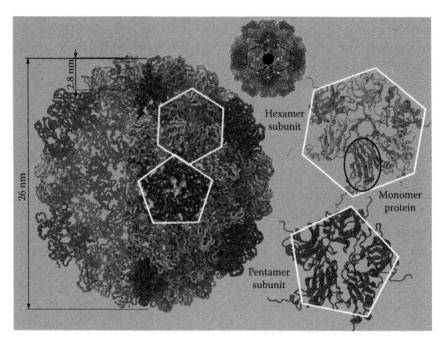

FIGURE 10.1   The crystal structure of the Cowpea chlorotic mottle virus (CCMV) from VIPERdb (PDB entry: 1CWP). Shown on the left is the side view of the CCMV shell. The protein domains forming pentamer capsomers are shown in blue, while the same protein domains in hexamer capsomers are shown in darker and lighter colors. The hexamer and pentamer capsomers composed of six and five copies of the same monomer protein domain (circled in the black ellipse), respectively, are magnified on the right. The CCMV capsid is a thick shell with an outer diameter of ~26 nm and a shell thickness of ~2.8 nm. The top structure displays the top view of the CCMV particle with the two-fold symmetry axis at the center of the common edge of adjacent hexamer subunits.

bis-phosphate carboxylase, along with carbonic anhydrase. It functions to concentrate CO_2 in its interior in close proximity to the enzymes to improve the overall enzymatic carbon fixation rate. Another type of bacterial protein nanocompartment is the family of native virus-like bioreactor structures called encapsulins (Figure 10.2). These small organelles contain peroxidases or ferritin-like proteins (Flps); both enzymes serve protective functions in oxidative stress pathways [4]. Higher organisms (eukaryotes) also possess nanocompartment protein reactors that incorporate ferritins (Fe(II) detoxifying enzymes) [5], or multienzyme complexes such as the pyruvate dehydrogenase enzyme complex [6], fungal fatty acid synthase [7], or lumazine synthase [8]. In addition to bringing the substrate into close proximity to the tethered enzymes, other macromolecules

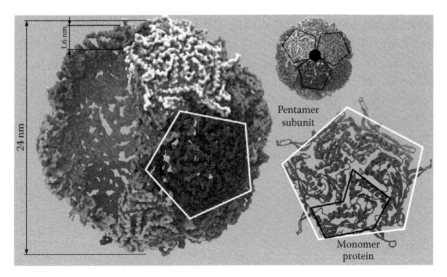

FIGURE 10.2   The structure of the empty *Thermotoga maritima* encapsulin constructed by replicating the structure of the pentamer subunit available from the Protein Data Bank (PDB entry: 3DKT). Shown on the left is the side view of the partially opened encapsulin nanocompartment. The pentamer subunit composed of five copies of the same single protein domain is magnified on the right. The secondary structure of the monomer protein domain (shown in the black polygon) is formed by the long α-helical stretches, extended β-sheets, and the E-loop. The encapsulin nanocompartment is a thin shell with an outer diameter of ~24 nm and with a shell thickness of ~1.6 nm. The top structure displays the top view of the encapsulin particle with the three-fold symmetry axis connecting the adjacent pentamer subunits.

are excluded from the interior volume, thereby improving the reaction rate.

Viruses that infect cells have biomedical and societal importance due to their roles as infectious agents. These viruses, plus the native cellular virus-like nanocompartment bioreactors, have potential applications in a number of areas. Indeed, viral capsids' convenient self-assembly, nanoscale size, environment-dependent swelling properties, and ability to be genetically altered to tailor their properties make them attractive candidates for biotechnological uses. In nanomedicine, they can function as gene delivery or drug delivery vehicles, as well as having biotechnological and industrial importance as nanoscale bioreactors. For example, unique bacterial nanocompartments such as encapsulin can act as efficient nanoreactors by packaging specific enzymes, making bacterial nanoparticles into prospective platforms for applications in nanotechnology [9]. Therefore, the study of the dynamics, energetics, and biomechanical properties of these viruses and native virus-like particles is of considerable importance.

One important way that these virus and virus-like structures have been characterized experimentally is using atomic force microscope (AFM)-based dynamic force spectroscopy measurements of their nanoindentations [10]. An idealized schematic of the AFM-based dynamic force measurement on a biological particle (virus) is presented in Figure 10.3. Although this is a relatively high-resolution technique, it suffers from a lack

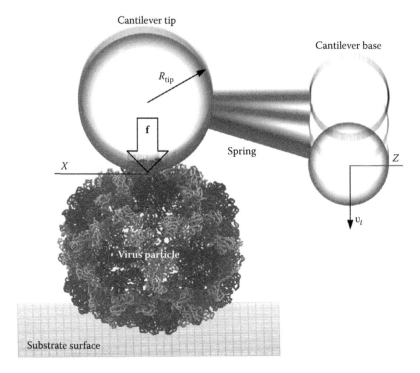

FIGURE 10.3 Highly idealized schematic of the setup used in dynamic force experiments *in vitro* or *in silico* (see Sections 10.5.1 and 10.5.2). The biological particle (virus shell) is positioned on the substrate surface. In the course of a measurement of the forward indentation, the cantilever base (virtual sphere) is moving in the direction perpendicular to the outer surface of the particle with the constant velocity v_f (dynamic force ramp). This creates a compressive force, which is transmitted to the cantilever tip (spherical object of radius R_{tip}) through the harmonic spring with the (cantilever) spring constant κ. The magnitude of compressive force (indicated by a large vertical arrow) ramps up linearly in magnitude with time, $f(t) = r_f t$, with the force-loading rate $r_f = \kappa v_f$, which mechanically loads the particle and produces particle indentations. The mechanical response of the particle can be measured by profiling the force of deformation (indentation force) F as a function of the cantilever base (piezo-) displacement Z (FZ curve) or as a function of the indentation depth X (FX curve).

of control in a number of ways: (1) there is no knowledge of the exact molecular position at which nanoindentation is occurring; (2) it lacks a fine structure molecular view of how the object (virus particle) is being deformed by the AFM tip; (3) as a result, energetics cannot easily be related to what is happening in the object during nanoindentation. The main challenge is that these properties could, in principle, be extracted from the AFM-based dynamic force measurements *in vitro* [10–12], whereby the forced compression of virus particles deforms their structure and ruptures the respective noncovalent protein–protein bonds. However, in practice, due to the multicapsomer nature of the viral capsids' structures (Figures 10.1 and 10.2), the molecular interpretations of the experimental force-indentation spectra at the level of single bonds and on the nanometer length scale is virtually impossible.

In this chapter, we feature our multiscale modeling approach, which we refer to as "nanoindentation *in silico.*" This method allows one to explore the biomechanics of virus particles in a computer. The method combines the all-atom molecular dynamics (MD) simulations of atomic structural models of virus particles available from the Protein Data Bank and Langevin simulations of their native topology-based coarse-grained models. Our coarse-grained description is based on the so-called self-organized polymer (SOP) model [13,14], which has been extensively used by several research groups to characterize the physico-chemical and biomechanical properties of biomolecules [15–22]. We shall show that our unique *in silico* "experiments" provide a complete and detailed simulation view of the entire process of compressive force induced capsid deformation, which can be compared with and used to inform the AFM experimental interpretation. These simulation studies can also be used to understand the process of capsid fatigue and structural collapse, and these processes cannot be easily understood by current experimental methodology. In this chapter, we demonstrate the utility of our approach to nanoindentation *in silico* by exploring the biomechanical properties of two specific examples of virus shell—the Cowpea chlorotic mottle virus (CCMV) capsid and the virus-like particle encapsulin.

10.2 VIRUSES AND VIRUS-LIKE PARTICLES

Viral capsids must meet the conflicting goals of being stable enough to protect their genomic material, yet being sufficiently unstable that they release their genome into the host cells during the process of infection.

Virus-like bacteriophage nanocompartments, too, should possess a broad range of evolved biomechanical characteristics to be able to regulate the influx of substrate and efflux of product and to efficiently facilitate the enzymatic catalysis in their interior. This makes the exploration of the physico-chemical and biomechanical properties of these important biological particles into a highly significant research objective. Neither bacterial nanocompartments nor virus shells contain any metabolic machinery. This suggests that the structure–mechanics–function relationship in these biology-inspired systems can be understood using the framework of equilibrium thermodynamics and nonequilibrium statistical mechanics. Therefore, we first provide some background for the upcoming discussion of these properties for the case of CCMV and encapsulin particles.

10.2.1 Cowpea Chlorotic Mottle Virus

CCMV is a member of the *Bromoviridae*, an important family of single-stranded RNA plant viruses, distributed worldwide, that infect a range of hosts and are the cause of some major crop epidemics [23]. The *Bromoviridae* family of plant viruses is composed of dozens of members, of which CCMV is perhaps the best studied representative (Figure 10.1) [24]. For over 35 years, it has been the subject of research into its biophysical and biochemical properties, ever since it was first reassembled *in vitro* from its constituent proteins and four RNA molecules. Four unique single-stranded RNA molecules comprise the viral genome; they are packaged by 180 identical single subunit capsid proteins. The 19.5 KDa (190 amino-acid sequence) capsid proteins self-assemble into 12 pentameric and 20 hexameric structures that then form an icosahedral virus capsid (triangulation number $T = 3$) which is 28.6 nm in diameter with an average shell thickness of 2.8 nm [24,25]. The shell is comprised of 60 trimer structural units and exhibits pentameric symmetry at the 12 vertices (pentamer capsomeres) and hexameric symmetry at the 20 faces (hexamer capsomeres) of the icosahedron (Figure 10.1).

CCMV has a number of interesting physico-chemical properties that make it an appealing target for study. The viral capsid can fully self-assemble in the absence of RNA, or with its genomic RNAs, to form a fully infectious viral particle. The x-ray structures have been determined for filled and empty CCMV capsids. In the case of filled capsids, RNA-capsid protein interactions exist at the RNA-capsid interface and these are visible in x-ray structures due to the RNA being locally ordered. However, most of the RNA located in the virion interior is not ordered sufficiently to be at all resolved. Unlike the double-stranded DNA bacteriophages, which can actively package their nucleic acid using adenosine triphosphate (ATP) to internal pressures up to 60 atm, the filled CCMV capsid is not packed under pressure. In contrast to filled capsids, empty CCMV capsids are simpler systems to study and ultimately understand completely the dynamics and energetics of the protein capsid. Experimentally, these viral capsids have been shown to undergo interesting cooperative swelling transitions. The native form of CCMV is stable in the environment of pH = 5. At higher pH ($\sim$7.5) and low ionic strength ($\mu = 0.1$), the capsid undergoes transition to a new stable swollen form, which is characterized by radial expansion ($\sim$10%) and the occurring of voids in the three-fold symmetry vertices [24]. This transition is reversible, when pH decreases back to pH = 5 or the concentration of ions (Ca^{2+} or Mg^{2+}) increases. The CCMV capsid is able to self-assemble *in vitro* (i.e., under well-defined conditions of pH, ionic strength, and divalent metal ion concentration), where CCMV capsid protein or capsid protein and RNA reassembles to form $T = 1$, $T = 3$, swollen $T = 3$, and $T = 7$ icosahedral particles, multishelled $T = 3$ and $T = 7$ particles, sheets, tubes, rosettes, and a variety of laminar structures [24].

Swelling transitions appear to be a general feature of plant viruses that may be related to their mechanism of infectivity. Therefore, studying CCMV and other viral capsid structures' dynamics, energetics, and biomechanical properties are of great interest. Unlike the case with animal viruses, cell surface receptors are not known to be involved in the plant virus infection process [26]. Studies have revealed that CCMV and more broadly the *Bromoviridae* virus family infect plant tissue only through damaged cell walls that can be produced by either mechanical or biological means [27]. Thus, in the case of physical damage to cells by tough plant outer structures, mechanical forces may be exerted directly upon virus particles by plant fibers, thereby enhancing their infectivity through a deformation process, which may function as an adjunct to infectivity by the common mechanism whereby environmental changes result in virus particle structural alterations, such as swelling.

10.2.2 Encapsulin

As mentioned in Section 10.1, viral capsids closely resemble, and may be the evolutionary origin of, a number of small native cellular organelles that serve a range of biological functions. Recently discovered bacterial nanocompartments, called encapsulins, belong to a family of virus-like bioreactor structures (Figure 10.2) [4,28]. Much like bacterial microcompartments, these are polyhedral protein cages that encapsulate enzymes to act as miniorganelles in prokaryotes [29–32]. Similar to virus particles, encapsulins are well known for being notoriously resilient structures, as they can withstand internal pressures of tens of atmospheres and possess the rigidity of hard plastics [33,34], which makes it interesting to investigate their materials properties. Bacterial nanocompartments also have icosahedral symmetry, but their resemblance to virus capsids is not limited to this icosahedral structure. The subunits of bacterial nanocompartments also adopt the HK97-fold, typical of the protein subunits of many bacteriophage capsids [4,28]. For example, the protein monomer of *Thermotoga maritima* encapsulin has structure homologous to capsid protein gp5 of bacteriophage HK97, which emphasizes the great similarity between viruses and bacterial nanocompartments.

The bacterial nanocompartment encapsulin from *T. maritima* is a protein shell, ~23–24 nm in diameter and ~1.6–2.0 nm in thickness, which consists of 60 copies of a single protein subunit, arranged in 12 pentamers with $T = 1$ icosahedral symmetry (Figure 10.2). Encapsulin naturally occurs in a variety of prokaryotes. In *Brevibacterium linens*, the native encapsulin is loaded with a dye-decolorizing peroxidase (DyP), whereas the *T. maritima* encapsulin accommodates a Flp. Both enzymes serve protective functions for the host as part of oxidative stress response pathways. The cargo enzyme of the bacterial nanocompartment can be either part of the same gene product, as in the *Pyrococcus furiosus* virus-like particle (*PfV*), or attached to the shell interior via a C-terminal anchoring sequence of the enzyme molecule, as in the encapsulin nanocompartment [4,28]. This naturally encoded mechanism for cargo loading makes encapsulin a particularly exciting scaffold as a nanocontainer. For both empty and DyP-loaded encapsulin shells, the dimer is a stable substructure and is likely to be the unit of assembly for encapsulin shells.

The monomers have particularly stable interactions across the two-fold symmetry axis of the shell (Figure 10.2), due to the extended E-loop of the protein subunits [4]. Based on the crystal structure of *T. maritima* encapsulin [4], the E-loop in one subunit makes direct contact to the neighboring subunit across the two-fold symmetry axis. This twisting motion pulls the adjacent subunits closer together. Cargo is proposed to bind to the same β-sheet of the encapsulin monomer as the interchain contacts of the E-loop [4], which explains the observed allosteric effect associated with cargo binding/unbinding. In the original report on the encapsulin structure, several pores were found in the shell wall that might be important for substrate flux in and out of the compartment [4]. This suggests a built-in mechanism, by which the cargo enzyme encapsulation might modulate the permeability of the shell for its substrates.

10.3 COMPUTATIONAL METHODS USED TO DESCRIBE VIRUS MECHANICS

10.3.1 Finite Element Analysis, Normal Mode Analysis, and Atomistic MD Simulations

There exist a number of theoretical methods, which have been used to describe the dynamics of virus capsids. Historically, the finite element analysis was the first method used to model the mechanical properties of biological assemblies [35,36]. This method was utilized to simulate nanoindentation of the capsid of native and swollen CCMV [35] and herpes simplex virus type 1 (HSV1) [36]. In this method, the protein shell is represented by a triangulated meshwork, where each node is described by the potential energy function. The nonlinear elasticity problem is approximated by interpolating the deformation among the deformed positions of the vertices in a polyhedral mesh, and solving for the nodal positions which best minimize the potential energy. This method is capable of describing the dynamic behavior of the capsid and resolving some nonlinear features of the capsid mechanical response. However, the finite element analysis has the following drawbacks: (1) the method does not take into account the discrete nature of the particle's structure, and (2) the particle's dynamics do not reflect the stochastic nature of mechanical deformation.

Another approach is based on the direct MD simulations of atomic structural models of biological particles. This approach has been used to perform the all-atom indentation study of southern bean mosaic virus (SBMV) [37,38]. However, an enormously large number of degrees of freedom ($\sim$1,000,000 atoms) limits the timescale of simulations to a few tens of nanoseconds. Consequently, when running MD simulations, one is forced to use rather unphysical force-loading rates that are 10^6–10^8-fold faster than the ones used in AFM experiments. Because of this, there emerges the fundamental question as to whether the deformation dynamics correspond to the diffusion-controlled limit. In addition, using fast loading rates makes it rather difficult to directly compare the results of experiment with simulations, which limits the predictive power of this method.

A variant of the normal mode analysis of equilibrium fluctuations of atomic positions has been used to study the mechanism of the swelling transition in CCMV [39] and equilibrium properties of viruses [40]. A modified version of this method was later developed, in which the particle's structure was represented by a network of interconnected Hookean springs (with C_α-atoms in nodes). The Hessian matrix is diagonalized, and the obtained matrix of eigenvectors (normal modes) defines low-frequency global motions of the system in the harmonic approximation. Although the amplitudes of equilibrium normal modes can predict the direction of global displacements, the normal mode analysis cannot describe "the unstable modes," which dominate the far-from-equilibrium dynamics of the compressive force induced buckling and collapse transitions. A similar method in spirit, based on elastic network models, was employed to study the low-frequency displacements of the HK97 capsid [41]. The method was extended (anisotropic network model) and was coupled to the elastic wave function equation to follow the dynamics of the shell along the main modes [42]. However, as we will show in the sequel, when tested mechanically, the virus shell does not respond to stress elastically, not to mention its behavior in the transition regime, in which the particle's structure fails mechanically.

These limitations of the equilibrium methods have prompted the development of alternative new approaches. An interesting method was proposed, which takes advantage of the roughly spherical symmetry of viruses and virus-like particles. The approach is based on the spherical harmonics expansion and it describes the virus particle deformation by a superposition of the first few modes of "out-of-plane" displacements [43]. Although this method provides information about the mechanical properties of the virus shell, it considers only the contribution to the deformation dynamics from the "out-of-plane" modes of motion while ignoring the orthogonal "in-plane" modes. However, as we will show, the in-plane modes play an important role in the transition regime, in which the structure transitioning from the native (undeformed) state to the collapsed (deformed) state is mediated by the strong coupling between the in-plane modes and out-of-plane modes.

10.3.2 Simplified (Coarse-Grained) Models

There seems to be no way of getting around the limitations described above other than simulating the deformation dynamics of the virus particle directly. To reduce the number of degrees of freedom, several researchers have proposed using a coarse-grained description of a polypeptide chain. In these methods, groups of atoms are replaced by one bead (unified center), and interactions between the beads are described by potential energy functions. The stochastic dynamics of the system are then obtained by propagating forward in time the Langevin equations of motion for each bead. A method based on Voronoi cells has been used to create the coarse-grained representation of the virus capsid [44,45]. Although this is a significant step forward, this method combines several amino acids into one bead, which lowers the structural resolution. To overcome this problem, a different C_α-atom-based coarse-graining approach was used, in which each amino acid was represented by its C_α-atom. This method was used to simulate nanoindentations of the CCMV capsid and Cowpea mosaic virus (CMV) capsid [46] by compressing the shells between the two plates. However, because in these simulations the force-loading rates are three orders of magnitude faster than their experimental counterparts, the results of experiment *in vitro* and *in silico* cannot be compared, which limits the interpretive capacity of the method.

10.4 NATIVE TOPOLOGY-BASED COARSE-GRAINED MODELING OF VIRUS MECHANICS

10.4.1 Nanoindentation *In Silico*: Rationale for Coarse-Grained Modeling of Virus Shells

Limitations of the previous methods exclude the study of a range of problems for which the dynamic force spectroscopic data are already available. We utilize the SOP model of a polypeptide chain to describe the mechanical properties of virus and virus-like particles (Figure 10.4). The SOP model is also based on the representation in terms of C_α-atoms, and it has proved to be very accurate at describing the physico-chemical properties of biomolecules [13–16,18,20,47] and biological assemblies [17,19, 21,48]. The simplified SOP model that we are using is based upon the notion that the unique features associated with native topology and local symmetry of capsomer arrangements, rather than their atomic details, govern the dynamic behavior of viruses and virus-like particles under mechanical stress (Figure 10.3). Furthermore, the additional ~200-fold+ computational speedup attained on a graphics processing unit (GPU) [47,48] brings us closer to the experimental conditions of force applications used in dynamic force assays, which makes the SOP model-based

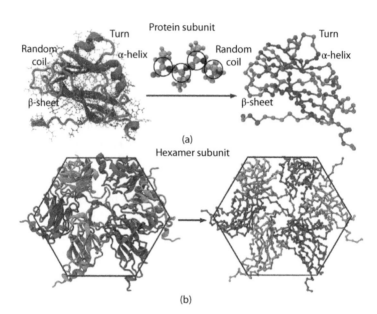

FIGURE 10.4 Schematic of the coarse-graining procedure involved in the construction of a self-organized polymer (SOP) model [13,14] of a polypeptide chain (see Section 10.4.2). Panel (a) exemplifies the coarse-graining procedure for the atomic structure of the protein subunit forming the pentamer and hexamer capsomers of the CCMV shell (see Figure 10.1). Each amino-acid residue is represented by a single interaction center (spherical bead) with the coordinates of the C_α-atom (represented by the black circles, four representative circles are indicated above the coarse-graining process arrow). Consequently, the protein backbone is replaced by a collection of the C_α–C_α covalent bonds with a bond distance of 3.8 Å. The potential energy function (molecular force field) given by Equation 10.1 describes the binary interactions between amino acids stabilizing the native folded state of the protein, chain collectivity, chain elongation due to stretching, and chain self-avoidance (see Equations 10.2 through 10.4). Because the SOP model is native-topology based, the coarse-graining procedure preserves the secondary structure of a protein: α-helices, β-strands and sheets, and random coil and turns. Panel (b) depicts the results of coarse graining of an isolated single hexamer capsomer. Six identical copies of the same protein monomer structure, coarse grained as described in panel (a), form a C_α-based model of the hexamer subunit. These and the pentamer subunits are then combined together to form a coarse-grained reconstruction of the full CCMV capsid. The SOP model preserves the tertiary structure of the virion particle and describes the geometry and 3D shape of the virion.

description of virus mechanics accelerated on a GPU (SOP-GPU software; see Section 10.4.4) into a unique computational modeling tool. In this "nanoindentation *in silico*" approach, the all-atom MD force fields are utilized in conjunction with coarse-grained models to resolve the atomic-level details of the intra- and inter-subunit protein–protein interactions. This information is then used to parameterize the coarse-grained SOP model of the virus particle. Importantly, in the framework of our multiscale modeling approach to nanoindentation *in silico*, there is no need for using either additional kinetic or thermodynamic information to carry out the multiscale modeling of the virus biomechanics.

10.4.2 SOP Model of Virus Particle

The SOP model was originally designed to address the protein forced unfolding problem. There is an impressive track record of success stories made possible by the application of this model to a variety of biological systems [13–22,47,48]. The SOP model of the polypeptide chain [13,14,47,48] can be used to describe each protein subunit forming a virus capsid (see Section 10.2). In the topology-based SOP model, each amino-acid residue is represented by a single interaction center described by the corresponding C_α-atom, and the protein backbone is represented by a collection of the C_α–C_α covalent bonds with the bond distance of $a = 3.8$ Å (peptide bond length; see Figure 10.4).

The potential energy function of the protein conformation U_{SOP} specified in terms of the coordinates of the C_α-atoms $\{r_i\} = r_1, r_2, \dots r_N$, where N is the total number of amino-acid residues, is given by

$$U_{SOP} = U_{FENE} + U_{NB}^{ATT} + U_{NB}^{REP} \tag{10.1}$$

In Equation 10.1, the first term is the finite extensible nonlinear elastic (FENE) potential:

$$U_{FENE} = -\sum_{i=1}^{N-1} \frac{k}{2} R_0^2 \log \left(1 - \frac{\left(r_{i,i+1} - r_{i,i+1}^0 \right)^2}{R_0^2} \right) \tag{10.2}$$

where $k = 14$ N/m is the spring constant, and the tolerance in the change of the covalent bond distance is $R_0 = 2$ Å. The FENE potential describes the backbone chain connectivity. The distance between the next-neighbor residues i and $i+1$ is $r_{i,i+1}$, and $r_{i,i+1}^0$ is its value in the native structure.

To account for the noncovalent (nonbonded) interactions that stabilize the native folded state, we use the Lennard–Jones potential:

$$U_{NB}^{ATT} = \sum_{i=1}^{N-3} \sum_{j=i+3}^{N} \varepsilon_h \left[\left(r_{ij}^0 / r_{ij} \right)^{12} - 2 \left(r_{ij}^0 / r_{ij} \right)^6 \right] \Delta_{ij} \tag{10.3}$$

In Equation 10.3, we assume that if the noncovalently linked residues i and j ($|i - j| > 2$) are within the cutoff distance of 8 Å in the native state, then $\Delta_{ij} = 1; \Delta_{ij} = 0$ otherwise. The value of ε_h quantifies the strength of the nonbonded interactions.

The nonnative (nonbonded) interactions are treated as repulsive:

$$U_{NB}^{REP} = \sum_{i=1}^{N-2} \varepsilon_l \left(\sigma_l / r_{i,i+1}\right)^6 + \sum_{i=1}^{N-3} \sum_{j=i+3}^{N} \varepsilon_l \left(r_{ij}^0 / r_{ij}\right)^6 \left(1 - \Delta_{ij}\right) \qquad (10.4)$$

In Equation 10.4, an additional constraint is imposed on the bond angle between the triplet of residues i, $i + 1$, and $i + 2$ by including the repulsive potential with parameters $\varepsilon_l = 1$ kcal/mol and $\sigma_l = 3.8$ Å. These define the strength and the range of the repulsion. In the SOP model, parameter ε_h sets the energy scale. This parameter is estimated based on the results of all-atom MD simulations of the virus particle at equilibrium (see Section 10.4.3).

The dynamics of the system are obtained by propagating forward in time numerically the Langevin equations of motion for each particle position r_i in the over-damped (Brownian) limit:

$$\eta \frac{dr_i}{dt} = -\frac{\partial U_i(r_i)}{\partial r_i} + g_i(t) \qquad (10.5)$$

In Equation 10.5, $U_i(r_i)$ is the total potential energy, which accounts for the biomolecular interactions (U_{SOP}) and interactions of particles with the indenting object—spherical tip (U_{tip}; see Equation 10.6). Also, in Equation 10.5, $g_i(t)$ is the Gaussian distributed zero-average random force, and η is the friction coefficient. To generate the Brownian dynamics, the equations of motion are propagated with the time step $\Delta t = 0.08\tau_H$, where $\tau_H = \zeta \varepsilon_h \tau_L / k_B T$ ($\Delta t = 20$ ps for CCMV and encapsulin). Here, $\tau_L = (ma^2/\varepsilon_h)^{1/2} = 3$ ps, $\zeta = 50.0$ is the dimensionless friction constant for an amino-acid residue in water ($\eta = \zeta m/\tau_L$), $m \approx 3 \times 10^{-22}$ g is the residue mass, and T is the absolute temperature [22,49]. To perform simulations of nanoindentation of a virus particle, we set T to room temperature and use the bulk water viscosity, which corresponds to the friction coefficient $\eta = 7.0 \times 10^5$ pN ps/nm.

10.4.3 Example: SOP Model Parameterization for CCMV Shell

For each virus system, numerical values of ε_h, which describe the strength of native interactions (see Section 10.4.2), can be determined from the all-atom MD simulations. To that end, we perform the all-atom MD simulations of the atomic structural model of the system in question using the crystal structures available. For example, for the CCMV shell at $T = 300$ K, we used the solvent accessible surface area (SASA) model of implicit solvation (CHARMM19 force field) [50]. To obtain accurate parameterization of the SOP model for CCMV, we used the crystal structure of the capsid (Viper Data Base; PDB code: 1CWP [24]) with $T = 3$ symmetry. First, we calculate the number of native contacts based on a standard cutoff distance between the C_α-atoms of 8.0 Å. The native contacts are divided into the interchain contacts and the intrachain contacts. For the CCMV shell (Figure 10.1), there were a total of $N_{intra} \approx 20,554$ intrachain contacts that stabilize the native state of the capsid protein, and a total of $N_{inter} \approx 3,405$ interchain contacts at the interfaces formed by capsid proteins. Next, we calculate the total energy of noncovalent interactions for each contact group. For CCMV, the total energy for the intrachain contacts is $E_{intra} = 25,898$ kcal/mol and the total energy for the interchain contacts is $E_{inter} = 3,746$ kcal/mol. Finally, to obtain the values of parameter ε_h, we divide the two numbers for each contact group. For CCMV, $\varepsilon_{intra} = E_{intra}/N_{intra} = 1.26$ kcal/mol and $\varepsilon_{inter} = E_{inter}/N_{inter} = 1.1$ kcal/mol. The atomic-level details of biomolecular interactions, contained in these parameters, are then exported into the SOP model-based reconstruction of a polypeptide chain of the proteins comprising the full CCMV particle.

10.4.4 SOP-GPU Software

Depending on the numerical algorithm and system size, most of the GPU-based calculations are from four-fold to 200-fold faster than heavily optimized CPU-based implementations. We developed a GPU-based computational methodology for: (1) the generation of pseudo-random numbers based on the Hybrid Taus, Lagged Fibonacci, and Mersenne Twister algorithms, (2) the calculation of molecular forces including the particle-based and interaction pair-based parallelization approaches, and (3) the numerical propagation of the Langevin equations of motion [47,48]. Pseudo-random numbers of high statistical quality are necessary to perform simulations in a stochastic thermostat, force calculations are required to describe the biomolecular interactions, and equations of motion are needed to generate the dynamics of the system in question. To fully optimize the computational resources, we also developed (1) the one-run-per-GPU approach and (2) the multiple-runs-per-GPU approach to generate concurrently several independent trajectories on a single GPU device [47]. The developed numerical algorithms on a GPU, in conjunction with the SOP model-based Langevin dynamics simulations, were incorporated into the SOP-GPU software package for the computational exploration of large-size biological assemblies [47,48]. The SOP-GPU package enables one to perform dynamic force measurements *in silico* (see Figure 10.3) in the experimental centisecond timescale [19–22,47,48].

10.5 NANOMANIPULATION *IN VITRO* AND *IN SILICO*

10.5.1 Single-Particle Dynamic Force Spectroscopy

Experimental single-molecule techniques, like AFM, have become available to explore the physico-chemical properties of biological particles [51,52]. Dynamic force measurements provide a unique methodology to deform select elements of a given structure, which opens a new direction of biophysical experimentation to probe the biomechanics of biological assemblies. AFM-based deformation experiments yield force-deformation (or *FX*) curves, which contain information on the spring constant, reversibility of deformation, forces required to reversibly deform and to irreversibly destroy the structures, etc. [25]. Several virus shells have been tested using the AFM-based approach, including the bacteriophages—Φ29, λ, and HK97 [33,53,54]; the human viruses—human immunodeficiency virus, norovirus, hepatitis B virus, adenovirus, and herpes simplex virus [55–61]; and other eukaryotic cell infecting viruses, including minute virus of mice, triatoma virus, and CCMV [62–64]. The AFM-based deformation of a particle can be performed in a regime of forward indentation, when the cantilever base moves with constant velocity (v_f) toward the particle, thereby gradually increasing the applied compressive force, or in a regime of force-quenched backward retraction, when the direction of cantilever movement is reversed and the force decreases to zero. However, as we pointed out, experimental force-deformation spectra, oftentimes, reveal a complex deformation mechanism, but the lack of structure-based information precludes detailed nanoscale interpretation and modeling of the spectral lineshapes [10–12,25].

10.5.2 Nanoindentation *In Silico* Method

Several limitations in experimental resolution call for the development of alternative research strategies. We recently pioneered the methodology of "nanoindentation *in silico*," which can be used to provide a structure-based understanding of the mechanisms of deformation and collapse of the virus shell (Figure 10.3) [21]. In this method, mechanical loading and unloading of a biological particle are carried out *in silico* (i.e., in a computer). Importantly, the nanoindentation measurements are performed under experimental conditions of dynamic force application $f(t) = r_f t$ (Figure 10.3), i.e., in our simulations, we use the experimental force-loading rates (r_f). Structural transitions can be resolved by examining the coordinates of amino-acid residues, and thermodynamic quantities and biomechanical characteristics can be gathered through detailed analysis of the energy output.

Our unique *in silico* "experiment" provides a complete and detailed simulation view of the entire process of particle deformation, relaxation, and collapse. For this reason, it can be used to provide a detailed mechanistic interpretation of the experimental force-deformation spectra. In our nanoindentation assays *in silico*, we are capable of maintaining full control over the simulation protocol, i.e., the amplitude and direction of applied force, exact location of the contact between the cantilever tip (indenter) and the biological particle (system), and surface-constrained particle residues to prevent the particle from rolling. As in experiments, we can also reverse the direction of applied force, decrease or increase the force amplitude, etc. Hence, the full control we have over the system during the entire deformation process can be used (1) to study nanoindentation at different symmetry points on the particle surface, and (2) to relate the energy changes observed during any point in the simulation to the specific molecular details observed in the particle's structure. In addition, we are capable of (3) switching back and forth from the forward indentation to the backward tip retraction [21]. The latter can be used to perform the simulations of repeated cycles of indentation and retraction, in order to understand the process of mechanical fatigue and structural collapse for the biological system in question.

In dynamic force measurements *in silico*, the cantilever base is represented by the virtual bead, connected to the spherical bead of radius R_{tip}, mimicking the cantilever tip (indenter), by a harmonic spring (Figure 10.3). The tip interacts with the biological particles via the repulsive Lennard–Jones potential:

$$U_{tip} = \sum_{i=1}^{N} \varepsilon_{tip} \left(\frac{\sigma_{tip}}{|r_i - r_{tip}| - R_{tip}} \right)^6 \tag{10.6}$$

thereby producing an indentation on the particle's outer surface. In Equation 10.6, r_i and r_{tip} are coordinates of the ith particle and the center of the tip, respectively; $\varepsilon_{tip} = 4.18$ kJ/mol and $\sigma_{tip} = 1.0$ Å are parameters of interaction, and the summation is performed over all the particles under (i.e., interacting with) the tip (or other indenting object).

For the cantilever tip (spherical bead in Figure 10.3), we solve numerically the following Langevin equation of motion:

$$\eta \frac{dr_{tip}}{dt} = -\frac{\partial U_{tip}(r_{tip})}{\partial r_{tip}} + \kappa \left(\left(r_{tip}^0 - v_f t \right) - r_{tip} \right) \tag{10.7}$$

In Equation 10.7, r_{tip}^0 is the initial position of spherical tip center (v_f is the cantilever base velocity; κ is the cantilever spring constant), and the friction coefficient $\eta = 7.0 \times 10^6$ pN ps/nm ($\zeta = 500$ is the dimensionless friction constant for a spherical tip in water). To generate the dynamics of the biological particle of interest tested mechanically, we solve Equations 10.1 through 10.5 numerically for the particle (see Section 10.4.2) and Equations 10.1 and 10.2 for the indenting object (spherical tip).

The moving with constant velocity (v_f) cantilever base (virtual bead in Figure 10.3) exerts through the cantilever tip, the time-dependent force $\mathbf{f}(t) = f(t)\mathbf{n}$ in the direction $\mathbf{n}$ perpendicular to the outer surface of the virus particle, which mechanically loads the particle. As a result, the magnitude of force $f(t) = r_f t$ exerted on the particle ramps up linearly in time t (force ramp) with the force-loading rate $r_f = \kappa v_f$. In the simulations of "forward indentation," the cantilever base (and spherical tip) is moving toward the virus capsid; in the simulations of "force-quenched retraction," the direction of cantilever base (tip) motion is reversed, which results in a gradual decrease of compressive force to zero (force-quench). We control the piezo (cantilever base) displacement Z, and the cantilever tip position X, which defines the indentation depth. The resisting force of deformation F from the virus particle, which corresponds to the experimentally measured indentation force, can be readily calculated using the energy output. Also, to prevent the capsid from rolling and to mimic adsorption of the particle on the substrate, we can always constrain the bottom portion of the particle by fixing several C_α-atoms contacting the substrate surface. Owing to the ~200-fold+ computational acceleration attained on contemporary graphics cards, in our simulations, we can afford to use the experimentally relevant values of cantilever velocity ($v_f = 0.1 - 1.0$ μm/s), which is a significant advantage of our method.

10.6 BIOMECHANICS OF CCMV AND ENCAPSULIN SHELLS

10.6.1 Equilibrium Properties of Virus Shells Are a Sum of Properties of Its Structure Elements

We carried out simulations of the isolated single pentamer and hexamer capsomers comprising the CCMV shell and the full CCMV capsid at equilibrium at room temperature. The simulation output for the residue coordinates was used to perform the normal mode analysis [65] of equilibrium fluctuations. In this method, one calculates the Hessian matrix for the displacements of the centers of mass of amino-acid residues (H_{IJ}). The eigenvalues $\{\lambda_I\}$ and eigenvectors $\{R_I\}$ are obtained numerically by solving the eigenvalue problem (diagonalization procedure) and are then used to calculate the spectrum of normal frequencies $\omega_I \propto \sqrt{\lambda_I}$ and normal modes $Q_I = \sum R_{IJ} q_I$, where q_I are the center-of-mass positions of residues ($I = 1, 2, \ldots$). We compared the spectra of equilibrium normal modes for the isolated single pentamer capsomers and hexamer capsomer, and for the full CCMV capsid. The normal mode spectra for the isolated single hexamer capsomer and for the full CCMV shell are presented in Figure 10.5. The normal mode spectrum for the pentamer capsomer shows a similar agreement with the CCMV shell spectrum (not shown in Figure 10.5). We see that the spectra of normal modes for the capsid fragment (hexamer) and for the entire capsid are practically indistinguishable, which implies that, at equilibrium, the mechanical characteristics of the whole system (CCMV capsid) can be represented by a sum of the mechanical properties of its structural components (i.e., capsomers). We also performed a comparative analysis of the normal modes of equilibrium displacements for the encapsulin nanocontainer and obtained similar results (data not shown).

10.6.2 Nanomanipulations *In Silico* Reveal Global Collapse Transitions in Virus Shells

We performed the dynamic force measurements *in silico* for the CCMV and encapsulin shells. We used the SOP model for the empty CCMV shell, obtained from the all-atomic structure of the capsid (PDB entry: 1CWP) after a 5-ns equilibrium run. For the encapsulin nanocontainer, we first constructed a complete shell from the all-atomic structure of a resolved pentameric subunit (PDB entry: 3GKT). Next, we performed a 5-ns equilibrium simulation run and used the final structure obtained at the end of the run to construct a SOP model of the encapsulin shell. We used spherical tips of radius $R_{\text{tip}} = 15$ and 20 nm for the CCMV shell and radius

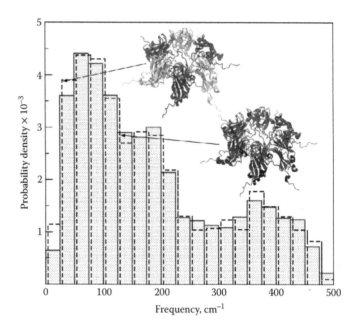

FIGURE 10.5 Equilibrium normal modes of motion of the CCMV shell. The spectrum is the profile of the probability density of normal (eigen-) frequencies as a function of frequency (histogram). Compared are the spectrum of eigen-frequencies obtained for an isolated single hexamer capsomer (white bars) and the spectrum of eigen-frequencies obtained for the entire CCMV shell (gray bars). Structures display the scale of normal displacements from the global modes, which include displacements of protein monomers and entire capsomers at the low-frequency end of the spectrum for frequencies <50 cm^{-1} (indicated in the left-hand structure), to the more local modes of motion of secondary structure elements such as α-helices and β-strands at 100–200 cm^{-1} (indicated in the right-hand structure).

R_{tip} = 10 and 15 nm for the encapsulin shell to compress each shell along the two-, three-, and five-fold symmetry axes. First, we collected the force-deformation curves for each study system by performing the simulations of forward indentations. Next, we carried out a few simulation runs, in which the forward indentations were interrupted by the backward tip retraction process. In the forward indentation (and backward tip retraction) measurements for the CCMV and encapsulin shells, the force magnitude $f(t) = r_f t$ increased (decreases) in time with the force-loading (force-quenching) rate $r_f = \kappa v_f$, where $v_f = 1.0$ μm/s and $\kappa = 50$ pN/nm. These are typical values of these quantities used in dynamic force experiments *in vitro* (dynamic force-ramp protocol). The resulting force-deformation spectra for the forward indentation and backward tip retraction for the CCMV and encapsulin shells are presented in Figures 10.6 and 10.7, respectively.

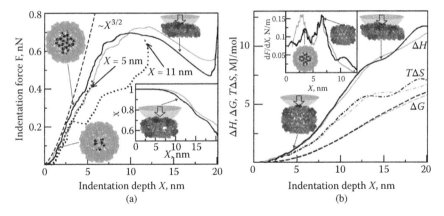

FIGURE 10.6 Nanoindentation *in silico* of the CCMV particle. Shown in black and gray color are two representative trajectories ($v_f = 1.0$ μm/s). The cantilever tip ($R_{tip} = 20$ nm) indents the capsid in the direction perpendicular to the capsid surface along the two-fold symmetry axis (see Figure 10.1). (a) The force-deformation spectra (*FX* curves) results for the forward deformation and backward retraction are represented by the solid and dotted black curves, respectively. The retraction simulations are performed using the structures of the deformed CCMV shell occurring at $X = 5$ and 11 nm. The retraction curves show that the 5 nm deformation can be retraced back almost reversibly (small hysteresis), whereas the 11 nm deformation is nearly irreversible (large hysteresis). The dashed line represents the fit of the initial *FX* curve in the elastic regime ($X < 3-5$ nm) by the nonlinear function $\sim X^{3/2}$. Structures on the left show the top view of the CCMV shell with the increasing tip-capsid surface contact area (blackened); the structure on the right shows the profile of the CCMV particle in the globally collapsed state (compressive force is shown by a large arrow). *The inset* is the profile of the structure overlap $\chi(X)$ (see Section 10.6.4), which decreases with the indentation depth X. The structure in *the inset* represents the CCMV particle right before the transition to the collapsed state (at $X \approx 10$ nm indentation). (b) The profiles of thermodynamic state functions: the enthalpy change ΔH, the Gibbs free energy change ΔG, and the entropy change $T\Delta S$ associated with the mechanical deformation and collapse of the CCMV particle. The structures show the profile of the CCMV particle in the elastic regime of mechanical deformation, where $\Delta H \approx T\Delta S$, and in the plastic regime, where $\Delta H > T\Delta S$. *The inset* shows the profile of the slope of the *FX* curve, i.e., dF/dX versus X, with two peaks which correspond to the mechanically activated (transition) states for the two types of transitions: (1) local curvature change in the tip-capsid surface contact area (first peak at $X \approx 3$ nm and the corresponding top view of CCMV particle); and (2) bending deformation of the portions of CCMV shell structure that are not in direct contact with the cantilever tip (second peak at $X \approx 7$ nm and the corresponding side view of CCMV particle).

We see that the force(F)-deformation(X) spectra (or *FX* curves) for both the CCMV and encapsulin shells are characterized by the initial steep increase in deformation force F for small deformation (or indentation) depth X. In this initial regime, the dependence of force F on deformation X is almost linear with the roughly constant slope, dF/dX, which

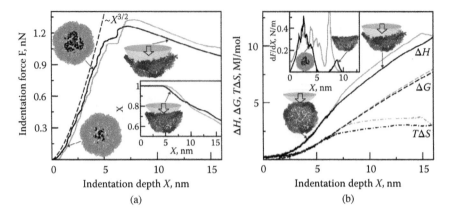

FIGURE 10.7 Nanoindentation *in silico* of the encapsulin nanocompartment represented by two trajectories shown in black and gray color (R_{tip} = 15 nm and v_f = 1.0 μm/s). Displayed are the same quantities as in Figure 10.6 but for the encapsulin nanocompartment indented along the three-fold symmetry axis (see Figure 10.2). (a) The FX curves for the forward deformation (retraction curves are not shown); the dashed line is the fit of the initial portion of the FX curve by the nonlinear function ~$X^{3/2}$. The left-hand structures show the top view of the encapsulin shell with the increasing tip-capsid surface contact area (blackened); the right-hand structure shows the profile of the shell in the globally collapsed state. *The inset* is the profile of the structure overlap $\chi(X)$ (see Section 10.6.4), which decreases with the indentation depth X. The structure in *the inset* is the encapsulin particle right before the transition to the collapsed state occurs (at X ≈ 7 nm). (b) The profiles of ΔH, ΔG, and $T\Delta S$ for the mechanical deformation and collapse of the encapsulin particle. The structures show the profile of the particle in the elastic deformation regime, where $\Delta H \approx T\Delta S$, and in the plastic regime, where $\Delta H > T\Delta S$. *The inset* shows the slope of the FX curve dF/dX as a function of X with two peaks: the first peak at X ≈ 3–4 nm (see the top view of encapsulin shell) and the second peak at X ≈ 7–8 nm (see the side view of encapsulin shell).

quantifies the mechanical compliance of the virus particle to the external compressive force (f). The average values of the slope for the CCMV and encapsulin shells are accumulated in Table 10.1. Interestingly, the initial pseudo-linear regime persists until the critical value of the deformation $X = X^*$ is reached, at which point the deformation force attains the critical (or maximum) force value $F = F^*$ (Figures 10.6 and 10.7). Hence, a pair of values, i.e., F^* and X^*, quantify the limits of mechanical stability of the virus particle. The average critical values of deformation force F^* and deformation depth X^* for the CCMV and encapsulin shells are given in Table 10.1. Beyond the critical point, i.e., for $F > F^*$ and $X > X^*$, the particle enters the dynamic regime, in which it undergoes a spontaneous shape-changing transition from the native (undeformed) state to the globally collapsed (fully deformed) state. This

TABLE 10.1 Biomechanical and Thermodynamic Properties of the CCMV Capsid (Figure 10.1) and Encapsulin Shell (Figure 10.2) from the Nanoindentation Measurements *In Silico* Performed along the Two-Fold Symmetry Axis (for CCMV) and Three-Fold Symmetry Axis (for Encapsulin)

System	F^*, nN	X^*, nm	k_{cap}, N/m	E, GPa	ΔG_{ind}, MJ/mol	ΔH_{ind}, MJ/mol	$T\Delta S_{ind}$, MJ/mol
CCMV	0.69 ± 0.02	11.7 ± 0.7	0.11 ± 0.01	0.18 ± 0.02	4.6 ± 0.4	11.9 ± 0.4	7.3 ± 0.8
Encapsulin	1.17 ± 0.06	6.7 ± 2.4	0.29 ± 0.05	1.36 ± 0.07	7.5 ± 0.4	11.5 ± 0.6	4.0 ± 0.2

Note: Summarized are the critical force F^* and critical deformation X^*, and the particle spring constant k_{cap} (calculated as the average slope dF/dX of the pseudo-linear portion of the FX curve). Also presented are the thermodynamic state functions: the Gibbs free energy change ΔG, the enthalpy change ΔH, and the entropy change $T\Delta S$. Theoretical estimates of all the quantities are obtained by averaging the results from three simulation runs obtained using the spherical tip with radius $R_{tip} = 20$ nm for CCMV (see Figure 10.6) and $R_{tip} = 15$ nm for encapsulin (see Figure 10.7), and the cantilever base velocity $\nu_f = 1.0$ μm/s. The values of ΔG, ΔH, and $T\Delta S$ are calculated using the native state ($X = 0$ nm) as the initial state and the globally collapsed state ($X = 20$ nm for CCMV and $X = 17$ nm for encapsulin) as the final state.

dynamic transition is also reflected in the sudden decrease in indentation force from the maximum value (F^*) to some lower values (see Figures 10.6 and 10.7). After attaining the minimum, the deformation force starts increasing again because the spherical tip starts indenting the lower portion of the collapsed virus structure attached to the substrate surface.

10.6.3 Force-Deformation Assays *In Silico* Allows One to Probe Elastic Properties of Virus Shells

We compared the simulated FX spectra obtained by indenting the CCMV and encapsulin shells along the two-, three-, and five-fold symmetry axes (data not shown). Interestingly, we found that the mechanical response of CCMV and encapsulin particles probed at different indentation points, i.e., when the shells are compressed along the two-, three-, and five-fold symmetry axes, exhibit small differences. These differences are fully reflected in the simulated FX spectra, the result implying that the mechanical response of the capsid depends on the symmetry of the local capsomer arrangement. Hence, our results are indicative of what might be a general property of virus particles, namely that the physical properties of the virus shells are dynamic but local characteristics of their structure and energetics. We profiled the dependence of the slope dF/dX of the FX curves for the CCMV and encapsulin shells on the indentation depth X. This quantity can be viewed as an X-dependent measure of mechanical compliance of a biological particle. Interestingly, we found that the slope dF/dX varies greatly with X and attains two maxima (see Figures 10.6 and 10.7). The peaks of the slope dF/dX correspond to the two transition states, i.e., mechanically activated states of the virus particle, for the global transitions: (1) from the native state to the intermediate structure (first peak of dF/dX) and (2) from the intermediate state to the globally collapsed state (second peak of dF/dX).

The above results for CCMV and encapsulin indicate that the profile of the instantaneous slope of the FX curve versus indentation depth X is a suitable and informative measure of the biomechanical properties of virus particles; it reveals that there is dynamic structure remodeling that takes place in virus shells during the indentation process, which makes the biomechanical properties of the virus shell dependent on the indentation depth X. Structure analysis of the simulation output revealed, both for the CCMV and encapsulin shells, that during the first transition, a change in the local curvature of the tip-capsid surface contact area occurs, which results in a subsequent increase of the contact area. Prior to the second transition, the side portions of the capsid undergo bending deformations, which lead eventually to the global collapse transition in the capsid structure (see structure snapshots in Figures 10.6 and 10.7).

For the CCMV and encapsulin shells, we also evaluated the average slope of the FX curves for the initial pseudo-linear (elastic) regime of deformation, which we refer to as the "spring constant" of the virus capsid, k_{cap}. We obtained values of $k_{cap} = 0.11$ N/m for the CCMV shell and $k_{cap} = 0.29$ N/m for the encapsulin shell (Table 10.1). A theory of thin shells provides a linear relationship connecting the particle spring constant k_{cap} and the Young modulus E, $k_{cap} = \lambda E h^2 / R$, where h and R are the capsid thickness and outer radius, respectively, and λ is a numerical prefactor ($\lambda \approx 1$). We used this relationship to estimate Young's modulus for the CCMV and encapsulin shells. We found that $E = 0.18$ GPa for CCMV and $E = 1.36$ GPa for encapsulin (Table 10.1). The value of Young's modulus for CCMV agrees with the experimental estimate of 0.14 GPa reported in Ref. [62]. Hence, our results indicate that the CCMV shell is a softer particle as compared to the encapsulin nanocompartment.

We also carried out simulations of nanoindentations of the CCMV and encapsulin particles by using a spherical tip of smaller size (15 nm for CCMV and 10 nm for encapsulin). We compared the FX curves for the CCMV and encapsulin shells generated using a spherical tip of smaller size (data not shown) with the FX curves obtained by using a spherical tip of larger size (Figures 10.6 and 10.7). We found that the FX spectra for both the CCMV and encapsulin shells exhibit a steeper slope and larger values of the critical force (F^*) and critical deformation (X^*) when indented with a larger tip-sphere. Hence, our results indicate that the physical properties of the CCMV particle depend on the geometry of mechanical perturbation, as the mechanical response changes with tip size.

10.6.4 Structure Collapse Transition in Virus Particles Is a Shape-Changing Process

To quantify the extent of structural similarity between a given conformation and a native (reference) state, we employed the structure overlap, $\chi(X) = (2N(N - 1))^{-1} \sum \Theta(|r_{ij}(X) - r_{ij}(0)| - \beta r_{ij}(0))$, where in the Heaviside step function Θ, $r_{ij}(X)$ and $r_{ij}(0)$ are the interparticle distances between the ith and jth residues in the transient structure obtained for the extent of deformation X and in the native (undeformed) state for which $X = 0$, respectively, and $\beta = 0.2$ is the tolerance for the interparticle distance change. This is an interresidue distance-based metric of the structural similarity, which ranges from $\chi = 0$ (for completely dissimilar structures) to $\chi = 1$ (for fully similar structures). Hence, changes in the structure overlap should indicate alterations at the secondary and tertiary structure level. The results of calculations of $\chi(X)$ for the CCMV and encapsulin shells are presented in Figures 10.6 and 10.7, respectively (the inset in panel a). The structure overlap starts deviating from unity for large deformations, i.e., for $X = 10$ nm deformation (for CCMV) and $X = 5$ nm deformation (for encapsulin). Interestingly, we see that both the CCMV and encapsulin particles in their fully collapsed states, which correspond to $X =$ nm and $X = 17$ nm indentation, respectively, remain 60%–70% similar to their native structures. Hence, unlike the protein force unfolding problem, transitioning to the collapsed state of a virion particle does not imply substantial structural changes at the level of its subunit secondary and tertiary structure. In a sense, the collapse transition and associated mechanical failure of the virus particle can be thought of as a spontaneous shape-changing process rather than a structural transition.

10.6.5 Nanoindentations *In Silico* Allow One to Probe Thermodynamics of Mechanical Transitions in Viruses

We studied the thermodynamics of mechanical deformation of the CCMV and encapsulin shells. For each system and indentation run, we evaluated the total work of indentation w by converting the *FZ* curve into the *FX* curve by integrating the area under the *FX* curve. Next, we repeated this procedure but for the retraction *FX* curve to evaluate the reversible part of work w_{rev} (see Figures 10.6 and 10.7). By evaluating the relative difference, $(w - w_{rev})/w$, we found that in the elastic regime of deformation ($X < 5$ nm for CCMV and $X < 3$ nm for encapsulin) $\sim$10%–12% of the total energy is dissipated. In the transition range (10 nm $\leq X \leq$ 15 nm for CCMV and 7 nm $\leq X \leq$ 10 nm for encapsulin), the relative difference can be as high as 25%–30%, which agrees with the large hysteresis observed in the *FX* curves (Figures 10.6 and 10.7).

Because in the *NPT* ensemble (N is the total number of residues, P is pressure, and T is temperature) $w_{rev} = \Delta G = \Delta H - T\Delta S$, where ΔG, ΔH, and ΔS are the Gibbs free energy change, enthalpy change, and entropy change, respectively, we are able to estimate ΔH and $T\Delta S$. We profiled and compared the dependence of these thermodynamic state functions on the indentation depth X. The results obtained for the CCMV and encapsulin particles are presented in Figures 10.6 and 10.7, respectively. We see that the mechanical deformation of the CCMV and encapsulin particles requires a large investment of energy. Indeed, the X-dependent changes for all three quantities ΔG, ΔH, and $T\Delta S$ are in the megajoule (MJ) range. We estimated the total changes in ΔG, ΔH, and $T\Delta S$ associated with the mechanical deformation of the CCMV and encapsulin particles and transitioning from their native states ($X = 0$ nm deformation) to the globally collapsed state ($X = 20$ nm for CCMV and $X = 17$ nm for encapsulin). The obtained values of ΔG_{ind}, ΔH_{ind}, and $T\Delta S_{ind}$ are accumulated in Table 10.1.

Interestingly, the enthalpy changes for CCMV and encapsulin are fairly similar (11.9 MJ/mol for CCMV vs. 11.5 MJ/mol for encapsulin), whereas the Gibbs free energy changes are almost two-fold different (4.6 MJ/mol for CCMV vs. 7.5 MJ/mol for encapsulin) owing to the difference in the entropy changes (7.3 MJ/mol for CCMV vs. 4.0 MJ/mol for encapsulin). Hence, the deformation and collapse processes are more spontaneous for the CCMV particle than for the encapsulin particle. The differences in ΔG_{ind} and $T\Delta S_{ind}$ can be understood by considering what we discussed earlier (in Section 10.6.3), namely that of the two particles the CCMV shell is a softer structure (compare, e.g., the values of k_{cap} and E in Table 10.1). Hence, during the indentation process, the encapsulin nanocompartment is less prone to structure remodeling, which explains the lower value of $T\Delta S_{ind}$ and the associated higher value of ΔG_{ind}.

For small deformations, i.e., for $X < 5$ nm (for CCMV) and for $X < 3$ nm (for encapsulin), ΔH and $T\Delta S$ have similar changes ($\Delta H \approx T\Delta S$), whereas for large deformations, i.e., for $X > 5$ nm (for CCMV) and for $X > 3$ nm (for encapsulin), changes in ΔH exceed those in $T\Delta S (\Delta H > T\Delta S)$. Hence, changes in the local curvature of the tip-capsid surface contact area for small deformations (first peak in the profiles of the dF/dX in Figures 10.6 and 10.7) are accompanied by small changes in ΔH as well as $T\Delta S$. However, bending deformations of the side portions of the capsid structure for large deformations (second peak in the profiles of the dF/dX) are accompanied by larger changes in ΔH but smaller changes in $T\Delta S$. Hence, our results indicate, at least for the CCMV and encapsulin shells, that (1) the entropy change and enthalpy change both contribute to the capsid stiffening for small deformations, which is reflected in the increase of the slope dF/dX (for $X < 3-4$ nm for CCMV and for $X < 2-3$ nm for encapsulin) whereas (2) the capsid transitioning to the collapsed state, reflected in the decrease of dF/dX to zero (mechanical failure), is driven mainly by the enthalpy change (for $X > 9-10$ nm for CCMV and for $X > 6-7$ nm for encapsulin).

10.6.6 Nanoindentations *In Silico* Illuminate Dynamic Interplay between the Out-of-Plane and In-Plane Displacements

The essential dynamics approach [66] allows one to calculate the most important displacements for a virus system tested mechanically, which reside in a subspace of just a few degrees of freedom. The remaining degrees of freedom represent less important fluctuations. Dynamic correlations between particle positions at time t, $\mathbf{X}(t) = \{X_1(t), X_2(t), \ldots, X_N(t)\}$, and the position in the reference (equilibrium average) structure $\mathbf{X}_0 = \{X_1(0), X_2(0), \ldots, X_N(0)\}$ can be expressed by considering the covariance matrix $\mathbf{C}(t) =< (\mathbf{X}(t) - \mathbf{X}_0)(\mathbf{X}(t) - \mathbf{X}_0)^T >$, where $< \ldots >$ denotes the ensemble averaging and the superscript T represents the transposed matrix. By construction, $\mathbf{C}$ is a symmetric matrix, which can be diagonalized by an orthogonal transformation $\mathbf{T}$, $\mathbf{C} = \mathbf{TLT}^T$, where $\mathbf{L}$ is the diagonal matrix of eigenvalues and $\mathbf{T}$ is the matrix of eigenvectors of $\mathbf{C}$. In the 3D space, there are $3N-6$ eigenvectors with nonzero eigenvalues for a system of N particles (excluding translations and rotations). The eigenvalue l_I in the center-of-mass representation ($I = 1, 2, \ldots, N$) corresponds to the amplitude of displacement $\mathbf{X}(t) - \mathbf{X}_0$ along the Ith eigenvector $\mathbf{t}_I$. The principal coordinates $P_I(t)$ are obtained by projecting the displacement $\mathbf{X}(t) - \mathbf{X}_0$ onto each eigenvector $P_I(t) = \mathbf{t}_I(\mathbf{X}(t) - \mathbf{X}_0)$. In the Cartesian basis, these projections are given by $\mathbf{X}(t) = P_I(t)\mathbf{t}_I + \mathbf{X}_0$ [65,66].

We performed the in-depth analysis of the simulation output for coordinates of amino acids obtained from the simulations of nanoindentations of the CCMV and encapsulin shells using the essential dynamics approach. For each system, we performed two separate analyses using the output for coordinates for the elastic deformation regime ($0 < X < 4.5$ nm for CCMV and $0 < X < 4$ nm for encapsulin) and for the pre-transition range ($6 < X < 10.5$ nm for CCMV and $4 < X < 7$ nm for encapsulin). Both for CCMV and encapsulin, we focused on the first and the second most important modes of essential dynamics, which capture ~85% of the dynamics of CCMV and ~70% of the dynamics of encapsulin. For CCMV, these modes are displayed in Figure 10.8. In the elastic regime, the first mode corresponds to the "out-of-plane" displacements of the pentamer and hexamer capsomers (77% of dynamics), whereas the second mode corresponds to the "in-plane" displacements of capsomers (8% of dynamics). However, in the transition regime, each of these modes of collec-

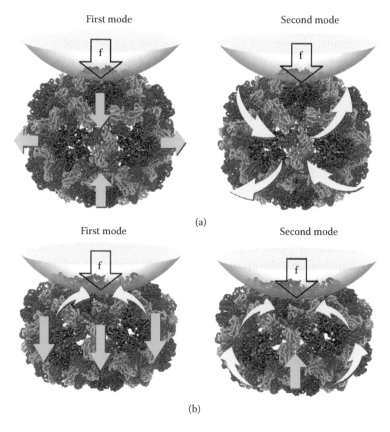

FIGURE 10.8 Nonequilibrium dynamics of the mechanical deformation of the CCMV shell in the "essential subspace" (essential dynamics). Shown are the displacements of pentamer capsomers (in blue) and hexamer capsomers (in red), which correspond to the most important first two modes of the collective excitations (first mode and second mode). These modes are projected along the reaction coordinate (deformation or indentation depth X) coinciding with the direction of applied compressive force $\mathbf{f}(t)$ (large arrow) in the initial elastic regime of deformation (a), which corresponds to the roughly linear portion of the FX curves, and in the regime of spontaneous shape-changing transition (b), which corresponds to the nonmonotonic portion of the FX curves with the force peaks (see Figures 10.6 and 10.7). The "out-of-plane" modes and "in-plane" modes of collective displacements are indicated by the straight green and curved yellow arrows, respectively.

tive motion exhibits strong coupling between the out-of-plane and the in-plane displacements (Figure 10.8). Hence, in the regime of mechanical deformation, which corresponds to the dynamic evolution of the CCMV shell right before the collapse transition, the out-of-plane and in-plane displacements are mixed. We obtained similar results for the encapsulin shell (data not shown). The only difference was the smaller amplitude of the in-plane displacements for encapsulin as compared to those for CCMV.

The results of essential dynamics analysis can be used to interpret our findings regarding the (ir)reversibility of mechanical deformation for the CCMV shell (Figure 10.6). The retraction curves show that, in the elastic regime ($X < 5$ nm), the deformation of CCMV is reversible (with small hysteresis in the FX curve), whereas in the transition regime (11 nm $< X <$ 15 nm), the deformation of CCMV is irreversible (with large hysteresis). These findings can now be rationalized using the physical picture of out-of-plane and in-plane displacements (Figure 10.8). In the elastic regime of deformation, the out-of-plane excitations dominate the near-equilibrium displacements of capsomers. Therefore, when a compressive force is quenched, as in the retraction experiments, the first mode (pure out-of-plane displacements) provides a mechanism for capsid restructuring, and the amount of energy dissipated is small. However, the out-of-plane and the in-plane displacements are strongly coupled in the far-from-equilibrium transition regime. Here, the capsid is capable of restoring its original shape (owing to the out-of-plane displacements), but this capsid restructuring comes at a cost of exciting additional degrees of freedom (i.e., the in-plane displacements) and, hence, a larger amount of dissipated energy.

10.6.7 (Ir)reversibility of Deformation and (In)Elastic Properties of Virus Shells Is Directly Correlated

The obtained results show that the thermodynamic properties, i.e., changes in the state functions ΔG, ΔH, and $T\Delta S$, and the dynamic properties of the virus particles, i.e., reversibility and irreversibility of the compressive force induced deformation, are directly correlated with their mechanical characteristics—that is, the initial elastic response for small deformations and the inelastic response for large deformations. The connection among the thermodynamic, dynamic, and biomechanical properties is shown in Figure 10.9 for the CCMV shell. In particular, our results indicate that

FIGURE 10.9 Summary of the physico-chemical properties of the CCMV particle based on the results of computational exploration of its indentation nanomechanics. This summary is superimposed upon an image of a CCMV infected leaf, where a representative area is enlarged in two successive stages to achieve a nanoscale molecular level view. The summary shows that the mechanical force-driven transformations in the CCMV shell structure are driven by a direct contact between the CCMV particle and the indenter, and so the virion's physical properties depend not only on the magnitude but also on the geometry of mechanical input (compressive force indicated by a large vertical arrow). The collapse transition and associated mechanical failure of the virus particle can be thought of as a spontaneous shape-changing transformation. In the elastic regime of mechanical deformation, which corresponds to the roughly linear initial portion of the force-deformation spectrum (*FX* curve), the out-of-plane displacements dominate the in-plane modes of collective excitations. In this regime, mechanical deformation can be retraced back reversibly owing to the complete capsid restructuring, and the associated energy losses due to dissipations are small (little or no hysteresis). In the transition regime, which corresponds to the nonmonotonic portion of the *FX* curve with the force maximum, the irreversible changes to the capsid structure and shape occur while the capsid is transitioning from the native-like state to the globally collapsed state. In this regime, the mechanical response of the virus particle is inelastic due to strong coupling between the out-of-plane and in-plane modes of collective mechanical excitations. This coupling provides a mechanism for significant energy dissipation and is the microscopic origin of the large hysteresis observed in the *FX* curves from the simulations of forward (forced) indentation and backward (force-quenched) retraction.

the extent of reversibility of mechanical deformation of virus particles, i.e., the extent to which their mechanical deformation can be retraced back reversibly, is determined by the dynamic interplay between the out-of-plane and the in-plane types of collective displacements. In the elastic regime (pseudo-linear portion of the FX curve), changes in ΔH and $T\Delta S$ are comparable (see Figures 10.6 and 10.7) and the main mode of collective motions is dominated by the "out-of-plane" displacements. These provide a direct pathway for the virus particle's full restructuring, which results in a reversible deformation as observed in the FX retraction curve (with small hysteresis). In the inelastic regime (nonmonotonic portion of the FX curve), changes in ΔH far exceed those in $T\Delta S$ (Figures 10.6 and 10.7) and the "out-of-plane" and "in-plane" displacements become mixed. This mixing is taking place already in the pre-transition range of mechanical deformation right before the transition to the collapsed state occurs. In this regime, there are neither pure "out-of-plane" nor "in-plane" displacements (see Figure 10.8b). The coupled nonequilibrium essential modes of deformation cannot be reconstructed using a linear combination of the out-of-plane modes and the in-plane displacements. The degree of this coupling (or mixing) and, hence, the extent to which the deformation is reversible, is determined by the indentation depth. For large deformations beyond critical (force peak in the FX curve), the virus capsid does not restructure completely, which results in an irreversible deformation (with large hysteresis in the FX curves). This is, in part, because some of the deformation energy that has been pumped into the virus particle by the indenter (tip-sphere) has been lost to excite the in-plane motion of pentamer and hexamer capsomers, including their translocation and rotation (see the insets in Figure 10.9) made possible by the dynamic coupling between the out-of-plane and in-plane modes of collective excitations.

10.7 MODELS OF MECHANICAL DEFORMATION OF VIRUS PARTICLES

10.7.1 Identifying Degrees of Freedom for Virus Particles' Deformation

The thin-shell theory is widely used to model the deformation of biological particles [10,11,25,33,54,62]. Let us briefly review the assumptions underlying this theory. In general, deformation of a shell can be described by the second-rank deformation tensor [67]:

$$\bar{\bar{u}} = \begin{pmatrix} u_{xx} & u_{xy} & u_{xz} \\ u_{yx} & u_{yy} & u_{yz} \\ u_{zx} & u_{zy} & u_{zz} \end{pmatrix} \tag{10.8}$$

The thin-shell theory is based on the assumption that when the particle is subjected to mechanical loading, there exist only two types of deformation: the "out-of plane" bending u_{zz} along the axis of force application (z-axis); and the "in-plane" stretching, described by tensor components u_{xx}, u_{yy}, u_{xy}, and u_{yx}. The theory assumes that the coupling (mixing) between these types of motions is negligible, which translates to the equality of the off-diagonal components u_{xz}, u_{yz}, u_{zx}, and u_{zy} to zero. Hence, in the thin-shell approximation the deformation tensor becomes [67]

$$\bar{\bar{u}} = \begin{pmatrix} u_{xx} & u_{xy} & 0 \\ u_{yx} & u_{yy} & 0 \\ 0 & 0 & u_{zz} \end{pmatrix} \tag{10.9}$$

The total energy of the deformed thin shell can then be expressed as a sum of the tensor components for the out-of-plane and the in-plane deformations, i.e.,

$$U = \int \left(U_1(\varsigma) + U_2(u_{\alpha\beta}) \right) dA \tag{10.10}$$

where $dA = dxdy$ is the element of the particle surface area. In the integrand in Equation 10.10, $U_1(\varsigma)$ is the potential energy due to pure bending of the shell, which depends on the out-of-plane shape change $\varsigma(x, y)$ of the shell. For small deformations, the potential energy for the out-of-plane bending is given by [67]

$$U_1 = \frac{Eh^3}{24(1-\sigma^2)} \left\{ \left(\frac{\partial^2 \varsigma}{\partial x^2} + \frac{\partial^2 \varsigma}{\partial y^2} \right)^2 + 2(1-\sigma) \left[\left(\frac{\partial^2 \varsigma}{\partial x \partial y} \right)^2 - \frac{\partial^2 \varsigma}{\partial x^2} \frac{\partial^2 \varsigma}{\partial y^2} \right] \right\} \tag{10.11}$$

where E is the particle Young's modulus, h is the particle thickness, and σ is Poisson's ratio. The second term in the integrand in Equation 10.10, $U_2(u_{\alpha\beta})$ is the potential energy due to stretching of the shell, which depends on the in-plane components of the deformation tensor $u_{\alpha\beta}$ (α, $\beta = x$ and y). For small deformations, the potential energy for the in-plane stretching is given by [68]

$$U_2 = \frac{Eh}{2(1 - \sigma^2)} \left\{ \left(u_{xx} + u_{yy} \right)^2 - 2(1 - \sigma) \left(u_{xx} u_{yy} - u_{xy}^2 \right) \right\} \quad (10.12)$$

The thin-shell theory has been used in the past to describe small deformation of thin protein shells of viruses and bacteriophages [25]. However, a large number of virus particles have thick shells that might experience large deformations. In these circumstances, the thin-shell theory cannot be used. In addition, our results of nanoindentation of the CCMV and encapsulin shells indicate that there is strong coupling between the in-plane and out-of-plane types of collective deformation, which plays an important role in the mechanics of virus particles especially during their transitioning to the collapsed state. This mixing of the in-plane and out-of-plane deformations breaks down the assumption of their independence underlying the theory of thin shells (see Equations 10.9 and 10.10).

This calls for the development of new models to describe the dynamic evolution of biological assemblies subject to mechanical stress. It is important to identify the relevant degrees of freedom, which contribute most to the particle deformation (indentation) depth X, which can be viewed as a reaction coordinate. The challenge is to build a model of virus mechanics that links the experimentally observed quantities, such as the slope of the force-deformation spectrum (FX curve), the critical force, and critical deformation for the collapse transition (amplitude F^* and position X^* of the force peak), with the unobserved physical parameters that are characteristic of the structure, local geometry, and overall shape of the particle. Also, it should be expected that the mechanical response of a cylindrical particle (i.e., microtubule) is different from the response of a spherical particle (CCMV or encapsulin shell). Hence, the model should also take into account the geometry of an indenting object.

Our results from *in silico* nanoindentation of CCMV show that there are (at least) two types of excitations relevant to the forced deformation and collapse of the biological assemblies: (1) the local curvature change in a small area of contact formed by an indenting object and a biological particle, and (2) the bending deformations of a larger portion of the biological particle's structure, which is not in direct contact with an indenter. These types of mechanical excitation for the case of nanoindentation of CCMV are visualized in Figure 10.10, which show small-amplitude deformations due to the local curvature change in the limited area on the CCMV

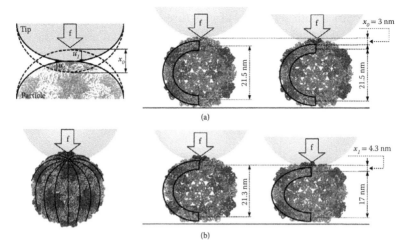

FIGURE 10.10 Identifying the most relevant degrees of freedom for theoretical modeling of the biomechanics of virus particles using the mechanical deformation of the CCMV capsid as an example. Panel (a) shows the physical contact between the indenter (spherical tip) and the particle (scheme on the left), which undergo normal displacements u_1 and u_2 (see Equation 10.14). Initially, the compressive force application (indicated by a large vertical arrow) results in the local curvature change in the portion of the virion outer surface under the tip and the subsequent increase in the tip-capsid surface contact area (dashed lines represent the contour lines of the tip-sphere and the particle in their undeformed states). This is the microscopic origin of initial deformation of the virus shell (x_0), which dominates the initial pseudo-linear portion of the FX spectra (see Figures 10.6 and 10.7 for the CCMV and encapsulin shells, respectively). The structures in the middle (native state) and on the right (partially deformed state) show the amplitude of this type of mechanical excitation, $x_0 \approx 3$ nm (see Equations 10.16 and 10.17). Panel (b) displays a different type of mechanical excitation associated with bending of the side portions of the virion structure that are not in direct contact with the indenter. This part of the structure can be partitioned into the bent elements (beams) of equal size (mixed top-side view of the CCMV shell on the left). For large values of the applied normal compressive force, these elements undergo gradual deformation, which results eventually in the capsid transitioning to the collapsed state for deformations beyond critical, i.e., for $X > X^* = 9-11$ nm for the CCMV shell and 7–8 nm for the encapsulin shell. Comparison of the structure in the middle (partially deformed state) and on the right (globally collapsed state) reveals the amplitude of this type of mechanical excitation, $x_1 \approx 4.3$ nm (see Equations 10.22 and 10.23).

outer surface under the tip (indenter), and the global bending transitions in the side portions of the CCMV shell. A model of virus mechanics based on these degrees of freedom should emphasize the connection between salient features of a force-deformation spectrum (FX curve), especially in the region of the spectrum around the collapse transition (critical values F^* and X^*), with the probability distribution of the strength, which describes the limits of mechanical stability of structure elements comprising the virus particle. To address this aspect of the problem, we can employ the statistics of extremes applied to the global transitions in biological matter. This approach might prove to be useful at describing the dynamic evolution of the structural damage in a biological particle subject to mechanical stress. Furthermore, our results also indicate that the *FX* spectra develop weak nonlinearity (see Figures 10.6 and 10.7).

10.7.2 Local Curvature Change of the Indenter-Particle Contact Area

We have performed a careful analysis of the structure changes for the CCMV and encapsulin particles. We found that in all the simulation runs that initially, (1) the mechanical loading due to the dynamic force ramp $f(t) = r_f t$ is amortized by the local curvature change in the tip-surface contact area; then, (2) the tip-surface contact area grows with force f (see Figures 10.6 and 10.7). Hence, the indentation depth X and indentation force F should involve a contribution from the local curvature change. To describe the restoring force (indentation force) of the biological particle resisting mechanical deformation F, due to local curvature change in the tip-surface contact area, we can adopt the Hertz model that describes the mechanical interaction of two solid bodies in contact [67,69].

According to the Hertz model, when the two spherical bodies, i.e., the biological particle and the cantilever tip, are pressed against each other (without friction and adhesion) with normal force F, the distribution of pressure created in their contact area is given by the following function of two variables x and y, which represent a point on the surface [69]:

$$P(x, y) = \frac{3F}{2\pi b^2} \sqrt{1 - \frac{x^2}{b^2} - \frac{y^2}{b^2}} \qquad (10.13)$$

where b is the radius of the (circular) contact area. When subjected to this pressure, the amino-acid residues of the particle and the tip forming the contact area undergo normal displacements u_1 and u_2, respectively [67]. This results in the local curvature change both in the particle and in the tip, and the total displacement (indentation) is given by the sum

$$x_0 = u_1 + u_2 = D_H \iint_A \frac{P(x, y)}{r} dA \qquad (10.14)$$

where A is the contact area, and $r = (x^2 + y^2)^{1/2}$ is the Euclidean distance from any point on the surface of the contact area (x, y) to the normal axis of compressive force. In Equation 10.14, D_H is a constant coefficient given by

$$D_H = \frac{3}{4}\left(\frac{1-\sigma_1^2}{E_1} + \frac{1-\sigma_2^2}{E_2}\right) \tag{10.15}$$

where E_1 and E_2 are the Young's moduli for the biological particle and the tip, and σ_1 and σ_2 are the Poisson's ratios for the particle and the tip, respectively. Equation 10.14 for two spherical bodies can be solved as described in Ref. [67]. Corresponding to the local curvature change, the indentation force F can be expressed as a function of the total indentation depth x_0 (Equation 10.14) as

$$F_H(x_0) = \frac{1}{D_H} \cdot \sqrt{\frac{R \cdot R_{\text{tip}}}{R + R_{\text{tip}}}} \cdot x_0^{3/2} \tag{10.16}$$

where R is the radius of the spherical particle (virus shell), and R_{tip} is the size of spherical indenter (cantilever tip). By performing the integration of Equation 10.16 over x_0, we can obtain the corresponding expression for the energy of indentation:

$$U_H(x_0) = \frac{2}{5D_H} \cdot \sqrt{\frac{R \cdot R_{\text{tip}}}{R + R_{\text{tip}}}} \cdot x_0^{5/2} \tag{10.17}$$

The radius of the contact area b formed by two interacting spherical bodies as a function of the indentation force F and the indentation depth x_0 is given by the following equations:

$$b(F) = \left(\frac{D_H \cdot R \cdot R_{\text{tip}}}{2(R + R_{\text{tip}})}\right)^{1/3} F^{1/3} \tag{10.18}$$

$$b(x_0) = \sqrt{\frac{R \cdot R_{\text{tip}}}{R + R_{\text{tip}}}} x_0^{1/2} \tag{10.19}$$

where D_H, entering Equation 10.18, is defined in Equation 10.15.

The Hertz model predicts that the indentation force F_H is a weak nonlinear function of the indentation depth due to the local curvature change x_0 (i.e., F_H scales as $x_0^{3/2}$; see Equation 10.16). In agreement with this prediction, the observed force-deformation spectra (FX curves) show some weak nonlinearity as a function of X. This is because F (indentation force) and X (indentation depth) involve contributions from F_H (Hertzian force) and x_0 (Hertzian deformation), respectively (see Figures 10.6 and 10.7). The observed effect, i.e., a nonlinear Hertzian increase in F as a function of the increasing deformation depth X, can be understood by considering the local curvature change under the tip and time evolution of contact area in the course of indentation (the insets in Figures 10.6 and 10.7). We see that the contact area (and the radius b) increases with force F and indentation depth X, which agrees with the Hertz model, because the dynamic force ramp loads mechanically an increasingly larger portion of the capsid structure. Hence, during the indentation process, the mechanical resistance is expected to gradually increase with the extent of loading, which results in the increase of the slope of the FX curve (Figures 10.6 and 10.7).

10.7.3 Global Mechanical Deformation of Virus Shell Structure

Structure analysis of the trajectories of nanoindentations of the CCMV and encapsulin shells showed that at the later stage of the indentation process, side portions of the biological particle that are not in direct contact with the indenter (spherical tip) also develop bending deformations (see Figure 10.10). This is because over longer time periods, the mechanical tension from the indenter (tip) propagates further by gradually engaging an increasingly larger portion of the virus particle. Hence, the indentation force F and indentation depth X also involve contributions from the bending deformations of the portions of the particle structure that are further away from the indenting object.

To describe the restoring force of the particle due to the bending deformations in response to the external mechanical loading, we can adopt a model which assumes that the side portions of the virus particle can be represented by a collection of roughly equal parallel vertical structure elements ("parallel beams"). The beams bend slightly, which results in their curvature change and the associated bending deformation x_1. Since the bending amplitude is small (see Figure 10.10), the energy change associated with this type of mechanical excitation can be written using the following harmonic approximation [68]:

$$U_B = \frac{EI}{2} \int_L \left(\frac{d^2 s(l)}{dl^2} - \frac{d^2 s_0(l)}{dl^2} \right)^2 dl \qquad (10.20)$$

In Equation 10.20, $s_0(l)$ and $s(l)$ are functions of the shape of the structure elements (vertical beams) undergoing bending deformations at the beginning and the end of deformation, respectively, EI is the flexural rigidity, which can be written as the product of the Young's modulus E and the moment of inertia I, and L is the total length of each element. We assume that all the structural elements are equal and that they are capable of withstanding the equal deformation x_1. Then, their shape can be expressed as

$$s(l) = (R + x_1/2)\sqrt{1 - \frac{l^2}{(R - x_1/2)^2}} \qquad (10.21)$$

where $0 \leq l \leq R - x_1/2$ is the internal parameter that describes the elliptic shape of the vertical element in the Cartesian coordinates, and R is the radius of the spherical particle (virus capsid). By substituting Equation 10.21 into Equation 10.20, and performing the integration over x, we can obtain the expression for the bending energy, $U_B(x_1)$. By taking the derivative of the expression for U_B with respect to x_1, we obtain the corresponding expression for the restoring (bending) force as a function of x_1:

$$F_B\left(x_1\right) = EI\frac{\pi x_1}{32R^5}\left(\frac{576R^6 + 16R^4x_1^2 + 28R^2x_1^4 - 5x_1^6}{\left(4R^2 - x_1^2\right)^2}\right) \qquad (10.22)$$

For a small deformation, x_1, we can expand Equation 10.22 in a Taylor series in powers of x_1 and retain only the linear term in the expansion, which reads

$$F_B(x_1) \cong \frac{9EI\pi}{8R^3} \cdot x_1 \qquad (10.23)$$

Owing to their small amplitude, the bending deformations of the vertical structure elements (parallel beams) x_1 contribute linearly to the total deformation of the virus particle X. Hence, these types of mechanical excitations experienced by the virus particles can be thought of as a collection of compressed harmonic springs. This also explains why the force-deformation spectra (FX curves in Figures 10.6 and 10.7) are only slightly nonlinear. Indeed, from a visual inspection of the force-deformation spectra, we see that the FX curves show some weak nonlinearity for small X values, i.e., in the initial deformation regime that corresponds to the Hertzian type of mechanical excitation ($X \approx x_0$) described in the previous section. For larger values of X, the FX profiles become more linear, especially in the pre-transition regime, when the indentation depth X is approaching the critical value X^* and the indentation force F is approaching the critical value F^* (see Figures 10.6 and 10.7); at that point the transition to the collapsed state occurs (force drop in the FX curves). Hence, the models of biomechanics of virus shells should also take into account the global bending deformations described above, which play an important role when describing the gradual transitioning of the virus shell to the collapsed state.

10.8 CONCLUDING REMARKS

Hierarchical supramolecular assemblies that spontaneously undergo self-assembly, disassembly, or shape change play fundamental roles in all biological systems. Detailed understanding of the microscopic structural origin underlying these unique properties of biological assemblies and the mechanisms of their response to external mechanical and molecular signaling factors represent currently a key research challenge. Although state-of-the-art experiments have become available to explore the unique physico-chemical properties of biological assemblies including viruses and virus-like particles and bacteriophage [53]–[64], owing to their high complexity and immense system size, these experiments often yield data that are nearly impossible to interpret accurately without some input from theoretical modeling [21]. In this chapter, we have presented our novel *in silico* nanoindentation approach, which combines multiscale modeling (all-atom and coarse-grained simulations) and high-performance (GPU) computing, to explore the physical chemistry and biomechanics of viruses (CCMV) and virus-like particles (encapsulin). In contrast to previous modeling efforts, this approach enables one to carry out large-scale simulations for the biological systems ($\sim 10^4$–10^5 residues) in the experimentally relevant (10–100 ms) timescale. It takes only 4–8 days of wall-clock time to complete one nanoindentation run for the encapsulin shell on a contemporary GPU (GeForce GTX 780) using our SOP-GPU software package.

In recent years, theoretical biophysicists and computational biologists have witnessed considerable performance gains associated with numerical modeling on a GPU. Hence, the computational methods that we have presented offer a powerful new modeling tool for the computational exploration of complex biological assemblies. We demonstrated that our *in silico* nanoindentation approach allows one to map out the free energy landscape underlying mechanical deformation of the CCMV and encapsulin shells. This information is not available from experiments or other computational methods. Therefore, our approach allows one to follow the dynamics of the biological particle in question beyond the near-equilibrium regime of elastic deformations into the transition regime, where the global shape change and structure collapse occur. Hence, our *in silico* nanoindentation methodology enables researchers to theoretically generate the entire force-deformation spectrum, which contains all the information about the biomechanics of the virus system in question. To the best of our knowledge, no other computational method has this capability. Furthermore, a direct comparison of the experimental force-deformation spectra with simulated *FX* curves enables researchers to uniquely relate dynamic structural features in virus capsids in the regime of their transitioning to the collapsed state with their physical chemistry and materials properties.

The virus capsids and the native cellular virus-like nanocompartment have potential applications in a number of areas. Indeed, biotechnological applications of nanomaterials range from catalysis in constrained environments in organic synthesis, to transport and delivery of substrates into cells in nanomedicine, and to form building blocks for nanotechnology [9]. In nanomedicine, native virus and virus-like particles can also function as gene delivery or drug delivery vehicles, as well as having biotechnological and industrial importance as nanoscale bioreactors. For these reasons, detailed knowledge about the dynamics, energetics, and biomechanical and materials properties of these biologically inspired nanoparticles is of considerable scientific interest. Consequently, the use of engineered virus-like particles for nanotechnology applications is becoming a rapidly growing field [70,71].

The computational methodology that we have developed for the multiscale modeling of virus particles and other protein assemblies can help a wide range of researchers: (1) to advance a conceptual understanding of the physical properties of closed protein shells, (2) to resolve the multiple dynamic modes of collective excitations leading to the emergence of mechanical stiffening and softening in the biological matter, (3) to characterize the (ir)reversibility of mechanical deformation of virus particles, and (4) to describe specific roles of the nonequilibrium collective modes of the capsomers' displacements and their connection to thermodynamic functions such as Gibbs free energy, enthalpy, and entropy. Because these properties are likely to be shared among different virion classes, these results obtained for the CCMV and encapsulin capsid are significant to understand the biomechanics of other protein shells at the nanoscale. When combined with dynamic force experiments *in vitro*, these nanoindentations *in silico* enable researchers to interpret the experimental forced indentation patterns in an unprecedented level of detail with regards to the thermodynamic and structural changes in the virus capsid in question in response to external mechanical deformation. Via a systematic study of the properties of different virion classes, it might become possible to gain insights into the physico-chemical taxonomy underlying all virions in nature. And, in a more practical sense, this understanding may provide biomedical researchers with molecular clues concerning the design of drugs to help prevent or cure viral infections.

The development of simple physical models of virus shells is important to understand their biomechanical and materials properties. Currently, there are no analytically tractable models that can be employed to describe the force-deformation spectral lineshapes available from single-particle dynamic force experiments. This impediment limits the potential information gains for a large number of virus particles, for which the experimental AFM data are already available. Our results of modeling of different types of mechanical excitations for the CCMV and encapsulin nanoshells show that although these biological particles are microscopic structures, their immense system size (many thousands of amino-acid residues) permits one to use the framework of continuous damage theory. Furthermore, the FX spectra for single virus particles differ from each other even when collected for the same indentation point or symmetry axis (see Figures 10.5 and 10.6). This implies that mechanical deformation is a stochastic process, and so theoretical modeling of the biomechanical properties of virus shells should be based on the theory of random processes and involve the description of the relevant biomolecular characteristics in terms of their distribution functions.

REFERENCES

1. Kausche, G.A., Pfankuch, E., and Ruska, H. (1939) Die Sichtbarmachung von pflanzlichem Virus im Übermikroskop. *Naturwissenschaften* **27**: 292–299.
2. Rossmann, M.G., Morais, M.C., Leiman, P.G., and Zhang, M. (2005) Combining X-ray crystallography and electron microscopy. *Structure* **13**: 355–362.
3. Cannon, G., Bradburne, C., Aldrich, H., Baker, S., Heinhorst, S., and Shively, J. (2001) Microcompartments in prokaryotes: Carboxysomes and related polyhedra. *Appl. Environ. Microbiol.* **67**: 5351–5361.
4. Sutter, M., Boehringer, D., Gutmann, S., Gunther, S., Prangishvili, D., Loessner, M.J., Stetter, K.O., Weber-Ban, E., and Ban, N. (2008) Structural basis of enzyme encapsulation into a bacterial nanocompartment. *Nat. Struct. Mol. Biol.* **15**: 939–947.

5. Smith, J.L. (2004) The physiological role of ferritin-like compounds in bacteria. *Crit. Rev. Microbiol.* **30**: 173–185.

6. Aevarsson, A., Seger, K., Turley, S., Sokatch, J.R., and Hol, W.G. (1999) Crystal structure of 2-oxoisovalerate and dehydrogenase and the architecture of 2-oxo acid dehydrogenase multienzyme complexes. *Nat. Struct. Biol.* **6**: 785–792.

7. Jenni, S., Leibundgut, M., Boehringer, D., Frick, C., Mikolásek, B., and Ban, N. (2007) Structure of fungal fatty acid synthase and implications for iterative substrate shuttling. *Science* **316**: 254–261.

8. Ritsert, K., Huber, R., Turk, D., Ladenstein, R., Schmidt-Bäse, K., and Bacher, A. (1995) Studies on the lumazine synthase/riboflavin synthase complex of *Bacillus subtilis.* Crystal structure analysis of reconstituted, icosahedral beta-subunit capsids with bound substrate analogue inhibitor at 2.4 A resolution. *J. Mol. Biol.* **253**: 151–167.

9. Fischlechner, M. and Donath, E. (2007) Viruses as building blocks for materials and devices. *Angew. Chem. Int. Ed. Engl.* **46**: 3184–3193.

10. Roos, W.H., Ivanovska, I.I., Evilevitch, A., and Wuite, G.J.L. (2007) Viral capsids: Mechanical characteristics, genome packaging and delivery mechanisms. *Cell. Mol. Life Sci.* **64**: 1484–1497.

11. de Pablo, P.J., Schaap, I.A.T., MacKintosh, F.C., and Schmidt, C.F. (2003) Deformation and collapse of microtubules on the nanometer scale. *Phys. Rev. Lett.* **91**: 98101–98104.

12. Schaap, I.A.T., Carrasco, C., de Pablo, P.J., MacKintosh, F.C., and Schmidt, C.F. (2006) Elastic response, buckling, and instability of microtubules under radial indentation. *Biophys. J.* **91**: 1521–1531.

13. Hyeon, C., Dima, R.I., and Thirumalai, D. (2006) Pathway and kinetic barriers in mechanical unfolding and refolding of RNA and proteins. *Structure* **14**:1633–1645.

14. Mickler, M., Dima, R.I., Dietz, H., Hyeon, C., Thirumalai, D., and Rief, M. (2007) Revealing the bifurcation in the unfolding pathways of GFP by using single-molecule experiments and simulations. *Proc. Natl. Acad. Sci. U S A* **104**:20268–20273.

15. Dima, R.I. and Joshi, H. (2008) Probing the origin of tubulin rigidity with molecular simulations. *Proc. Natl. Acad. Sci. U S A* **105**: 15743–15748.

16. Lin, J., Hyeon C., and Thirumalai, D. (2008) Relative stability of helices determines the folding landscape of adenine riboswitch aptamers. *J. Am. Chem. Soc.* **130**: 14080–14084.

17. Zhang, Z. and Thirumalai, D. (2012) Dissecting the kinematics of the kinesin step. *Structure* **20**: 628–640.

18. Hyeon, C. and Onuchic, J.N. (2007) Mechanical control of the directional stepping dynamics of the kinesin motor. *Proc. Natl. Acad. Sci. U S A* **104**: 17382–17387.

19. Theisen, K.E., Zhmurov, A., Newberry, M.E., Barsegov, V., and Dima, R.I. (2012) Multiscale modeling of the nanomechanics of microtubule protofilaments. *J. Phys. Chem. B* **116**: 8545–8555.

20. Zhmurov, A., Brown, A.E.X., Litvinov, R.I., Dima, R.I., Weisel, J.W., and Barsegov, V. (2011) Mechanism of fibrin(ogen) forced unfolding. *Structure* **19**: 1615–1624.

21. Kononova, O., Snijder, J., Brasch, M., Cornelissen, J.J.L.M., Dima, R.I., Marx, K.A., Wuite, G.J.L., Roos, W.H., and Barsegov, V. (2013) Structural transitions and energy landscape for cowpea chlorotic mottle virus capsid mechanics from nanomanipulation *in vitro* and *in silico*. *Biophys. J.* **105**: 1893–1903.

22. Kononova, O., Jones, L., and Barsegov, V. (2013) Order statistics inference for describing topological coupling and mechanical symmetry breaking in multidomain proteins. *J. Chem. Phys.* **139**: 121913–121925.

23. Scott, S.W. (2006) Bromoviridae and allies. In *Encyclopedia of Life Sciences*. Chichester: John Wiley and Sons.

24. Speir, J.A., Munshi, S., Wang, G., Baker, T.S., and Johnson, J.E. (1995) Structure of the native and swollen forms of cowpea chlorotic mottle virus determined by X-ray crystallography and cryo-electron microscopy. *Structure* **3**: 63–78.

25. Roos, W.H., Bruinsma, R., and Wuite, G.J.L. (2010) Physical virology. *Nat. Phys.* **6**:733–743.

26. Roenhorst, J.W., van Lent, J.W.M., and Verduin, B.J.M. (1988) Binding of cowpea chlorotic mottle virus to cowpea protoplasts and relation of binding to virus entry and infection. *Virology* **164**: 91–98.

27. Milne, J.R. and Walter, G.H. (2003) The coincidence of thrips and dispersed pollen in PNRSV-infected stonefruit orchards: A precondition for thrips-mediated transmission via infected pollen. *Ann. Appl. Biol.* **142**: 291–298.

28. Akita, A., Chong, K.T., Tanaka, H., Yamashita, R., Miyazaki, N., Nakaishi, Y., Suzuki, M., Namba, K., Ono, Y., Tsukihara, T., and Nakagawa, A. (2007) The crystal structure of a virus-like particle from the hyperthermophilic archaeon *Pyrococcus furiosus* provides insight into the evolution of viruses. *J. Mol. Biol.* **368**: 1469–1483.

29. Yeates, T.O., Crowley, C.S., and Tanaka, S. (2010) Bacterial microcompartment organelles: Protein shell structure and evolution. *Ann. Rev. Biophys.* **39**: 185–205.

30. Yeates, T.O., Thompson, M.C., and Bobik, T.A. (2011) The protein shells of bacterial microcompartment organelles. *Curr. Opin. Struct. Biol.* **21**: 223–231.

31. Kerfeld, C.A., Heinhorst, S., and Cannon, G.C. (2010) Bacterial microcompartments. *Ann. Rev. Microbiol.* **64**: 391–408.

32. Kang, S. and Douglas, T. (2010) Some enzymes just need a space of their own. *Science* **327**: 42–43.

33. Ivanovska, I.L., de Pablo, P.J., Ibarra, B., Sgalari, G., MacKintosh, F.C., Carrascosa, J.L., Schmidt, C.F., and Wuite, G.J. (2004) Bacteriophage capsids: Tough nanoshells with complex elastic properties. *Proc. Natl. Acad. Sci. U S A* **101**: 7600–7605.

34. Hernando-Perez, M., Miranda, R. Aznar, M., Carrascosa, J.L., Schaap, I.A.T., Reguera, D., and de Pablo, P.J. (2012) Direct measurement of phage phi29 stiffness provides evidence of internal pressure. *Small* **8**: 2366–2370.

35. Gibbons, M.M. and Klug, W.B. (2008) Influence of nonuniform geometry on nanoindentation of viral capsids. *Biophys. J.* **95**: 3640–3649.

36. Klug, W.S., Roos, W.H., and Wuite, G.J.L. (2012) Unlocking internal prestress from protein nanoshells. *Phys. Rev. Lett.* **109**: 168104.

37. Zink, M. and Grubmuller, H. (2009) Mechanical properties of the icosahedral shell of southern bean mosaic virus: A molecular dynamics study. *Biophys. J.* **96**: 1350–1363.

38. Zink, M. and Grubmuller, H. (2010) Primary changes of the mechanical properties of southern bean mosaic virus upon calcium removal. *Biophys. J.* **98**: 687–695.

39. Tama, F. and Brooks, C.L. III. (2002) The mechanism and pathway of pH induced swelling in cowpea chlorotic mottle virus. *J. Mol. Biol.* **318**: 733–747.

40. Tama, F. and Brooks, C.L. III. (2005) Diversity and identity of mechanical properties of icosahedral viral capsids studied with elastic network normal mode analysis. *J. Mol. Biol.* **345**: 299–314.

41. Rader, A.J., Vlad, D.H., and Bahar, I. (2005) Maturation dynamics of bacteriophage HK97 capsid. *Structure* **13**: 413–421.

42. Yang, Z., Bahar, I., and Widom, M. (2009) Vibrational dynamics of icosahedrally symmetric biomolecular assemblies compared with predictions based on continuum elasticity. *Biophys. J.* **96**: 4438–4448.

43. May, E.R., Aggarwal, A., Klug, W.S., and Brooks, C.L. III. (2011) Viral capsid equilibrium dynamics reveals nonuniform elastic properties. *Biophys. J.* **100**: L59–L61.

44. Arkhipov, A., Freddolino, P.L., and Schulten, K. (2006) Stability and dynamics of virus capsids described by coarse-grained modeling. *Structure* **14**: 1767–1777.

45. Arkhipov, A., Roos, W.H., Wuite, G.J.L., and Schulten, K. (2009) Elucidating the mechanism behind irreversible deformation of viral capsids. *Biophys. J.* **97**: 2061–2069.

46. Cieplak, M. and Robbins, M.O. (2010) Nanoindentation of virus capsids in a molecular model. *J. Chem. Phys.* **132**: 015101.

47. Zhmurov, A., Dima, R.I., Kholodov, Y., and Barsegov, V. (2010) SOP-GPU: Accelerating biomolecular simulations in the centisecond timescale using graphic processors. *Proteins* **78**: 2984–2999.

48. Zhmurov, A., Rybnikov, K., Kholodov, Y., and Barsegov, V. (2011) Generation of random numbers on graphics processors: Forced indentation in silico of the bacteriophage HK97. *J. Phys. Chem. B* **115**: 5278–5288.

49. Barsegov, V., Klimov, D., and Thirumalai, D. (2006). Mapping the energy landscape of biomolecules using single molecule force correlation spectroscopy (FCS): Theory and applications. *Biophys. J.* **90**: 3827–3841.

50. Ferrara, P., Apostolakis, J., and Caflisch, A. (2002) Evaluation of a fast implicit solvent model for molecular dynamics simulations. *Proteins* **46**: 24–33.

51. Kasas, S. and Dietler, G. (2008) Probing nanomechanical properties from biomolecules to living cells. *Pflug. Arch. Eur. J. Phys.* **456**: 13–27.
52. Engel, A. and Müller, G.J. (2000) Observing single biomolecules at work with the atomic force microscope. *Nat. Struct. Biol.* **7**: 715–718.
53. Ivanovska, I., Wuite, G., Joensson, B., and Evilevitch, A. (2007) Internal DNA pressure modifies stability of WT phage. *Proc. Natl. Acad. Sci. U S A*, **104**: 9603–9608.
54. Roos, W.H., Gertsman, I., May, E.R., Brooks, C.L. III, Johnson, J.E., and Wuite, G.J.L. (2012) Mechanics of bacteriophage maturation. *Proc. Natl. Acad. Sci. U S A* **109**: 2342–2347.
55. Kol, M., Shi, Y., Tsvitov, M., Barlam, D., Shneck, R.Z., Kay, M.S., and Rousso, I. (2007) A stiffness switch in human immunodeficiency virus. *Biophys. J.* **92**: 1777–1783.
56. Baclayon, M., Shoemaker, G.K., Uetrecht, C., Crawford, S.E., Estes, M.K., Prasad, B.V., Heck, A.J.R., Wuite, G.J.L., and Roos, W.H. (2011) Prestress strengthens the shell of Norwalk virus nanoparticles. *Nano Lett.* **11**: 4865–4869.
57. Liashkovich, I., Hafezi, W., Kühn, J.E., Oberleithner, H., Kramer, A., and Shahin, V. (2008) Exceptional mechanical and structural stability of HSV-1 unveiled with fluid atomic force microscopy. *J. Cell. Sci.* **121**: 2287–2292.
58. Pérez-BernÁ, A.J., Ortega-Esteban, A., Menendez-Conejero, R., Winkler, D.C., Menendez, M., Steven, A.C., Flint, S.J., de Pablo, P.J., and San Martin, C. (2012) The role of capsid maturation on adenovirus priming for sequential uncoating. *J. Biol. Chem.* **287**: 31582–31595.
59. Roos, W.H., Radtke, K., Kniesmeijer, E., Geertsema, H., Sodeik, B., and Wuite, G.J.L. (2009) Scaffold expulsion and genome packaging trigger stabilization of herpes simplex virus capsid. *Proc. Natl. Acad. Sci. U S A* **106**: 9673–9678.
60. Roos, W.H., Gibbons, M.M., Arkhipov, A., Uetrecht, C., Watts, N.R., Wingfield, P.T., Steven, A.C., Heck, A.J.R., Schulten, K., Klug, W.S., and Wuite, G.J.L. (2010) Squeezing protein shells: How continuum elastic models, molecular dynamics simulation and experiments coalesce at the nanoscale. *Biophys. J.* **99**: 1175–1181.
61. Snijder, J., Reddy, V.S., May, E.R., Roos, W.H., Nemerow, G.R., and Wuite, G.J.L. (2013) Integrin and defensin modulate the mechanical properties of adenovirus. *J. Virol.* **87**: 2756.
62. Michel, J.P., Ivanovska, I.L., Gibbons, M.M., Klug, W.S., Knobler, C.M., Wuite, G.J.L., and Schmidt, C.F. (2006) Nanoindentation studies of full and empty viral capsids and the effects of capsid protein mutations on elasticity and strength. *Proc. Natl. Acad. Sci. U S A* **103**: 6184–6189.
63. Carrasco, C., Carreira, A., Schaap, I.A.T., Serena, P.A., Gómez-Herrero, J., Mateu, M.G., and de Pablo, P.J. (2006) DNA-mediated anisotropic mechanical reinforcement of a virus. *Proc. Natl. Acad. Sci. U S A* **103**: 13706–13711.
64. Snijder, J., Uetrecht, C., Rose, R., Sanchez, R., Marti, G., Agirre, J., Guérin, D.M., Wuite, G.J.L., Heck, A.J.R., and Roos, W.H. (2013) Probing the

biophysical interplay between a viral genome and its capsid. *Nat. Chem.* **5**: 502–509.

65. Hayward, S. and de Groot, B.L. (2008) Normal modes and essential dynamics. *Methods Mol. Biol.* **443**: 89–106.

66. Amadei, A., Linssen, A.B.M., and Berendsen, H.J.C. (1993) Essential dynamics of proteins. *Proteins Struct. Funct. Bioinform.* **17**: 412–425.

67. Landau, L.D. and Lifshitz, E.M. (1986) *Theory of Elasticity (Theoretical Physics)*, 3rd edition, vol. 7, Oxford, UK: Butterworth-Heinemann.

68. Timoshenko, S.P. (1961) *Theory of Elastic Stability*, 2nd edition, New York: McGraw-Hill Book Company.

69. Johnson, K.L. (1985) *Contact Mechanics*, 1st edition, Cambridge, UK: Cambridge University Press.

70. O'Neil, A., Reichhardt, C., Johnson, B., Prevelige, P.E., and Douglas, T. (2011) Genetically programmed in vivo packaging of protein cargo and its controlled release from bacteriophage P22. *Angew. Chem. Int. Ed.* **50**: 7425–7428.

71. Patterson, D.P., Prevelige, P.E., and Douglas, T. (2012) Nanoreactors by programmed enzyme encapsulation inside the capsid of the bacteriophage P22. *ACS Nano* **6**: 5000–5009.

Index

Acidic islet, 327
Adaptive Poisson–Boltzmann
    solver (APBS), 197–198
Advanced CG NCP model, 316–318
All-atom molecular dynamics (AA
    MD) simulations, 276
D-Amino-acid residues treatment,
    85–86
Amyloid formation, 93
Anharmonic potentials, 194
APBS, *see* Adaptive
    Poisson–Boltzmann solver
Apomyoglobin, 174
Associative memory Hamiltonian
    (AMH) family of
    potentials, 140
Associative memory, water
    mediated, structure and
    energy model (AWSEM),
    124
Atomistic simulations, 29
AWSEM, *see* Associative memory,
    water mediated, structure
    and energy model
AWSEM Hamiltonian, neural
    network origin
  associative memory
    Hamiltonian family of
    potentials, 140
  biological computation, 145
  charges, 139

Desulfovibrio vulgaris
    rubredoxin, 141
  folding, 145
  fragment memory, 144
  Hopfield network model,
    136–138
  physical–chemical
    interactions, 144–145
  potentials of mean force, 143
  protein chain length, 140
  protein structure prediction
    algorithms, 142
  recurrent artificial neural
    networks, 136
  residue pair-dependent
    hydrogen bonding, 143
AWSEM-MD
  apomyoglobin, 174
  combined protein–DNA
    potentials, 176
  cytoplasmic polyadenylation
    element binding, 179–180
  Debye–Hückel electrostatic
    interactions, 178–179
  docking algorithm, 175
  HJURP chaperone, 177–178
  homologous protein
    structures, 173–174
  "homologues excluded", 173
  LAMMPS open source code,
    176

nucleosomes, 176–177
protein aggregation, 179
protein structure prediction,
172
root-mean-square deviation,
173
structure predictions, 174–175
two-dimensional free energy
profiles, 175–176

Backbone-electrostatic and
correlation terms, 80–84
Backbone models, RNA 3D
structure
one-bead model, 230–231
two-bead model, 231–233
Backbone–nucleobase hybrid
models, RNA
five-bead model, 238–240
seven-bead model, 240–242
three-bead model, 236–238
Backbone potential, *see also*
AWSEM-MD
AWSEM,
three-beads-per-residue
backbone model, 147
chirality potential, 148
connectivity and chain
potentials, 147
excluded volume interaction
potential, 150
harmonic springs, 145, 147
protein-like stereochemistry,
146
Ramachandran plot, 148–149
steric repulsions, 150
Base–base interaction, 100
B-DNA bending, 206, 207
Bendability, 303

Bin/Ampiphysin/Rvs (BAR),
94, 95
Biomolecules, elastic models of
application
ENMs, normal modes, and
structural transitions,
198–201
FEMs and mechanical
responses, 201–208
membranes, elastic models
for, 208–211
computational models
elastic network models,
192–194
membrane, elastic models
for, 196–198
proteins and DNA,
continuum mechanics
models, 194–196
Bottom-up and top-down models
biological macromolecules, 33
CG units, interaction
potential, 31
Henderson's theorem, 32
iterative Boltzmann inversion
procedure, 32
nuclear magnetic resonance
relaxation experiments, 33
numerical optimization, 32
polymer structure and
dynamics, 32
Bottom-up
one-bead-per-nucleotide
DNA force field, 274–279

Car–Parrinello molecular dynamics
(CPMD), 19
CCMV, *see* Cowpea chlorotic
mottle virus

Center-of-mass fictitious sites, 38
CG DNA models, 17–19
CG lipid model, 13–17
Chemical nodes, 195–196, 206
Chirality potential, 148
Chromatin
    CG mesoscale, 321
    coarse-graining of, 308–312
    discrete surface charge
        optimization approach,
        323–326
    electrostatic environment,
        287
    electrostatic sphere-bead
        model, 326–329
    eukaryotic, 298–299
    fiber folding, 287
    modeling, 309–312
    organization, 300–302
    physical and structural
        properties, 302–308
    "six-angle" model, 322–323
Clash score, 257–258
Coarse-grained energy function,
        70–71
Coarse-graining (CG) models, 28
    atomistic simulations, 29
    diffusion coefficient, 31
    internal energy, 31
    protein dynamics, 31
    thermodynamic and
        dynamical
        inconsistencies, 30
Combined protein–DNA
        potentials, 176
Connectivity and chain potentials,
        147
Consistency principle, 344

Cowpea chlorotic mottle virus
        (CCMV), 374–375,
        see also viruses
    nanomanipulations *in silico*,
        387–392
    shells, equilibrium properties,
        387
CPEB, *see* Cytoplasmic
        polyadenylation element
        binding
CPMD, *see* Car–Parrinello
        molecular dynamics
Critical assessment of protein
        structure prediction
        (CASP) experiment, 258
Cytoplasmic polyadenylation
        element binding (CPEB),
        179–180

Debye–Hückel electrostatic
        interactions, 178–179
Debye–Hückel model, 206
Debye–Huckel potentials, 287
Debye–Waller factors, 201
Deformation index (DI), 254, 257
Denaturant, 127
Density functional theory (DFT)
        functionals, 19
Differential-scanning calorimetry
        (DSC), 101
Diploid genomes, 359–362
DiSCO approach, *see* Discrete
        surface charge
        optimization approach
Discrete molecular dynamics
        (DMD) sampling
    HPR data as restraints,
        251–253
    interactions and collisions, 244
    long-range restraints, 248

replica exchange, 245
RNA structure refinement, 252
SHAPE data as restraints, 247
Discrete nonlinear Schrodinger
        (DNLS) equation, 91–92
Discrete surface charge
        optimization (DiSCO)
        approach, 323–326
Disordered phases types, 127
Disulfide bonds treatment, 85
DNA
    bottom-up
        one-bead-per-nucleotide
        DNA force field, 274–279
    CG models for, 273
    cruciforms, 286
    nano tweezers, 286
    oxDNA, 282–286
    3SPN.2c, 279–282
Docking algorithm, 175
Duplex solutions, 286
Dynamics, in coarse-grained
        models
    calibration curve, 57
    degree of polymerization,
        57–58
    diffusion coefficient, 57
    dissipation of energy, 57
    local degrees of freedom, 57
    mean-square displacement,
        56–57
    molecular surface exposure, 57
    polyethylene chain, 58–59

Electrostatic sphere-bead model,
        326–329
Encapsulins, 370, 375–377
Energy landscape theory, 123–124
Eukaryotic chromatin, 298–299

Fine-graining approaches, 35–36
Finite extensible nonlinear elastic
        (FENE) potential, 41–42,
        352, 381
Force matching procedure,
        structural distribution, 37
Fraction of native contacts, 132
Fragment assembly RNA (FARNA),
        233–236
Free-energy differences, 123
Free-energy profiles (FEPs), 90
Free-energy-type potential, 28
Full Atom Refinement FARNA
        (FARFAR), 235–236

Gaussian network model (GNM),
        192
Genomes
    chromatin-conformation
        capture (3C)-based
        method, 342
    consistency principle, 344
    diploid, 343, 359–362
    dynamical genome hypothesis,
        342
    Gō-like potentials, 344
    interphase budding yeast,
        345–359
    in interphase budding yeast,
        345–354
    minimum frustration
        principle, 344
    native interactions, 344
    nuclear magnetic resonance
        measurement, 344
    topologically associating
        domains, 341–342
Gibbs free energy, 49–50

Global distance test total score (GDT_TS), 89–90
Globule-to-native transition, 128
Graphics processing unit (GPU), 379–380

Helfrich model, 196
Helmholtz free energy, 43, 92
Henderson's theorem, 32
Hessian matrix, 199, 378
HJURP chaperone, 177–178
"Homologues excluded" (HE), 173
Hoogsteen ring, 228
Hopfield network model, 136–138
HRP, *see* Hydroxyl radical probing
Hsp70 molecular chaperone, 96–97
Hydrogen bond potentials
    alpha helices *vs.* beta sheets, 158
    angular dependence, 158
    in AWSEM, 160–161
    liquid-crystal potential, 158
    types, 160–161
Hydroxyl radical probing (HRP), 246, 247, 251–253

IBI procedure, structural distribution, 37
INF, *see* Interaction network fidelity
Information entropy, 37
Integral equation coarse-grained (IECG) model, 29
Integral equation theory, 37–38
Interaction network fidelity (INF), 257
Interarray aggregation, 300–301
Intermolecular atomistic potential, 41

Internal-loop/bubble formation, 103
Interphase budding yeast, fluctuating genome structure in
    gene expression activity, 358–359
    genome movement, 354
    probability distributions, two-dimensional representation, 355
    Rabl-like configuration, 355
    telomeres and nuclear envelope distances, 357
    telomere–telomere distance distributions, 355–356
Interphase budding yeast, genome model
    chromosomes and envelope of, 345, 346
    equation of motion, 347–348
    genome-wide Hi-C data, 345
    knowledge-based potential, 353
    nuclear constraints, potential representing, 348–350
    simulation, 354
    spindle pole body, 345
    3D modeling, 346
    worm-like chain potential, 350–353
Inverse Monte Carlo (IMC) methods
    applications, 11–21
    approximation, 5
    Car–Parrinello molecular dynamics, 19
    CG DNA models, 17–19

CG lipid model, 13–17
density functional theory
    functionals, 19
electrolytes, 11–13
interaction potentials, 21
LiO potential, 20
mixed lipid–cholesterol bilayer,
    19
N-body potential of mean
    force, 4–5
Newton inversion, 6–7
pair potentials from RDFs,
    7–10
potential energy function, 4
software, 10–11
two-dimensional RDFs, 19
Iron–sulfur binding (Isu1), 97
Isothermal compressibility, 43
Iterative Boltzmann inversion (IBI)
    procedure, 32

Kissing hairpins, 286
Kubo cluster cumulant, 106–108

LAMMPS open source code, 176
Langevin equation for protein
    dynamics (LE4PD), 59–60
Lennard-Jones potential, 29, 41–42
LE4PD, *see* Langevin equation for
    protein dynamics
Levinthal paradox, 130, 131
Lysozyme protein, 192, 193

MARTINI
    and continuum models, 208
    force field, 71
    free energy profile, 212
    fusion pore stimulation, 208,
        209

Mean spherical approximation
    (MSA), 40
"Melted polymer" model, *in vivo*
    chromatin, 308
Messenger RNA (mRNA), 223
Minimum frustration principle, 344
Molecular chaperones, 95–97
Molecular dynamics simulations,
    28
Molecular renormalization
    coarse-graining
    (MRG-CG) technique,
    274, 277–278
    applications, 280
Molecular stripping, 176
Molten globule phase, proteins, 130
Monomer–monomer interaction,
    41–42
Mori–Zwanzig projection operator
    technique, 28
MREMD algorithm, 87
MUCA, *see* Multicanonical
    algorithm
MUCAREM, *see* Multicanonical
    replica exchange method
Multiblob coarse-grained model
    blob size, 38
    direct correlation function, 39
    fictitious interacting sites, 38
    intermolecular atomistic
        potential, 41
    intramolecular distributions,
        39
    mean spherical approximation,
        40
    monomer–monomer
        interaction, 41–42
    Ornstein–Zernike relation, 39

scaling behavior, 40–41
soft-sphere representation,
38–39
Multicanonical algorithm
(MUCA), 87–88
Multicanonical replica exchange
method (MUCAREM), 88
Multiplexed replica exchange
molecular dynamics
(MREMD) simulations, 79

Nanomanipulation
nanoindentation *in silico*
method, 384–386
single-particle dynamic force
spectroscopy, 384
Nano stars, 286
NARES-2P
DNA hybridization and, 101
force field, 101
MREMD simulations, 101, 102
NAST, *see* Nucleic acid simulation
tool
Native interactions, 344
Native topology-based
coarse-grained modeling
CCMV shell, SOP model
parameterization,
382–383
nanoindentation *in silico*,
379–381
SOP-GPU software, 383–384
virus particle, SOP model,
381–382
N-body potential of mean force,
4–5
NCP, *see* Nucleosome core particle
Newton inversion, 6–7

Nonnative (nonbonded)
interactions, 382
Nuclear magnetic resonance
(NMR) relaxation
experiments, 33
Nuclear Overhauser effect (NOE)
data, 60
Nuclear Overhauser effect
spectroscopy (NOESY)
cross peaks, 246
Nucleic acids, coarse-grained
simulations of
DNA folding, thermodynamics
simulations, 101, 102
DNAs and RNAs, folding
pathways simulations, 101
force field, calibration of, 100
model and effective energy
function, 98
parameterization, 98–100
premelting transition, 103
Nucleic acid simulation tool
(NAST), 230–231
Nucleobase models, RNA
energy function, 234
fragment assembly RNA,
233–236
Monte Carlo simulation
process, 233
protein data bank identifier,
233–234
rootmean-square distance,
234–236
Nucleosome arrays, 299, 321
Nucleosome core particle (NCP),
299
advanced CG NCP model,
316–318

attraction, 309
multiscale CG model, 319–320
spherical-bead NCP model,
 314, 315
Nucleosome–nucleosome stacking
 interaction, 305–307
Nucleosome repeat length (NRL),
 307
Nucleosomes
 advanced CG NCP model,
  316–318
 advanced nucleosome core
  particle model, 288
 AWSEM-MD, 176–177
 chaperon-assisted *in vivo*
  assembly of, 303, 304
 computational biophysics
  community, 287
 discrete charge optimization
  algorithm, 287
 electron microscopy
  observations, 289
 NCP-NCP attraction, 309
 poly-nucleosomal arrays, 286
 poyl-electroyle model, 288
 single NCP, multiscale CG
  model, 319–320
 small-angle x-ray
  scatteringspectra, 289
 sphere models with grafted
  histone tail chains,
  313–316
 x-ray diffraction, 289
Nucleotide binding domain (NBD),
 96

Odijk–Skonick–Fixman (OSF)
 theory, 208
Oligomerization, 301

Ornstein–Zernike integral
 equation, 33, 34
Ornstein–Zernike relation, 39
"Out-of-plane" displacements,
 378
OxDNA, 282–286

Pairwise and FM-based contact
 potentials
 coarse-grained energy
  function, 153
 distance comparison function,
  151
 Hopfield-type neural networks,
  154
 hydrophobic residues,
  attraction between, 152
 PMF-type interactions, 151
 residue pair-dependent
  interactions strengths,
  153
Parameter optimization, protein
 folding funnels, *see also*
 AWSEM-MD
 accuracy, 163–164
 algorithm, 164
 AWSEM force field
  parameters, 163
 configurational entropy,
  165–166
 energy landscape sculpting
  approach, 162
 glass transition temperatures,
  165, 167
 Hamiltonian parameters, 162
 Lagrange multiplier, 168, 169
 least-square approach, 170
 mean-field theory, 168
 "problematic" proteins, 170

random energy model, 165
self-consistent, 171
spin-glass model, 163
Parameter optimization strategy,
    125
Phosphate–phosphate/
    phosphate-base
    interaction, 100
PICK1–BAR interactions, 94, 95
PMF, coarse-grained system and
    cluster-cumulant
    expansion, 71–73
Poisson–Boltzmann model, 206
Polymer melts, variable-level
    coarse-grained
    representation
    Carnahan–Starling expression,
        45
    compressibility parameter c_0
        evaluation, 47–48
    packing fraction for UA
        simulations, 46
    polyethylene, atomistic
        simulations, 46–47
    soft-sphere model, 45
Poly-nucleosomal arrays, 287
Polypeptide chains, UNRES model,
    73, 74
Position-specific iterative basic
    local alignment search
    tool (PSI-BLAST) server,
    88
Potentials of mean force (PMFs), 71
Principle of minimum frustration,
    131
PRISM approach, 37
Proline residue, trans–cis
    isomerization of, 86

Protein
    chain length, 140
    and DNA, continuum
        mechanics models,
        194–196
    dynamics, coarse-grained
        method, 59–60
    and kinks with UNRES model,
        91–93
    structure prediction
        algorithms, 142
Protein data bank identifier
    (PDB-ID), 233–234
Protein–DNA interactions,
    103–104
Protein-folding pathways and FELs,
    90, 91
Protein interacting with C kinase 1
    (PICK1), 94, 95
Protein physics, energy landscape
    theory
    analogy of folding, 128
    in AWSEM, 134–135
    backbone stereochemistry, 129
    computational biology, 126
    denaturant, 127
    disordered phases types, 127
    folding funnel, 132, 134
    folding *in vivo*, 126
    folding problem, 125–126
    fraction of native contacts, 132
    free energy profile, 132–134
    globular protein, 132, 133
    globule-to-native transition,
        128
    "intrinsically disordered"
        proteins, 129–130
    Levinthal paradox, 130, 131

molten globule phase, 130
principle of minimum
    frustration, 131
Ramachandran plot, 130
random energy model, 132

Quantum chemical calculation,
    122, 158

Rabl-like configuration, 355
Radial distribution functions
    (RDFs), 316
Ramachandran plot, 130, 148–149
Random energy model (REM), 132
Real monomer sites, 38
Recurrent artificial neural
    networks, 136
REMD, see Replica exchange
    molecular dynamics
REMUCA, see Replica exchange
    multicanonical algorithm
REMUCAREM, see Replica
    exchange multicanonical
    with replica exchange
    method
Renormalization group (RG)
    theory, 274
Replica exchange molecular
    dynamics (REMD), 87
Replica exchange multicanonical
    algorithm (REMUCA), 88
Replica exchange multicanonical
    with replica exchange
    method (REMUCAREM),
    88
Representability and transferability,
    CG models, 36–37
Residue pair-dependent hydrogen
    bonding, 143
Ribosomal RNA (rRNA), 223

RNA
    atomistic models, 225
    backbone models, 230–233
    backbone–nucleobase hybrid
        models, 236–242
    groove triplex, 228–229
    HRP chemistry reactivity
        profile, 251–253
    as messenger molecule, 222
    methods to derive distance
        restraints, 245–247
    monomer, 226
    nucleobase models, 233–236
    nucleotides, 2D
        representations, 226
    pseudoknots, 228
    radiolabeling experiments, 222
    RNA-puzzles, 253–261
    sampling methods, 242–245
    SHAPE chemistry, 247–251
    stem loops, 228
    tetraloops, 228
    three-dimensional structures,
        224
    types, 222–223
    Watson–Crick base pairing,
        227–228
    X-ray crystallography, 199,
        224, 286, 369
Root-mean-square deviation
    (RMSD)
    AWSEM-MD, 173
    nucleobase models, RNA,
        234–236

SBMV, see Southern bean mosaic
    virus
SC–SC interaction potentials,
    76–77

SC torsional potentials, 78–80
Selective 2'-hydroxyl acylation analyzed by primer extension (SHAPE) method, 246–251
Self-organized polymer (SOP) model, 373
  CCMV shell, parameterization for, 382–383
  polypeptide chain, 381
  virus particle, 381–382
Serial replica exchange method (SREM), 88
"Six-angle" model, 322–323
Small nuclear ribonucleic particles (snRNPs), 223
Soft-sphere representation, 38–39
Solvent accessible surface area (SASA) model, 383
Southern bean mosaic virus (SBMV), 377
Sphere models with grafted histone tail chains
  cation-induced aggregation, 315
  in DH approximation, 313
  harmonic bonds, 316
  nucleosome core particles, 313, 314
  polyelectrolyte behavior, 316
  radial distribution functions, 314, 316
  spherical-bead NCP model, 314, 315
SREM, *see* Serial replica exchange method
Steric repulsions, 150
Strand displacement reactions, 286

Synthetic polymers modeling
  atomic scale simulations, 34
  CG lengthscale, 34
  coarse-graining approach, 34, 35
  fine-graining approaches, 35–36
  Ornstein–Zernike integral equation, 33, 34
  polymer systems, multiscale simulations, 34

Tail-mediated nucleosomal aggregation, 287
Target T0663, 89
Tethered hydroxyl radical probing (t-HRP), 248
Theoretical predictions, coarse-grained model against atomistic simulations
  degree of polymerization, 49
  entropy, 52–53
  Gibbs free energy, 49–50
  Helmholtz free energy, 48–49, 50
  potential energy, 51–52
Thermodynamically consistent coarse-grained model
  advantages, 53–54
  IBI procedure, 54
  Lennard–Jones liquid, 56
  total correlation function, 54–56
Thermodynamics
  free energy and pressure, 44–45
  Helmholtz free energy, 43, 92
  internal energy per chain, 43

isothermal compressibility, 43
phase transition, 44
potential energy, multiblob
 description, 44
Thermostat algorithm, 87
3SPN.2c, 279–282
T-HRP, *see* Tethered hydroxyl
 radical probing
Tip effect, 198
Transfer RNA (tRNA), 222–223
Two-dimensional free energy
 profiles, 175–176

United atom (UA) potential, 29
UNited-RESidue (UNRES) force
 field, 70
UNRES model
 D-amino-acid residues
  treatment, 85–86
 backbone-electrostatic and
  correlation terms, 80–84
 biological applications
  amyloid formation, 93
  molecular chaperones,
   95–97
  PICK1–BAR interactions,
   94, 95
 in blind prediction of protein
  structures, 88–90
 disulfide bonds treatment, 85
 energy function, 73–76
  calibration, 84–85
 implementation, 78
 local potentials, 78
 MD simulations, 86–88
 proline residue, trans–cis
  isomerization of, 86
 protein and kinks with, 91–93
 protein-folding pathways and
  FELs, 90, 91

SC–SC interaction potentials,
 76–77
SC torsional potentials, 78–80

Vfold, 231–233
Viruses
 CCMV shell, SOP model
  parameterization,
  382–383
 equilibrium properties, 387
 force-deformation assays
  in silico, 392–393
 irreversibility and inelastic
  properties, 398–400
 mechanical deformation
  models
  degrees of freedom, 400–404
  global mechanical
   deformation, 406–408
  local curvature change,
   404–406
 mechanical transitions in,
  thermodynamics, 394–396
 mechanics, computational
  methods for
  direct MD simulations, 377
  finite element analysis, 377
  Hessian matrix, 199, 378
  normal mode analysis, 378
  "out-of-plane"
   displacements, 378
  simplified (coarse-grained)
   models, 378–379
 nanoindentation *in silico*,
  379–381
 nanomanipulation *in vitro* and
  in silico, 384–392
 out-of-plane and in-plane
  displacements, dynamics
  approach, 396–398

particle, SOP model, 381–382
SOP-GPU software, 383–384
structure collapse transition,
    393–394
and virus-like particles,
    373–377
Virus-like particles
    Cowpea chlorotic mottle virus,
        374–375
    encapsulin, 375–377

Water-mediated interactions,
        see also AWSEM-MD
    binding predictions, 157
    contact potentials, 154

electrostatic repulsion, 156
hydrophobic effect, 154
Miyazawa–Jernigan type
    contact potentials,
    154–155
Monte Carlo sampling, 156
protein monomer folding, 155,
    156
protein's core, 157–158
Weighted histogram analysis
    method (WHAM), 79

X-ray crystallography, 199, 224,
    245–246, 286, 369

Printed and bound by CPI Group (UK) Ltd, Croydon, CR0 4YY

01/11/2024

01782623-0012